GODS OF THE UNIVERSE™

Design Patent
Trademark ®

Written by
NICHOLAS JAMES KOUSHIAFES/NIKO

Published by
GODS OF THE UNIVERSE™

GODS OF THE UNIVERSE™

First published on March 21, 1996

Printed in the United States of America

Published by
GODS OF THE UNIVERSE™
P.O. Box 1543
Highland, Indiana 46322
U.S.A.

International Standard Book Number : 0-9607228-6-6
Library of Congress Catalog Card Number : 95-94055

DEDICATION

First of all, this book is dedicated to the astronauts/supreme beings, whom our ancient ancestors called Gods, GODS OF THE UNIVERSE,™ for without them, mankind, as we know ourselves today, would not have been possible. For it was the Gods who thought enough of the planet Earth to stop here, and create mankind, and then later relate our creation and early history to ancient mankind, who must have eagerly scratched pictographs of the story into the dirt, or onto clay tablets, surely not fully understanding it at that instant. And when the Gods again return, perhaps mankind's technological state of development and understanding will be advanced far enough to completely lift the so-called veil of mystery, our lack of comprehension, for the more advanced we become, the more common and understandable the mysterious events appear to us. But that is only if the Gods elect to return to an Earth whereon mankind threatens their own existence, and the very existence of the planet. For as caretakers of the Earth, mankind truly shall not be viewed upon as performing an outstanding and commendable task. What will the judgment of the Gods be this time? You be the critic.

Secondly, this book is dedicated to all of the scholars who have endeavored to increasingly lift the veil of mystery which surrounds the transliterations of the clay tablets even further, and give us insight. From King Ashurbanipal, who in circa 645 B.C. gathered the clay tablets in his famous library at Nineveh, to a priest of Babylon named Berossus, who in circa 290 B.C. included as one of his three part books about the Babylonians, called Babyloniaka, the flood story derived from the clay tablets, to George Smith, who in 1872 announced he had deciphered and discovered a Babylonian version of the Biblical flood story on clay tablet fragments at the British Museum, and many others who have no doubt spent many countless hours locating, accumulating, deciphering, and trying to understand the specific meaning of ancient words, thoughts, and stories: allegorically and metaphorically veiled truths. But it is not our lack of understanding that has prevented our seeing the ancient relationship between the Gods and mankind, it is our inability to project our minds into the future, beyond the boundaries of our comprehension, into a highly advanced technological state of development of the Gods, and then place that advanced technology back in time, to our distant past, next to the primitive technology of early mankind. The contrast is very great, almost unimaginable. This is what has happened in the past on Earth, as witnessed by the clay tablet accounts, and shall undoubtedly reoccur in the future.

And finally, the challenge and measure of each of us is that we make a contribution, and leave this place called Earth, and its inhabitants, mankind, better-off when we depart, than when we arrived. Therefore, I consider the writing of this book as my contribution to mankind, and dedicate it to all mankind, and especially to those individuals who have the dark recesses of their minds illuminated through reading it, and embrace the same perceptions and ideals which I hold.

CONTENTS

PREFACE

Not since the time the Babylonian priest and scholar Berossus, who in circa 290 B.C., nearly 2,300 years ago, wrote in the Hellenic language, prevalant during that period, a three book rendition called Babyloniaka, which is now lost, and presented the Babylonian history to the Hellenistic culture, has a composite version of the creation and early history of mankind through the great flood been rendered, to my knowledge, except for excerpts and quotes.

This is perhaps the oldest story every told by mankind, and very possibly could be the first religion practiced on Earth: the source, the wellspring and forerunner of all our present day religions, and because of the numerous clay tablet sources from which this account was drawn, all of which contain a common fabric, perhaps this account is as close as we can come to the original story, handed down to us by our distant ancestors. The clay tablet story is an older version of the Biblical Genesis, and wherever the clay tablet narrative has been unavailable or missing, I have relied upon the parallel Biblical account, primarily Chapter 1 through Chapter 9 of Genesis. However, we may never have a full and more accurate story from any of our sources today, unless a complete set of clay tablets is found, or a complete volume of Babyloniaka is located. We may have to wait for time to pass, until the original storytellers, our creators, the Gods, space travelers from an advanced civilization, return and reiterate the story to us in its quintessential form.

Upon my desk, I have a constant reminder taken from a clay tablet, the Epilogue[1] of the clay tablet derived book Enuma Elish, which refers to the fifty epithets, probably characterizations of manifestations or effects upon the Earth during various stages/periods of its orbit, the planet Nibiru,

---

[1] J. Page 111; (see bibliography letter identification on page 383 for footnote reference books. Wherever possible, missing words on the clay tablets are rendered through perceptive interpretive research, and are underscored with dots in the footnotes by this author; and words enclosed in brackets without underscoring dots are restorations of the original text by their authors; and helpful words, not in the original text, are enclosed in parentheses).

here called Marduk, is known by, and it exhorts mankind to continue to remember the names. It states:

> Let them be held in remembrance, and let the first man proclaim them;
> Let the wise and the understanding consider them together!
> Let the father repeat them and teach them to his son;
> Let them be in the ears of the pastor and the shepherd!

Finally, the name of the hero of the flood story has been translated into several different names by the succeeding generations, among which are, moving forward in time: Sumerian — Ziusudra, Akkadian — Utnapishtim, Babylonian — Atra-hasis, Hebrew — Noah, and Hellenic — Deucalion, Xisuthros, and Sisithros; but make no mistake about it, they are all one and the same.

INTRODUCTION

Briefly, the synopsis of the story is as follows: very old clay tablets inscribed with cuneiform characters have recently been deciphered, and reveal a story of the genealogy of all mankind, through aliens, technologically mature and genetically complete beings, supreme beings, or as our ancient ancestors called them, Divine Beings, Lofty Ones, and later deities, Gods, whom today we call extraterrestrials, astronauts, who came to the Earth thousands of years ago. In search of an ideal replacement worker for their rebellious subordinate work force, and through laboratory experiments, they create various mutates, using the existing quadruped animals. These animal mutates are not successful, and the Gods resort to creating prototypes of early mankind, human beings, by mixing the genes from an arbitrarily evolving ape-like Earth creature, Homo erectus, and one of their own subordinate workers. This eventually results in the perfection of Homo sapiens, pre-Neanderthal-like mankind, early mankind, except that we were purposely made sterile, like their subordinate workers. Thus, we become the ideal programmed worker. Later, through the benevolent intervention of their chief scientist, who gives mankind the ability to reproduce, the gift of knowledge, mankind prospers and proliferates, and then finally, at a much later date, the transition is made to Cro-Magnon mankind, recent mankind, basically, mankind as we know ourselves today. Then, mankind prospers, and becomes independent and assertive, and the sons of the Gods begin to mate with the daughters of mankind, causing the genes of the interbred offspring of the Gods to degenerate, and a diminution of their pure bred perfected alien race. This evil conduct, among other offenses by mankind, angers the administrating God, who calls for a meeting, at which the Gods decide to subdue this threat from mankind, through disciplinary measures of spreading pestilence and famine upon all mankind. This sequence culminates in the story of the original "Noah", Ziusudra, and the chief scientist God who surreptitiously engineers the escape of Ziusudra and his family, friends, and some animals and birds from the great flood; after which the Gods agree that mankind was worth saving, and shall be allowed to conditionally resume propagating upon the Earth. Then, Ziusudra and his wife are given the gift of immortality for saving the seed of mankind upon the Earth, a genetic trait the Gods have perfected for themselves, and they are both taken to the abode of the Gods to dwell.[2]

This clay tablet story has many similarities to the Genesis story in the Bible, except it is much, much older, and it identifies the Gods as astronauts; but it should not be interpreted as a traditional Biblical story, for it is not. Also, it is not a science fiction story. It is a substantive and very topical story in which the controversial issue between evolutionist and creationist is firmly addressed, a story of the creation and early history of mankind on Earth, whose express purpose was to work for and to serve the astronauts, whom early mankind called Lofty Ones, and eventually Gods. A story of astronauts/supreme beings coming to Earth, and their exploits, all of which were pressed onto soft clay tablets by their creation, mankind/human beings, thousands of years ago. Additionally, the Gods could be considered our genetic parents, and our actual earthling mothers and fathers our biological parents; and this realization we should all acknowledge and distinguish.

In writing this story through perceptive interpretive research, which I explain within 'Influence' in my Epilogue, I have carefully followed the deciphered clay tablets, which were initially found and excavated in the mid-1800's and thereafter to this day, and figuratively, I have combined the many scattered fragments and reconstructed some parts of the story and words, from the framework and the direction of its sentence outline, into a composite and comprehensive work for the first time in over 2,300 years, and throughout, my script closely adheres to the original version, which has been passed down to us from our most distant ancestors. Additionally, upon the clay tablet account of the creation and early history of mankind, which was written through the perspective of the primitive technology of early mankind, I have superimposed the space-age technology of the astronauts. I have even tried to capture the laconic style of writing employed by the ancients, expressing much with few words, including repeating words and blending segments of sentences, as we shall see, who wrote as though they

---

[2]M. Page 1 through page 25, introduction; and U. frontispiece through page 14; and E. page 1 through upper page 3.

were visually experiencing the story, as if it appeared to them in a dream, as opposed to exposure through oral or written mediums, including abruptly shifting the topic from sentence to sentence, which I compare presently to our seeing one picture after another flashed upon a screen; or as we ourselves picture events unfolding in our own dreams, which is professionally referred to as hypnogogic imagery, as I have personally experienced many times very early in the morning, while writing this book, and then awakening with fresh visions and insights, and with a high pitched sound of total concentration within my ears, and then quickly transferring my visions into words, which can best be described as a thinking and writing frenzy, what I call perceptive interpretative research.

Although I did not explicitly endeavor to discover the true shape of the boat of the great flood, comparatively recently referred to as "Noah's" ark, or the boat of Ziusudra, serendipitously as a byproduct, I claim and believe I have discovered the shape of the boat, which is not a rectangle, and can prove it by at least three different methods: Babylonian, Hebrew, and Hellenic, for all of the seemingly conflicting descriptions of these diverse languages are correct, our interpretations of them have been in error. I do expect some disagreement and criticism regarding the text of the book, due to the interpretations I put forth, but regarding the shape of the boat/ark, after the proof within my Comments, Diagrams, and Illustrations has been seriously considered, there is not much room for criticism.

The discovery of the shape of the boat/ark is of great historical significance, a major historical discovery, and this announcement is the pre-eminent highlight and claim of my work, and it contradicts and will shatter the accepted traditional religious beliefs pertaining to the shape of the boat, and the technology of the astronaut/God who gave the design of this unique boat/ark to Ziusudra/Noah, and saved mankind from the great flood. The significance of the discovery: the boat/ark is from an advanced technology not indigenous to Earth, their advanced technology, and not the primitive technology of early mankind; therefore, the supreme being who gave the design to Ziusudra/Noah was what today we call an astronaut. This may be one of the most important revelations in our lifetime.

Furthermore, this is the first time the shape of an object from an extraterrestrial advanced technology has been identified, and is proof and thereby gives credence to reports that supreme beings have visited the

Earth, which is confirmation of their existence, and that the ancient Gods were astronauts. And if the flood story is correct, and a cataclysmic flood did occur, as I believe, then other so-called mythological stories are also correct, and did occur. The Voyager II Spacecraft in August 1989 recently confirmed the hypothesis I offered in my copyrighted 1982 advance copy of this book, as I explain within my Epilogue, regarding rings on all of the outer planets: especially including Neptune, discovered in 1846, upon which no other person, to my knowledge, had ever put forth the published thought of an encircling ring; and if I may, I now refer to the rings of Neptune as the NIKO rings. This is no longer a hypothesis, it is now an accepted fact. Eventually, the design of the boat will also be an accepted fact. And regarding the discovery of the shape of the boat/ark, as among future discoveries, I believe when the concealed or hidden truth becomes apparent and visible, it will be very difficult for most people to understand and accept, for it will be totally different from what they had perceived and expected; for the truth is not always what we had envisioned it to be.

For centuries, scholars have been trying to identify the shape of the boat, and I believe I have put my finger on it and finally found it; and for those who will still be in doubt, I believe they will eventually affirm and herald my discovery, once their eyes are opened after seeing the proof and evidence I give for the unique shape of the boat/ark, for it should not only be known as the boat/ark of Ziusudra/Noah, it should also and foremostly be known as the boat/ark of the astronaut/God who gave the design to Ziusudra/Noah.

Although the ancient texts from which this book was drawn are both religious and historical in substance, the book was written through and should be viewed through the predominant context of the historical perspective: the creation and early history of mankind. However, if you wish to view the book through a stronger context of the religious perspective, you may do so with ease, a circumstance I also find myself in many times, for as you read, it is frequently difficult to distinguish between the two views.

After reading this book, you may come to the same conclusion that I arrived at after writing it: the religious doctrines contained in the clay tablet story presented serves a perpetuating purpose, while my contributions thereto are a sincere commitment to a revolutionary religious concept;

to actually revive, re-establish, and didactically espouse the ancient religious belief of polytheism, the belief in many Gods, which the clay tablets identify as GODS OF THE UNIVERSE,™ for the express purpose of preparing mankind/human beings for the reappearance of the astronauts/supreme beings, whom our ancient ancestors called Divine Beings/Lofty Ones/Gods. This can be achieved within the present religious institutions, by expanding the parameters of the monotheistic belief which prevails today to include the polytheistic believe which prevailed throughout ancient times; because monotheism and polytheism are the same, they are both theisms, a belief in a supreme being, the only aspect which distinguishes them is the number of supreme beings acknowledged. We believe in more than one Angel, why not believe in more than one supreme being; which is actually the same belief that we hold today, a belief in a supreme being, except a belief in more of them. And if Angels are also considered Gods/aliens, from the same source, which they certainly are, but a specific category of God/alien, the astronaut category, as they are called within the clay tablets, for they are also designated as Gods, which is the correct way to view them, then everyone who believes in Angels, whether they realize it or not, already believes in many Gods/aliens, polytheism; for our ancient ancestors had first-hand knowledge, a real life experience by witnessing them and the events surrounding them, and thus a better understanding of who they were, and thereby accurately identifying them as astronaut Gods/aliens.

Furthermore, the religion of our ancient ancestors is also the "space-age" religion of today, for we now have also seen astronauts and hold them in high esteem, our own astronauts; and imagine if extraterrestrial astronauts were to suddenly appear, I am certain our high esteem would immediately transcend into a reverence, for the two experiences are identical: old is new, new is old. We must always be cognizant that we live in a cyclical world and universe.

Finally, it is time for mankind to awaken, and become aware and enlightened, and not doubt our ancient ancestors eyewitness accounts, we should try to understand them, and begin to correctly interpret the distorted so-called mythological ancient writings, through a new perspective, in a new and different light, the light of space travel and genetic engineering, now that we are on the threshold of exploring these enigmas ourselves.[3]

[3]X. Page 302 through upper page 307.

PROLOGUE

The following is a composite and interpretive story of the creation and early history of mankind on Earth, whose express purpose was to work for and to serve their creators. The story was primarily derived from fragments of scattered clay tablets, now in museums, most of which were found in the ancient land generally called Babylonia, and are essentially of Sumerian origin. The Sumerian literary compositions, from which the basis of this comprehensive work was taken, were written in cuneiform characters, on soft clay tablets, some dating back to approximately 4,000 B.C., and possibly beyond. Their origin and oral traditions, however, as we shall see, were much, much older, reaching back to a pre-Sumerian period, and into the past, to a legendary time; acknowledging that perhaps portions of the story may be only remnants of very remote accounts of concurrent events from diverse lands, all of which I believe are linked, and could include or be accounts of Atlantis, and of the sinking of Atlantis in approximately 8,800 B.C.. Due to missing pieces, and the fragmentary condition of the tablets, some parts have been added for continuity, but the main story closely adheres to the original version, which has been passed down to us from our most distant ancestors.

*

We observe the outside of a twentieth century museum.

PRESENT

Inside the museum, we see the curators facilities, a large room with large rectangular wood tables, all filled with various artifacts in different stages of assortment. We scan over the museum sorting tables, and after a brief moment, stop on one particular table. Lying upon the table, there are many light brown colored clay tablet fragments, some of which are joined, but most of which are not joined. They appear as pieces of a mosaic. Also lying upon the table, we view a map of the Persian Gulf area, and hear the sounds of a flying doves wings flapping in the air, and hear the dove singing its song. The map has names of current countries and bodies of water. Iraq stands out among the names, and we begin to zoom into south central Iraq, and superimposed on the map, we see the regressing years of a calendar fade in.

*

As we continue to zoom in, but before the superimposed regressing years of the calendar reach 1898 A.D., the years fade away, and the map blends into the rays of the Sun shooting over the horizon, as the Sun rises over the flat desert land on an early morning in mid July. We look at the Sun for a moment, as it becomes brighter and rises into the clear blue sky to the 10 a.m. position. Then we look down from the Sun, and follow a dove as it enters our field of vision. We follow the flight of the dove, and it leads us to the muddy looking Euphrates River,[4] in which there are round, pitch coated reed gufa boats[5] near the banks of the river, and Arabs poling in them wearing white turban-like hats. We notice the curious shape of the boats, as we zoom in for a close up view. Then we scan to the river bank, where an Arab is applying pitch

---

[4]B. Page 24.
[5]B. Page 25; and L. lower page 104.

to the outside of his gufa boat, which he has pulled up onto the bank. Again, the dove crosses our vision, and we follow it as it flies into the distance. From the doves view in the air, we observe mounds of ancient ruins[6] in the distance. From the ancient ruins, the dove flies over us and alights upon the ruins. It is now very quiet.

RUINS OF ANCIENT NIPPUR 1898 A.D.

We scan over the ruins,[7] and begin to hear the muffled sounds of excavating amidst conversations. As we move over the top of one of the mounds, the excavating and conversation sounds grow and become clear, and suddenly we are looking down at an archaeological dig site,[8] with approximately 60 university students and 30 local Arabs scattered throughout the mounds, away from us, within the bright sunlight. The students are excavating the ruins, scraping and brushing, while digging into the mounds.[9] There are many reed baskets around the site, some filled with soil, while others are filled with unearthed clay tablet fragments.

Our eyes stop on a group of 12 students and 4 professors, who with the aid of the local Arabs are carefully excavating the ruins. They are digging with their trowels, and using their brushes to clear the loosened soil. The Arabs are assisting by carrying out reed baskets of excavated soil, which is a combination of sand, crumbled plaster, and crumbled unbaked, sun-dried clay building bricks. They carry the baskets on their back, one at a time, using a rope support, which they grasp at their chest.[10]

We notice a student uncovering and brushing soil off of a clay tablet, which only has one corner missing. Many tablets are partially visible in the soil. The full size of the average tablet is 3/4 by 3 1/2 by 5 3/4 inches long. The student excitedly calls for the attention of a professor, who is about 100 feet away, "Professor, professor, look at how many clay tablets we've found today." Then

---

[6]B. Page 28, and page 122; and L. opposite page 64.
[7]G. Frontispiece, and page 5.
[8]K. Page 8, plate 1.
[9]B. Page 19.
[10]Q. Illustration, Workmen in a Quarry; and U. frontispiece, lower register.

the student points to ten baskets full of clay tablet fragments, and continues, "You were right, this must be the ancient temple library[11] we've dug into," as the student places the tablet into a basket. Two professors quickly walk to the student's site, and look down at the partially excavated tablets.

The first professor declares, "Yes, this confirms my guess. This was the site of the temple library. This area should yield a very large number of important clay tablets with significant inscriptions, as they were most valuable, and were collected and kept in the library," and we see a close up view of the ten baskets full of the clay tablet fragments. Then another student gives each professor a fragment of a tablet that has just been unearthed. Both are about one-quarter to one-half broken off. The second professor adds, "After the previous expeditions by our university, it's good that this archeological expedition can participate in this most important find. These past nine years digging at Nippur have been very rewarding."

The first professor asserts to the second professor, "These cuneiform clay tablets lay buried, undisturbed, describing events that occurred thousands of years ago. The ancient writings have come down to us directly, and have not been altered since they were inscribed on the moist, soft clay tablets." The second professor adds, "The inscriptions on them are truly remarkable, listen to this." He holds up his fragment and looks at it closely, and we see a close up view of the cuneiform clay tablet fragment, as the professor begins to slowly decipher and read it. He pauses on some of the words, for they are difficult to decipher upon the partially soil filled fragment, as he continues, "After Heaven had been moved away from Earth, after Earth had been separated from Heaven."[12]

The first professor responds, "It sounds like many of the other clay tablets, like the Biblical creation in Genesis. Except these tablets are much, much older than the creation version in the Bible. Some of the tablets even describe events that compare to the Biblical day of the Lord, such as a trembling of the Earth. I believe these tablets could be some of the ancient writings from which the Bible was taken." The second professor reflects, "Since the original discovery of similar tablets in eighteen forty-two, fifty-six years ago, there

---

[11]G. Page 1 through page 29.
[12]K. Lower page 37.

has been much controversy about the Biblical parallels. But I agree with you, I think your assumption is correct. The Biblical day of the Lord, and the clay tablet trembling of the Earth, are describing the same event; and the Anunnaki, another clay tablet name for the Gods, sound like a description of the Angels in the Bible." The first professor suggests, "Gods and Angels, there is a strong similarity between the clay tablets and the Bible, where each is found." Again, we see a close up view of the same tablet, as the second professor again begins to slowly decipher and read it, "Let me read you more." He pauses, and then continues, "After the name of man had been fixed[13]. . . ," and simultaneously the professor's voice and the clay tablet fade away, and we do not hear the balance of the deciphered clay tablet, and now we view a black void.

*

As we view the black void, it begins to lighten and appear as a black sky, and then the black sky begins to fill with bright twinkling stars as we look into deep space, and among the twinkling stars, approaching toward us quickly, is a brighter object, which appears as a cross in space.

THE PLANET NIBIRU[14]
440,800 B.C.

We see our solar system, with the planets orbiting counter clockwise, as they appear from a northern point in outer space, and off in the distance, approaching counter clockwise, and orbiting toward them from a southern

---

[13]K. Lower page 37.

[14]J. Page 109, line 109 "Let his name be Nibiru, 'the Seizer of the Midst'!"; and T. page 215 Planet NIBIRU:, page 218 The Heaven (our solar system) he divides in half (orbits through the center, near the Asteroid Belt) and stands (is visible in its orbit) as Nibiru, and (the land will overflow with plenty) When Nibiru culminates . . . . (reaches its perigee), and page 222 When standing in (visibly orbiting through) the place of the celestial battle (near the Asteroid Belt): Nibiru (shall be its name)., and upper page 361 It is the god (planet) Nibiru ["Planet of Crossing"]; and X. middle page 99, including note 12 Pallas of the Greeks (Hellenes) was another name for Typhon (Nibiru); also Seth of the Egyptians was an equivalent of Typhon (Nibiru); it means 'overmastering' and 'overpowering', and in very many instances 'turning back', and again 'overpassing' (Nibiru may be what some today loosely refer to as "Nemesis" or the "Death Star").

plane, a different plane of orbit, similar to the 17 degree angle of Pluto, except extremely wider and elliptical in orbit, to eventually orbit away nearest the northern axis of the planets, we observe the bright radiating planet; Nibiru. It is approximately 1-1/4 times the size of Jupiter,[15] or approximately 14 times the diameter of the Earth at the equator. It has a bright, sunset red color,[16] which is radiating[17] and glowing from the planet. It has four[18] large, Mercury size moons orbiting around it, with two of the moons orbiting at 90 degrees to the other two, within two concentric rings of clouds composed of water molecules, a vertical ring within a slightly larger horizontal ring, which give a cross-like appearance to the planet. The two rings appear to be bisecting each other, but they are not. We zoom toward the planet, then pause, which gives us a close up view, and then we continue to zoom to the planet.

---

[15]T. Lower page 199 He was the loftiest of the gods (widest orbit), surpassing was his height (largest planet); His members were enormous (large moons), he was exceedingly tall (largest system of planet and moons in size), and upper page 360 Who by the planet Jupiter cannot be distracted (for it is larger than Jupiter, which has been reported to contain 90% of the mass of the known planetary system of the Sun), and upper page 363 Supreme, Supreme, Anointed; Lord whose shining crown (halo of satellites) with terror is laden. Supreme planet: (largest planet); and J. page 59, line 5 "O Marduk (Nibiru), thou art chiefest (largest) among the great gods (the planets)," and page 61, line 28 They rejoiced, and they did homage (unto him, saying), "Marduk (Nibiru) is king (largest planet)!", and page 93, line 5 O Asaru-alim-nuna (an epithet for Marduk/Nibiru), "the mighty one (planet with strongest forces)," and page 95, line 14 None among the gods (the other planets) can [rival him] (are as large as Marduk/Nibiru in size)!, and page 105 "[Who] among the gods (all the planets) is gloriously supreme (largest)!", and page 207, line 12 O mighty lord (Marduk/Nibiru), to whom there is no rival (no other planet possess your traits) in the assembly (in the solar system) of the great gods (of the planets)!; and X. middle page 27 Between the Earth and Jupiter there revolves a small planet, Mars, a tenth part of the Earth in mass, where, according to the scheme, a planet (Nibiru) ten to fifty times as large as the Earth should be expected.

[16]T. Page 218 The great planet: At his appearance, dark red, and upper page 363 Daily within the Lion he is afire (appeared daily within this period as a fireball); and X. page 109, note 13 The Babylonian Talmud, Tractate Shabbat 146a. According to Midrash Shir (15a-15b) the pharaoh warned the Israelites not to leave Egypt, because they would meet the bloody star Ra (in Hebrew "Evil"). (I believe this took place at a much later time; as I further explain under 'Most Recent Orbit' on page 317 in my epilogue), and page 116 (according to ancient Chinese annals) "a brilliant star issued (came forth) from the constellation Yin."

[17]X. Upper page 175, including note 17 "Sometimes there are (wavy) hairs attached to the planets," (etc.), but hair or coma (particles "blowing in the solar wind") is a characteristic of comets, and in fact "comet" is derived from the Greek (Hellenic) word for "hair" (the comet-like radiating attributes of Nibiru, including its comet-like orbit, were symbolically expressed by our ancient ancestors as wavy lines/hairs; as I illustrate on page 308 and page 309 in my comments).

[18]T. Upper page 361 It is the Hero, the planet with four heads (four moons/satellites).

We scan the interior of a large, dome shaped, futuristic assembly hall on the planet Nibiru. The walls are a silvery-white metallic color, and are self luminous. There is a circular seating area, with white pew-like molded circular seats, with divided segments to permit access. There is a special, slightly raised section with four seats, which is a part of the circular seating area. In the center of the assembly hall, there is a low circular ring-like base, with a dome viewing monitor upon it, which is presently dark. The dome viewing monitor is a three dimensional hologram-like television, 6 feet in diameter, and 3 feet high.

We notice a group of Gods,[19] 34 Gods and 16 Goddesses, entering the hall and sitting, the highest and most supreme Gods, also called great Gods, who as astronauts, are also called Anunnaki Gods, among which are seven who are the highest in authority, who are also called the seven great Anunnaki Gods. There are a total of 50 Gods,[20] a full assembly, gathered in the hall, waiting for the assembly to begin.

All the highest Gods have very fair white pigment skin, with tight-curly white hair, like sheep's wool;[21] medium length on the Gods, and shoulder length on the Goddesses. Some Gods have tight-curly white natural shaped beards of medium length, and some Gods have green eyes, while others have blue eyes. The highest and most supreme Gods are dressed in full length red robes, from elbow to ankle, with a 5 inch wide cummerbund-like gold waist

---

[19]T. Prologue through Chapter 6 (dingir/DIN.GIR/Divine Beings and i-lu or ilu/I.LU/Lofty Ones (shepherders) = supreme beings = Gods; and because of our somewhat enlightened understanding today, we could also refer to these beings as extraterrestrials or aliens, therefore Gods = aliens).

[20]L. Upper page 115.

[21]V. Revelation 1:14 His head (skin) and his hairs were white like wool, as white as snow, and Daniel 7:9 and the hair of his head like the pure wool (similar to these. In Revelation 1:14, John clearly identifies the tight-curly form of hair, as well as the skin and hair pigmentation/color of the supreme being as being white, and since Daniel also was a very keen observer and precise reporter, and also clearly identifies the tight-curly form of hair and its white color, I believe he would have mentioned the skin color of the supreme being if it was different from his own Caucasian skin color; and because he did not, we must presume that the skin color of the supreme being was the same or similar to his own, and not noticeably different. Also, because the two reports are similar, even though the description of Daniel is not as complete as that of John, I believe they are both descriptions of characteristics of the same family/genus of supreme being. Therefore, it must be presumed that the omitted portions of Daniel must also be similar to the more complete description of John).

band, and gold wedge-like sandals[22] with straps. Their robes have 3 inches of gold trim at the bottom, and 1 inch of gold trim at the neck and elbow. There are four main Gods among the highest Gods, the four highest Gods, who in addition to the above attire, have a 4 inch diameter identifying gold symbol pendant,[23] hanging around their neck on a gold chain necklace,[23] and a gold watch-like bracelet instrument on their wrist,[24] and wear a gold winged sphere symbol[25] of Nibiru on their uniform, above the left breast, as do all the Anunnaki Gods. All the highest Gods appear to be approximately 35 to 45 years of age.

There is an air flow door, a curtain of moving opaque air, within the doorway. We hear the soft sounds of quiet conversations among the assembly members. Our eyes stop on the four highest Gods, sitting next to each other at the slightly raised section with four seats, who are also four of the 7 great Anunnaki Gods. From in front of, from left to right, are sitting; God Enlil, God Anu, God Enki, and Goddess Ninhursag.[26]

The left middle God, God Anu, rises, and quickly there is silence, and all the Gods bow their heads to him. Then, in a warm tone, God Anu addresses the assembly, "In this assembly of Gods, I, God Anu, president of the assembly,[27] your leader, have brought you to my presence," and we see the previously dark viewing monitor light up, and during the remaining address by God Anu, we alternate between the assembly, the viewing monitor, and

---

[22]T. Page 137, fig. 72; and E. fig. 7 right.

[23]T. Page 137, fig. 72, and page 235, fig. 117; and E. fig. 4 (see my illustration on page 308 in my comments).

[24]Q. Illustration, Winged Figure.

[25]Q. Illustration, King Assurbanipal; and E. fig. 2; and T. page 219, fig. 112 (see my illustrations on page 309 and page 311 in my comments).

[26]M. Page 143, line 143 An (Anu), Enlil, Enki (and) Ninhursag (whose names mean: An/Anu = Heaven, Enlil = Lord (of the) Command, Enki = Lord (of) Earth, Ninhursag = Lady (of the) Mountainpeak. More common names are: An/Anu = Ouranos in Hellenic, Uranus in Roman; Enlil = Zeus in Hellenic, Jupiter in Roman; Enki = Kronos/Poseidon in Hellenic, Neptune in Roman; Ninhursag = Athena in Hellenic, Minerva in Roman); and T. page 88 through page 102; and U. page 15 through page 16; and L. page 118 through page 122; and I. page 95 through page 143, Chapter 4.

[27]M. Page 49, line 101 Anu, king (administrator) of heaven (space), was present, and page 97, line 51 Where has Anu the president gone, and page 99, line 39 Where has Anu the president gone (the highest in authority: see my illustrations under 'Creation Of Planets, Satellites, And Rings' on page 326 in my epilogue, for better understanding as you read the following address by God Anu).

God Anu, for the viewing monitor shows the topic of the spoken words. Then God Anu continues, "As you know, in the beginning, when this solar system was forming; when in the height, Heaven had not been named, and beneath, firm ground had not been called, there only existed the primeval Sun. From out of the Sun, first formed was the small planet Mercury, and the second formed was Tiamat, the large water planet. The waters of the Sun and Tiamat were mingled together, and in their midst, Mars and Venus were brought forth. Before they had grown in age and size, Saturn and Jupiter were formed, surpassing Mars and Venus in size. Then later, as lengthened the days and multiplied the years, from Saturn and Jupiter was formed Uranus, a rival of his ancestors, the older planets; and second formed from Saturn was Pluto, which became a satellite of Saturn. Then Uranus begot Neptune, as its equal and in its image."

"The planets then began to develop erratic orbits, and they surged back and forth, threatening and disturbing the Sun, but primarily the stability of Tiamat. The Sun could not lessen their troublesome ways, and as the effects of the Sun prevailed, it appeared that the Sun would destroy all the planets. But the forces of Neptune neutralized and calmed the Sun, as though pouring sleep upon the Sun, and Mercury was taken captive and also neutralized. Then, as a result of the Sun and Mercury being neutralized by Neptune, Tiamat roared and glistened, and from its own midst, due to its increasingly rapid rotation, caused eleven satellites and accompanying smaller debris to be formed and orbit around Tiamat, all fully electrically charged and fierce as monsters,[28a] the first of which is the Moon, which is the largest and has the widest orbit. Then, with this furious, fuming and raging host of eleven satellites, as they banded together and advanced, Tiamat threatened and disturbed the other planets."

"Then, after our planet, Nibiru, was created in the heart of deep space, our newborn enormous planet, the avenger,[28b] proceeding in its own orbit,

---

[28a] J. Page 53, line 94 "After this fashion, huge of stature, hath she (Tiamat/TI.AMAT, which means "Maiden of Life") made eleven (monsters)"; and X. lower page 27, and upper page 28 (the Moon).

[28b] J. Page 13, line 86, and page 33, line (105), and page 37, line (134) "If I, your avenger," and page 39, line 10, and page 47, line 58, and page 55, line 116, and page 57, line 138, and page 59, line 13 "O Marduk (Nibiru), thou art our avenger!", and page 91, line 143.

was gravitationally attracted into the solar system by the outermost planet, Neptune, causing a ring and eight major satellites to be formed from Neptune, with all eight satellites staying in orbit around Neptune, as Nibiru is guided into a new orbit by Neptune, toward Tiamat, in the same orbital direction, but beneath the other outer planets, each of which shall impart electromagnetic force to Nibiru as it passes them. Then, as Nibiru orbits past Uranus, which perfects the orbit of Nibiru toward Tiamat, a ring and nineteen major satellites are formed from Uranus, and four of the satellites, called Winds, and accompanying smaller debris are gravitationally attracted by, and orbit around Nibiru. Then, as Nibiru passes Saturn and Jupiter, rings are formed on each of them, and thirty-nine major satellites are collectively formed from them, and three of the satellites, also called Winds, and accompanying smaller debris are gravitationally attracted by, and orbit around Nibiru, as Saturn and Jupiter further perfect the orbit of Nibiru toward Tiamat; and the satellite Pluto is loosened from orbit around Saturn."

"As Nibiru, now fully charged with electricity and with its seven satellites, the Winds,[29] approached Tiamat from beneath and behind, and drew progressively closer towards converging and juxtaposing on the initial pass, each began to release their forces upon the other, and the Moon developed an erratic orbit, and then the forces of the Moon were partially neutralized by Nibiru, and then the other ten satellites of Tiamat became erratic, and then the rotation of Tiamat was abruptly slowed, due to the prevailing forces of Nibiru, and Tiamat began to fissure. Then one of the satellites of Nibiru, one of the Winds, was driven into the fissuring Tiamat, and then two additional satellites, Winds, were driven into the fissure, increasing and opening the fissure wider. Then Nibiru released an enormous electrical bolt[30] into the fissure, and Tiamat was neutralized, its

---

[29]J. Page 63, line 42, 43, 45, and line 46, and page 65, line 47 He (Marduk/Nibiru) sent forth the winds which he had created, the seven of them.

[30]X. Upper page 100 The head of the comet (Nibiru) did not crash into the earth (Tiamat), but exchanged major electrical discharges with it. A tremendous spark sprang forth (from Nibiru into the fissured depths of Tiamat) at the moment of the nearest approach of the comet (Nibiru, after 3 of its 7 satellites/Winds were driven into the fissuring Tiamat. This sequence is correct), and middle page 272 if for some reason (a close orbit of Nibiru) the charge of the ionosphere, the electrified layer of the upper atmosphere, should be sufficiently increased, a discharge would occur between the upper atmosphere and the ground; and E. fig. 7 (which I believe depicts Nibiru approaching monstrous Tiamat from behind, confirming a similar/counter clockwise orbit, preparing to release its forces upon her).

forces extinguished. Then ten of the eleven monstrous satellites of Tiamat, all except the Moon, were gathered, shattered into larger and smaller debris, and scattered into deep space in diverse elongated orbits by Nibiru, with the larger debris becoming asteroids and comets, and the smaller debris becoming meteoroids, and then the Moon was fully neutralized and stopped rotating, and Nibiru returned to deep space."

"Then, after completing the first full orbit, our planet, Nibiru, returned to the fissured Tiamat and its fully neutralized satellite, the Moon, and the forces of Nibiru crushed the lower half and totally cracked the upper half, and then severed the molten mass of the remaining core and split Tiamat into two halves; and the forces of one of the remaining four satellites of Nibiru, the North Wind,[31] drove the upper half of Tiamat, along with the Moon, into its present orbit. We now call this upper half of Tiamat, the planet Earth. Thus, the Earth was created."

"Then, on the second orbit, Nibiru drove the crushed remains of the lower half of Tiamat into a new confined orbit, creating a great band of asteroids, as a bracelet, forming the Asteroid Belt."[32]

"Then, on the third orbit, the forces of Nibiru drove the satellite Pluto from its loosened orbit around Saturn, to its present and outermost orbit, facing deep space, thus Pluto became a planet. And then, due to the increased rate of our own programmed evolution, advanced life developed quickly on Nibiru, sooner than it did on the other planets. That is how this solar system was created,[33]

---

[31] J. Page 75, line 132 And he (Marduk/Nibiru) made the North wind bear it away into secret places (as I further explain under 'Quantum Mechanics' on page 324 in my epilogue).

[32] X. Middle page 56 The large number of asteroids (which are also orbiting counter clockwise) between the orbits of the planets Mars and Jupiter suggests that at some unknown time another planet (Tiamat) revolved there; now only these meteorites (confined orbit asteroids) follow approximately the path along which the (1/2) destroyed planet (Tiamat) circled the sun (and probably some of the diverse orbit asteroids, comets, and meteoroids from the ten shattered satellites of Tiamat somewhat follow the path of the planet, Nibiru, which caused the destruction). Possibly a comet (or a planet, Nibiru) ran (released its forces) into it and shattered (1/2) it, and page 285, note 1 Between Mars and Jupiter are over a thousand (50,000) asteroids that have been thought to have once been (1/2) a planet.

[33] T. Chapter 7 (in which Nibiru) is referred to as Marduk and the 12th planet, and especially page 212, fig. 109, which depicts Marduk/Nibiru approaching Tiamat from behind, which supports my belief that their orbital direction was the same); and J. preface through page 115 (in which Nibiru is referred to as Marduk. However in J, the story is allegorically rendered); and E. page 3 through page 60, and lower page 75 through page 81 (it is very possible that the Biblical account of the creation of Heaven and Earth in 7 days, and the Sumerian/Babylonian account of the 7 satellites/Winds of Marduk/Nibiru triumphing over the 11 satellites/monsters of Tiamat, are a report of the same event, with the newer Biblical account being a transliterated version, thus connecting the two).

a long, long time ago."

"Now, as our heavenly planet, Nibiru, which crosses through the solar system of planets, and whose symbol is the cross,"[34] God Anu pauses, and the viewing monitor shows the symbol of Nibiru, the cross, as it appears on God Anu, and he continues, "travels through the deep space of the universe, in our normal orbit around the star we call the Sun, again we approach the main body of planets, and the life sustaining water planet, Earth, which is the remaining upper half of the original water planet, Tiamat, which our planet, Nibiru, split long ago." God Anu points to the viewing monitor, and it instantly zooms toward and shows the Earth, as seen from space, and then it resumes showing the topic of the spoken words.

At this point, Nibiru is still outside the orbital path of Pluto, and on its normal southern to northern orbital trajectory, which will eventually, at perihelion, take it between Mars and Jupiter, and more specifically, between Mars and the Asteroid Belt,[35] then God Anu continues, "One of our years on Nibiru, called a Shar year, is equal to three thousand six hundred Earth years, for it takes our planet Nibiru, three thousand six hundred Earth years to complete each orbit around the Sun.[36] The Earth is not yet within our force fields, but normally, continuing in our orbit, as our planet Nibiru, the largest planet in this solar system, orbits near the Earth, which is a small planet, Nibiru shall begin to exert its gravitational and electromagnetic forces upon the Earth, and because of this imbalance, the fluid parts of the Earth,[37] the molten core, the waters, and the air, are slightly disturbed and drawn toward Nibiru. This normally causes the Earth to tremble, stimulating moderate tremors with fissures, and volcanic eruptions which emit lava, and thick clouds of smoke, ash, and steam, and temperatures of both the surface of the Earth and air begin to increase. Also, all the waters, both above and below the surface of the Earth, are displaced upon the Earth, causing mild

---

[34]T. Page 218, and page 230, fig. 115, and upper page 361 It is the god Nibiru ["Planet of Crossing"]; and J. page 109, line 109 (interpreted); and O. page 65, ILL. 40, i and l, and upper page 78 illustrations, and lower page 97, and page 109, ILL. 95, a right; and N. page 111, ILL. 74 copper disk (perhaps), and upper page 114 'sun-disks'.

[35]T. Upper page 360 Within Tiamat's battle place reposes, and upper page 363 Supreme planet: a seat he has set up, Facing the confined orbit of the red planet [Mars].

[36]T. Chapter 8; and P. lower page 141 The Sumerian sign šár (Shar) has, among others, the meaning "universe" or the like (cosmic year?). As a number word it represents 3600 (etc.), special meaning of 3600 years, "Saros" (which renders/means sweep in Hellenic, orbits/sweeps through Heaven/across Earth) is used by Berossos (circa 290 B.C.); and M. lower page 20 unlike 7 or 3,600.

[37]X. Middle page 385 the fluid parts, the air and the water of the oceans, would certainly have their angular velocity disrupted.

flooding, and the disturbed air forms winds that begin to mildly blow. Simultaneously, because of the increased frictional heat generated by the molten interior core being drawn toward the exterior of the Earth, the surface waters of the Earth begin to excessively evaporate, and our force fields attract this pure water, causing it to mist up and appear as a rainbow, and then condense and fall back to the Earth as profuse moderate rain, which waters the whole face of the Earth.[38] These are the normal effects of our orbit, or passing, but depending on the proximity of the planets of the solar system at the time of passing, the forces of Nibiru could drastically influence and disturb the planets, especially the Earth and nearby planets, being closest to our orbital path."[39]

"Since our last orbit and passing, three thousand six hundred Earth years ago, the ice that has covered a large part of the Earth is receding, due to the warming temperatures. From our surveillance scanners, we can detect that animal life is continuing to slowly arbitrarily evolve and develop from the initial seeding by Nibiru,[40] long ago.[41] We can identify, from our mineral spectrum analysis, that gold, silver, platinum, cobalt, and many other precious metals, are abundantly present upon the Earth. Our mineral supplies are depleting, and are now very low, due to our increasing technological applications. This orbit would be an excellent opportunity to begin our mineral recovery operation."

"My son," God Anu indicates with his left arm, to his left, and we see a close up view of God Enki, as God Anu continues, "God Enki,[42] our chief scientist, shall lead and be in charge of the exploration force, which will consist of 6 assistant scientists," and we look at the viewing monitor, and see 4 Gods

---

[38]V. Genesis 2:6 But there went up a mist (water vapor/steam = fog) from the earth, and (condensed and fell as rain, which) watered the whole face (surface) of the ground (Earth, similar to this. This process, which we simply describe as rain, is obviously a form of distillation).

[39]T. Page 359 through page 362.

[40]T. Page 231 this means that the life that was on Earth so soon after Earth was born was itself a descendant of some previous life form, and not the result of the combination of lifeless chemicals and gases (the "primordial soup" theory), and "life on Earth may have sprung from tiny organisms from a distant planet" (probably not the only seeding, or source; as I further explain under 'Evolution' on page 356 in my epilogue).

[41]V. Genesis 2:5 And every plant of the field before it was in the earth, and every herb of the field before it grew (upon the Earth: the initial seeds of plants and herbs were from outside the Earth, similar to this).

[42]M. Page 43, line 15 [The bolt], the bar (control) of the sea (waters of Earth), and line 16 [They (the assembly) had given] to Enki, the prince (leader).

and 2 Goddesses, all wearing red robes, as God Anu continues, "along with a group of forty three other members of our astronaut corps, our astronauts, whom we call Anunnaki Gods, our miltiary[43] engineers, including officers and their commander: a total exploration force of fifty[44] Anunnaki Gods, shall descend to the planet Earth, and establish a base. Along with them, they shall take an adequate number of the Igigi[45] Gods, our subservient workers, our regular military, who shall provide all the heavy labor," and we look at the viewing monitor, and see 6 Igigi Gods, 4 males and 2 females, the lower beings, whose age and features are like the highest Gods. They are uniformed in tight blue attire, from elbow to boots; with 1 inch trim at the neck, elbow length gloves, calf length boots, tight briefs, and although they are not considered astronauts, Anunnaki Gods, they are permitted to wear the winged sphere symbol, all of gold-like material,[46] as God Anu continues, "And shall also provide and care for the day to day needs of our Anunnaki Gods. They shall be supervised by our Anunnaki God officers," and again we look at the viewing monitor, and see 6 Anunnaki God officers, 4 males and 2 females. They are uniformed just like the Igigi, except their attire is green, and they also wear the winged sphere symbol of gold-like material.

God Anu continues, "The destinies of all the forces," and we look at the viewing monitor, and see the two groups together, 6 Igigi Gods and 6 Anunnaki God officers, as God Anu continues, "shall be under the command of the seven great Anunnaki Gods, the commanders of our military." God Anu indicates with both arms outstretched, to his right and left, and we see a close

---

[43]C. Page 241, note 25, and page 293, note 20, and page 294, note 21, 22, and note 23, and page 295, note 18 (although military terminology is used to describe the Anunnaki in human terms, I believe these terms are for comparative purposes only, and the Anunnaki organizational structure is similar to the social order of a beehive colony, with the Anunnaki at the top; a kind of quasi-military order, an astronaut corps, peaceful and passive, and without bellicose intent).

[44]T. Upper page 244 The Anunnaki who are fifty, and page 294 An-nun-na-ki (a variant of Anunnaki), readily yields the meaning "fifty who from Heaven (space) went to Earth," (what today we would refer to as astronauts, therefore Anunnaki = astronaut).

[45]L. Upper page 123; and C. page 285 through page 295 "a category of angels", and "a class of angels" (returning to the analogy of a beehive colony, the Igigi, although a different group and somewhat different being, as the worker bee, at the bottom of the social order; utilized by the astronaut corps, but not considered astronauts).

[46]V. Daniel 10:5 and behold a certain man (space man) clothed in linen, whose loins were girded with (what appeared as) fine gold of Ū´–phăz (wore gold-like briefs), and 10:6 His body (suit) also was like the beryl (blue, or bluish green), (etc.), and his arms (gloves) and his feet (boots) like in colour to polished brass (gold-like, similar to these).

up view of the 7 great Anunnaki Gods, with the four highest Gods sitting next to each other at the slightly raised section, and three other highest Gods sitting directly behind them, 5 males and 2 females, as God Anu continues, "At this base, God Enki shall immediately proceed to build the first Earth city. Then, they shall begin to explore and secure the entire planet, and locate specific mining sites. To achieve this initial exploration, it shall take ten of our Shar years, thirty-six thousand Earth years."

"After this is accomplished, my son," God Anu indicates with his right arm, to his right, and we see a close up view of God Enlil, as God Anu continues, "God Enlil, our chief projects manager, shall descend to the planet, and shall be the administrator, the ruler, and shall have supreme command of the entire Earth project. God Enlil shall start the mining operation. With God Enlil, shall descend my daughter," God Anu indicates with his left arm, to his left, and we see a close up view of Goddess Ninhursag, as God Anu continues, "Goddess Ninhursag, our chief doctor. Together, you shall establish support cities to process the minerals, and supply Nibiru with a source of minerals, to assure our continued technological advancement."

"With you, to Earth, you shall take six hundred[47] of our Anunnaki Gods, along with adequate numbers of the Igigi Gods, who shall provide the labor for working the mines and building the support cities. The Igigi Gods, whom we have created especially to labor for the Gods, as subordinates, to act as our workers, and to serve our everyday needs, we have skillfully created.[48] We have given the Igigi Gods immortality, like ourselves, by using the tree of life genetic computer," God Anu indicates with his right arm, to the viewing

---

[47]T. Page 293 And on Earth, Six hundred he (God Anu) made reside. After he (God Anu) all their instructions had ordered, to the Anunnaki (astronauts) of Heaven (space) and of Earth he (God Anu) allotted their assignments, and page 294.

[48]M. Page 43, upper line 7 Anu, their (the Igigi Gods) father (creator: had the Igigi Gods created by decree), was the king; and K. page xvi, and page 72 An (the heaven-god) had caused the Anunnaki (the astronauts) to be born (commissioned: decreed the creation of their space exploration/astronaut program. In J. page XXXIII, An/Anu is also known in Hellenic as ᾿Ανὸν/ ᾿Ανός, therefore, An/Anu = ᾿Ανός/Οὐρανός in Hellenic = Ouranos: could also be, o = the, ur = Sumerian, Anos = An; or, our/ur = Sumerian, Anos = An. Anunnaki/An-nun-na-ki, literally means: 'Those Who From Heaven (An) to Earth (ki) Came' (astronauts), which I believe is the root word for the Hellenic word Ἄγγελος/Angelos (messenger/emissary), and our word Angel, all of which describe the same being. The common element is the first syllable, An, which represents Heaven (space) and Anu (God Anu), and the meaning of all of the words could be interpreted as: emissaries of An/Anu; and named after him because he authorized them. All Anunnaki/astronauts are also Gods/aliens, but not all Gods/aliens are Anunnaki/astronauts).

monitor, and we see a computer on the viewing monitor, with a 6 inch square viewing screen, as God Anu continues, "to alter their genes and suspend their aging process.[49a] But we have not given the Igigi Gods the ability to reproduce, using the tree of knowledge genetic computer. Thus they are, and shall remain, sterile."[49b] God Anu indicates with his right arm, to the viewing monitor, and we see another computer on the viewing monitor, also with a 6 inch square viewing screen, as God Anu continues, "As we have retained the reproductive function for ourselves."

God Anu pauses, and then he continues, "Also, I shall immediately dispatch one of our support satellites," God Anu indicates with his right arm, to the viewing monitor, and we see the special satellite on the viewing monitor. It is very large, and is flat cylinder shaped in design, its width equal to one-quarter of its diameter, as God Anu continues, "The Duku,[50] and place it in orbit around the Earth, to defend and assist the operation, as we have done on other planets. Our satellite and craft maintenance crews," God Anu indicates with his right arm, to the viewing monitor, and we see the viewing monitor, and another group of 6 Igigi Gods, 4 males and 2 females, as God Anu continues, "shall remain on the support facility to serve our defense and alternate exploration force, composed of our Anunnaki Gods, three hundred[51] in number, who shall staff the Duku satellite. I shall accompany God Enki to the Duku satellite, and from there, I shall return to our heavenly[52] planet, Nibiru, and leave the planet Earth to my subjects."[53]

[49a]L. Lower page 117 the gods were believed to be immortal.

[49b](The Igigi, which literally means 'Those Who See and Observe', are surrogate beings, whom the highest Gods have fashioned, possibly genetically altered higher Gods, or possibly a hybrid, or possibly a hybrid of the highest Gods themselves, a sterile worker, and although they are created, and develop and appear as adult male and female, they are sexless: genetically suspended in a pre-puberty state. For practical purposes, I refer to them as male and female, but they will never generatively reach that level of development. They were probably considered as appendages or implements, and certainly not as equal beings).

[50]K. Lower page 53 Let us cause them to descend from the Dulkug (I interpret Dulkug as equivalent to Duku), and upper page 54.

[51]T. Page 293 Assigned to Anu, to heed his instructions, Three hundred (Anunnaki) in the heavens (space) he stations as a guard (on the Duku satellite), the ways (operation) of (on) Earth to defend from the Heaven (space. This implies that there could be a threat to the operation, either from within the Earth assigned ranks themselves, especially the Igigi, for perhaps there was some question of their allegiance because of their subservient/lowest rank, or from other sources from Nibiru, or from sources other than Nibiru), and page 294.

[52]M. Page 43, line 13 Anu had gone up to heaven (Nibiru).

[53]M. Page 43, line 14 Leaving the earth to his subjects (those under his authority).

"On Earth, all the Anunnaki Gods shall wear the following equipment, as needed, while outside the protective environment of their craft," and again we look at the viewing monitor, and see the communication helmet,[54] and the ear, clip on, electromagnetic body regulator,[55] as they are mentioned by God Anu. Both appear as a gold-like metallic material, as God Anu continues, "The communication helmet, which has been specifically adjusted to Earth's electromagnetic wave lengths; and the ear, clip on, electromagnetic body regulator, to maintain a proper balance of electrons within the body."

God Anu pauses, and looks around the assembly, and then he continues, "As the assembly members have requested and I have proposed, I now command that we assign allotments[56] and establish divided[57] authority over the Earth," and again we look at the viewing monitor, and again we see the Earth. Then we see an overall view of the assembly, as all the members nod their head in agreement.

Then God Anu continues, "The assembly has spoken. With the entire assembly in agreement, the project is affirmed. Let us clasp hands together,[58] and begin this mission," and again we see an overall view of the assembly, as all the Gods rise and clasp hands in a circle, and then God Enlil demands, "Our oath to God Anu."

Together, all the Gods declare, "The decree of God Anu is binding."

*

From space, we observe Nibiru approaching the Mars area, and passing between Mars and Jupiter, on a plane of orbit that is similar to the 17 degree angle of Pluto, and then we see a close up view of Nibiru, in all its glory.

---

[54] T. Page 125, fig. 60.

[55] Q. Illustration, King Assurbanipal, and Winged Figure (similar to acupuncture procedure effects upon the ear lobe: see my illustration on page 311 in my comments).

[56] M. Page 43, line 12 (the Gods) Had cast lots (etc., assigned allotments).

[57] M. Page 43, line 12 Had cast lots and had divided (assigned control of the Earth among the Gods, thus the Gods portioned the Earth; and according to Egyptian priests to Solon in circa 600 B.C. via Plato in his circa 400 B.C. 'Timaeus' and 'Critias', prior to the Babylonian accounts of Berossus in circa 290 B.C.) "in speaking of the allotments of the gods, that they distributed the whole earth into portions differing in extent" (see footnote 47).

[58] M. Page 43, line 11 The gods had clasped hands together (although I interpret this literally, meaning they were in agreement, it may be a figure of speech, with the same meaning).

Then we see the Earth from space, and we zoom in for a close up view from the air, close to the Earth, and we notice the Earth slightly shaking.

From upon the Earth, we observe Nibiru, which is clearly visible against the star-filled sky, radiating and glowing in the northern dawn sky. Its two rings of clouds, within which orbit its four moons, give Nibiru the appearance of a cross in the sky, bathed in a sunset red glow. We look down to the Earth, and we notice moderate Earth tremors with fissures, and volcanic eruptions with lava and thick clouds of smoke, ash, and steam, causing frightened animals to run in all directions, and increasing temperatures of the ground and air, and water and steam bubbling up out of the ground, both in the wilderness and in the desert, and surface waters evaporating into a mist, with a rainbow forming in the sunlit distance, and it appears rain is imminent, and mild flooding of fog laden seas and rivers, and mildly blowing winds, causing tree branches to gently sway.[59]

From space, we view the Duku satellite moving into orbit near the Earth, and then from the Earth, we observe Nibiru growing smaller as it speeds away. Again, we look down to the Earth, and the moderate Earth tremors with fissures and volcanic eruptions and profuse moderate rain are diminishing, and temperatures are decreasing, with a rainbow forming and then slowly fading away in the sunlit distance, and the mild flood waters are receding, and the winds are decreasing, and the animals are returning to a calm state of disposition, and the Earth is returning to normal, as the effects of Nibiru subside.

From space, we observe Nibiru speeding out of sight, twinkling as it disappears into the star-filled space of our universe, and then from space, near the Earth, we view the Duku satellite orbiting close to the Earth. Then four saucer shaped shuttle craft[60] descend from the lower axis, each craft being 75 feet in diameter, glistening in the bright sunlight, and in formation they accelerate toward the Earth. From the underside of each craft, they emit a brilliant glowing light, which is not from a flame or burning process, but is like that of a firefly.

We observe God Enki in his craft, standing in front of and carefully watching a wide array of computers with flashing display lights, and with him

---

[59]T. Lower page 360 through page 362; and V. Isaiah 35:6 for in the wilderness shall waters break out, and streams in the desert (similar to this).

[60]A. Page 151, figure 16 (similar to this. Logically, the Gods would favor normal Earth conditions, and prefer to descend either before or after the perturbation period on Earth, depending on severity).

are 8 Anunnaki God engineers, and 10 Igigi Gods, who are seated off to a side, waiting to land, and 2 Anunnaki God officers, who are seated in front of the computer consoles, busy watching the flashing display lights and touching the heat sensitive numbers which control them. There is also one of the 7 great Anunnaki Gods, their commander, who is standing next to them watching the consoles, and there are 6 assistant scientists, who are seated off to another side. All are wearing loose fitting flight attire of gold-like material, except the Igigi Gods, who are not considered astronauts, Anunnaki Gods. Then God Enki touches a heat sensitive number on a computer console, and a small version of the dome viewing monitor on Nibiru rises from the floor, and shows a close up view of the scene below. They all look at the Earth, growing larger, as they descend toward it. From inside the craft, still high over the Earth,[61] we look down at the Persian Gulf area, and it appears to quickly move up toward us, as we zoom toward where the Persian Gulf water meets the land, the northern most tip of the gulf, the Tigris and Euphrates River delta area.

From the ground, we watch the four craft hover for a moment, glistening in the bright sunlight, descend and prepare to land, as the brilliant glowing firefly-like light emitting from the underside of each craft begins to diminish, and 4 pods like cattle hoofs[62] extend 6 feet down, and then the craft land, as the glowing light of each craft has now been reduced to a very low level, with the craft each resting on its pods. Then the glowing light further diminishes, and then completely expires, and after a moment, a side of each craft extends down, forming a 4 feet wide ramp and doorway. Then, God Enki emerges, wearing a communication helmet, as he walks through the air flow door, a curtain of moving opaque air, which prevents any outside elements from entering the craft, and down the ramp, followed by the members of his craft, all wearing helmets, and some wearing body regulators, except the Igigi Gods, who are not permitted, because of their subservient status, to wear helmets or body regulators on Earth. We scan and see the land, which is mostly flat, with some trees and short grass. We are near a marshy area, with tall reeds and the Euphrates River in the close background, to the east, with clear water in it. All

---

[61]T. Chapter 9.

[62]V. Ezekiel 1:7 and the sole of their feet was like the sole of a calf's foot (similar to this); and A. page 14, fig. 1, and page 15, fig. 2, and page 23, fig. 4, and lower page 35 through upper page 37, and page 56, Verse 7.

the Gods with God Enki have just scanned the area, as have the members of the other three craft, and they all gather into one group. There is God Enki; one great Anunnaki God commander, a male; 6 assistant scientists, 4 Gods and 2 Goddesses; 8 Anunnaki God officers, 6 males and 2 females; 34 Anunnaki God engineers, 26 males and 8 females: a total of 50 Anunnaki Gods. Additionally, there are 42 Igigi Gods, regular military, 35 males and 7 females. God Enki looks at the group, and declares, "I, God Enki, leader of the exploration force, proclaim myself the first ruler[63] of planet Earth. On this base site, I shall build the first city on Earth. It shall be called the 'Far Away City', Eridu.[64] From Eridu, I shall explore the Earth, and follow the decrees of God Anu."

*

It is bright daylight, and from a distance, we see the compound of Eridu, the city of God Enki, the first Earth city, with the Euphrates River to the east, close to the city, with clear water in it, and a few irrigation canals near the marshes.

ERIDU
.11 Shar Years Later
or
400 Earth Years Later
440,400 B.C.

The base is completed, and in its glory. Approaching, we observe the beautiful oval walled compound,[65] with no structures outside the wall. The wall is 12 feet high, and it envelopes a colorful 300 foot high central 7 tiered

---

[63]H. Page 71, line 3, and line 4 (A-lulim (ak) = God Enki).

[64]M. Page 141, line 88 [When the decree] (of God Anu) of kingship (rule of God Enlil) had come down from heaven (space), and line 93 The first of these cities, Eridu, he (God Enlil) gave to the leader (leader of initial exploration) Nudimmud (an epithet for God Enki, which means 'He Who Fashions Things'); and T. page 259 E.RI.DU ("house in faraway built"); and K. page 62, and page 63; and H. page 70, note 5, and upper page 71, line 1, and line 2, and page 37, note 86.

[65]I. Illustration opposite page 77 (similar to this, except with no structures outside the wall).

ziggurat,[66] composed of solid stages, platforms, or tiers, progressively reduced in height and width, and each coated with a different color plaster, white, black, purple, blue, vermilion, and silver, and gold colored temple with antennas atop it, which is a communication and control center, being nearly fully visible from our approach angle. Both structures are made of unbaked, sun-dried clay bricks, and while the ziggurat is colorful, the wall is coated with cream colored plaster. The oval walled compound is approximately 900 feet by 600 feet, and has 12 feet high double main gate doors of aluminum-like metal, which are composed of solid one foot square panels. As we near the city, we notice 10 Igigi Gods digging irrigation canals near the marshes, using only a plastic-like spade, and hauling the clay away from the canals, to be shaped into bricks, using a plastic-like basket, which they carry on their back, and using a plastic-like rope support, which they grasp at their chest. They are providing the heavy labor, and are somewhat sweating, slightly dusty and dirty, and slightly tired, and the pace is normal. Three Anunnaki God officers, all wearing helmets and two wearing body regulators, who are all crisp and clean, are supervising the Igigi Gods, who are also serving the highest Gods and performing the lighter and relatively easy domestic work for the highest Gods. Also, there are 4 Igigi Gods, carrying their spades, being led by one Anunnaki God officer, wearing a helmet and body regulator, marching in orderly formation to the work site. We now move closer to the city, and we notice 5 Igigi Gods, tending a small flock of sheep, and an area of crops in the small fields.

We now go through the gate and look inside the oval wall, and attached to the wall near the gate, we notice a small rectangular guard house, and then we see a full view of the colorful central ziggurat, with its 7 tiers, the only structure inside the wall, with a base of 300 feet by 300 feet, which we had not been

---

[66] T. Page 35, fig. 11, and page 276 through page 279 not only a perfect square (300 feet long by 300 feet wide) but a perfect cube (300 feet high) as well; and B. page 123 (for colors, see the book Atlantis - The Antediluvian World by Ignatius Donnelly, page 347, who quotes Ancient History of the East by Lenormant, vol. i., page 463. I believe the ziggurat also served as a navigational beacon/landmark, and as well as being symbolic of the Gods, the 7 tiers could be symbolic and represent the Earth, the 7th planet, counted from the perspective of an outside observer, an astronaut, who would enter our known solar system from beyond it, and pass Pluto, Neptune, Uranus, Saturn, Jupiter, Mars, and then encounter Earth).

able to fully see while we approached. The height in feet of each tier, beginning with the base, is: 110, 60, 20, 20, 20, 20, and 50 for the temple. The 1st and 2nd tiers have one front and two diagonal ramp-like stairs, built upon the face and converging at the center as they lead upwards to the 1st and 2nd tier. The 3rd, 4th, 5th, and 6th tier each have a ramp, built upon the face, connecting each of these tiers. We notice the gold colored temple at the top, as the seventh tier, the temple of God Enki, his communication and control center, with antennas, composed of 2 separate long rods piercing several rings, atop the temple. There is one Igigi God tending a small sheepfold nearby, which is composed of a low sun-dried clay brick wall, coated with cream colored plaster, and 2 Igigi Gods working in the courtyard, and pampering the ziggurat. All are somewhat sweating, slightly dusty and dirty, and slightly tired, with one Anunnaki God officer, who is wearing a helmet, and who is crisp and clean, supervising them, and the pace is normal. Four craft are on the ground near the ziggurat, and one is being maintained by 3 Igigi Gods. They have removed a side panel, and are in the process of replacing an electric circuit-like board. One of the other craft is preparing to ascend from within the walled city. After a moment, the ramp retracts to form a side of the craft, the underside begins glowing, the craft rises, the 4 pods like cattle hoofs retract, the glowing increases and becomes a brilliant glowing firefly-like light as it ascends, and the craft glistens in the bright sunlight, and we follow it until it quickly disappears into the distant bright blue sky.

From inside the craft, high over the land, we scan from Eridu to southeastern Africa, as we follow the rapid flight of the craft. The craft is now moving more slowly above the land, over an open area and toward a forest, filled with lush greenery of tall thick bushes and trees, and with gently sloping hills in the background.

From the ground, gazelles and other small animals are being frightened by, and running away from the craft, as it glistens in the bright sunlight, with its brilliant glowing firefly-like light emitting from the underside.

Inside the craft, we observe God Enki with 2 assistant scientists, 1 God and 1 Goddess, standing in front of the computers, and 2 Anunnaki God officers seated in front of the computer consoles, and 4 Igigi Gods seated off to a side, waiting to labor, and one great Anunnaki God commander standing next to them. They are all looking at the flashing display lights on the computers; and at the small viewing monitor, which shows a close up view of the scene below, with the running animals, and lush greenery and trees everywhere. Then God Enki announces, "The computers indicate large deposits of gold in this area. We shall begin our mining exploration here. We shall call this area the 'Land of Precious Metals', Abzu."[67]

From the air, we see the craft slowly continuing, at above tree top levels, moving over the forest, toward another open area.

From the ground, we watch the craft slowing moving above the trees, with rays of sunlight filtering through the trees. We look down from the tops of the trees, and follow the rays of filtered sunlight downward, and they fall onto a group of 16 prehistoric ape-man creatures. They have slowly arbitrarily evolved to a stage approaching early Homo erectus, whom for convenience we shall call ape-like creatures, for they look more like apes than man, and they have a brain size of approximately 600 cubic centimeters. They have brown skin, brown eyes, thick straight black-brown body hair, approximately 1-1/2 inches long, which looks shaggy because it is unkempt, and long straight black-brown hair on their heads, which is also unkempt. They are naked. They have far more animal than man characteristics. There are 8 adults, 4 males and 4 females, and 8 children, 4 males and 4 females. We scan the group and see a close up view of the faces of the ape-like creatures. Their eyes are wide open, and their expressions are filled with amazement. They are frozen in wonderment, as they stare up from beneath the trees, and silently watch the craft very slowly moving over their heads, above the trees.

From the air, we see the craft slowly moving horizontally toward the open area, drift over the open area, hover over the clearing, descend and prepare to land, as the brilliant glowing firefly-like light emitting from the underside begins to diminish, and the 4 pods like cattle hoofs extend 6 feet down, and then the craft begins to land.

From the ground, the ape-like creatures silently watch from behind the trees, as the craft lands nearby, and the glowing light further diminishes, and then completely expires. They watch it on the ground for a moment, and then

---

[67]M. Page 43, line 17 [After Anu] had gone up to heaven (returned to Nibiru), 18 [And Enki] had gone down to the Apsû, lower line 1, 2, 3, and lower line 4; and K. page 65.

a side of the craft extends down, forming a 4 feet wide ramp and doorway. Then their amazement turns into fright, fearing the strange flying creature has opened its mouth to devour them, and they turn and run away into the forest.[68]

*

It is mid afternoon on a bright day, and we see one craft in the clearing, and the forest of tall and thick bushes and trees in the background.

ABZU
.14 Shar Years Later
or
500 Earth Years Later
439,900 B.C.

Inside the craft, we scan over laboratory-like equipment, as we are in the biological laboratory craft of God Enki, the chief scientist of the Gods. There is a plentiful array of electronic equipment, consisting of a variety of computers with flashing display lights, and metal spectrum computers which test the various rock samples. There are many glass-like flasks and beakers containing bubbling chemicals of different colors, which additionally test the samples of rock, that are next to the flasks and beakers. Also, there is a microwave-like furnace, and within it, gold samples are being instantaneously melted down, producing a brilliant liquid gold color.

Then we scan to God Enki and an assistant scientist, a Goddess. They are seated at a laboratory table, and God Enki remarks to her, "Here in Abzu, our mineral exploration assignment on Earth is proving to be very successful,"[69] and the assistant scientist replies, "Yes God Enki, we have located areas of large concentrations of gold and other minerals. They are in abundance here, far more than we had expected." God Enki adds, "From beneath the lands,

---

[68]A. Foreword through page 174 (similar to this).
[69]T. Page 285, fig. 142, and page 293, fig. 146.

and beneath the waters of the seas,[70] we shall soon begin to extract large quantities of each precious metal, and transport them to our heavenly planet, Nibiru. That shall certainly please God Anu." The assistant scientist jokingly asks, "You have even begun to experiment with the creatures of the Earth. I have heard that you call the mutates your plants. Is it true?"

Again, we scan over the laboratory flasks and beakers, and across to the other side of the craft, where there are four small cages, two of which contain rabbits, and two contain dogs. They have been mutated, and one rabbit has two heads, and the other rabbit has four front legs and the head of a dog. One dog has eight legs, four front and four rear, and the other dog has two large muscular front legs. God Enki responds, while we observe the mutates, "Yes, I have a special nickname for the creatures. I call the results of my experiments, my plants, because I consider the entire planet Earth as my garden, and comparatively speaking, my creations, which I genetically develop and nurture, I endearingly refer to as my plants, placed within my garden to grow and fulfill their genetic programming. This is similar to the method I use to genetically alter and cultivate our hybrid species of foodstuff: our plants. In this way, my creations are similar to plants." Then the assistant scientist asks, "Why have you begun the experiments?" God Enki replies, "I mutated the genes to determine if I could produce a more adaptable creature, as I have done on other planets, that would be helpful and aid the Igigi Gods, as they toil and labor. These also did not produce a favorable result. I shall attempt other genetic formulas in the future." As our eyes examine them, we see a close up view of the mutates.

*

We see a bright blue daylight sky, as seen from upon the Earth. We look down, and from a distance, we observe the outside of the oval walled compound of Eridu, the city of God Enki, with its colorful central 7 tiered ziggurat, and temple with antennas atop it, as the seventh tier, still with no structures outside the wall. The outside irrigation canals are complete, and are veining through the area from the clear Euphrates River to the east, and no one is working outside the oval wall.

---

[70]T. Lower page 294.

ERIDU
9.75 Shar Years Later
or
35,100 Earth Years Later[71]
404,800 B.C.

We notice a craft descending from the sky toward Eridu, glistening in the bright sunlight, with its brilliant glowing firefly-like light emitting from the underside, hover, descend and prepare to land within the oval wall, as the glowing light begins to diminish, and the 4 pods like cattle hoofs extend 6 feet down, and then the craft begins to land.

From inside the oval wall, we watch the craft as it lands near the ziggurat, still the only structure inside the wall, and near four craft on the ground, and the glowing light further diminishes, and then completely expires. Beyond the craft, we notice the beautiful compound of Eridu: it looks fresh and clean, just as it had when it was built, with 3 Igigi Gods tending the small sheepfold nearby, and 5 Igigi Gods working in the courtyard, and pampering the ziggurat, who are still somewhat sweating, slightly dusty and dirty, and slightly tired, with one Anunnaki God officer, who is wearing a helmet, and who is crisp and clean, supervising them, and the pace is still normal. A short distance away, we notice God Enki standing with a group of Gods, looking at the craft. There are 6 assistant scientists, 4 Gods and 2 Goddesses, and 3 Anunnaki God officers, and one great Anunnaki God commander. They are waiting to greet the newly arrived Gods, and all are wearing helmets and some wear body regulators.

Then, after a moment, a side of the craft extends down, forming a 4 feet wide ramp and doorway, and walking through the air flow door, a curtain of moving opaque air, and down the ramp, we see God Anu leading the way, followed by God Enlil and Goddess Ninhursag, 4 Anunnaki God officers, 4 Igigi Gods, and one great Anunnaki God commander. All wearing their usual attire, and all, except the Igigi Gods, are wearing helmets and some wear body regulators. God Enki greets God Anu, "Welcome God Anu, my father, to planet Earth, and the first Earth city, Eridu." There are no hand shakes, only spoken

---

[71]H. Page 71, note 8 (400 + 500 + 35,100 = 36,000 years per Berossus, which represents the total time God Enki spent on the Earth until God Enlil arrived).

words and a bow of the head, for the Gods are always highly conscious of unnecessarily transmitting germs. As he gazes at the compound and the ziggurat, God Anu responds, "You have done well God Enki, my son," and God Enki[72] greets God Enlil[73] and Goddess Ninhursag, "Welcome God Enlil, my brother, and Goddess Ninhursag, my sister." God Enlil responds, "It is good to see you in person, God Enki," and Goddess Ninhursag adds, "Thank you God Enki. We acknowledge your welcome," while all the other Gods exchange welcome and thank you words, as they bow their heads. Then God Enlil reflects, "We have all been observing your progress on Earth, through the viewing monitor on Nibiru, for the past ten Shar years, thirty-six thousand Earth years."

God Anu, first looking at the group with God Enki, and then quickly again at the ziggurat, and then at God Enki, declares, "God Enki, my son, upon your report that there are no dangers, and that the mining sites have been selected, secured and thoroughly explored, I, God Anu[74] decree that supreme command and rule of the Earth,"[75] and God Anu now looks at God Enlil, and continues, "be given to God Enlil, my son, and that you immediately begin the mining operations. Also, you should construct a base to process the metals, and then other cities,[76] to additionally support the mining operations. The

---

[72]H. Page 70, note 6 [á]-lulim, or á-lulim-ak, "horn of the stag" (could refer to the obvious fact that a horn is in front of a stag, the first part of the stag to arrive, or "leads" the stag, the main part of the body, as God Enki was the first to arrive and leader of the exploration forces, followed by the main part of that body, the regular forces from Nibiru; therefore horn of the stag means first to arrive or leader) = Ἀλωρος (in Hellenic = God Enki; who is also known in Hellenic, in J. page XXXIII, as Ἀόν/Ἀός).

[73]H. Page 71, note 10 [á]-làl-gar = Ἀλαπαρος (in Hellenic = God Enlil; who is also known in Hellenic, in J. page XXXIII, as Ἴλλινον/ Ἴλλινος, which appears to be a variant, especially when considered phonetically, of the ancient word Ἕλληνες, which is the word from which Hellēnes is derived).

[74]M. Page 141, line 89 After the lofty crown (bestowed authority of God Anu) and the throne of kingship (rule of God Enlil) had come down from heaven.

[75]M. Page 141, line 90 Enlil perfected Earth (perhaps this is the source of monotheism, the belief in "one" God: "one" God/supreme being selected/designated to administer the Earth and command the other Gods/supreme beings, and later human beings, upon the Earth); and H, page 71, line 5 Alalg̃ar (= God Enlil. It is my observation that [á]-làl-gar/Alalg̃ar could be the root word for Allah, the Arabic name for the Supreme Being of Islam in the Koran, reported to be a compound of al = the, and ilah = God; or taking the widely accepted progression, probably from the same source, we have i-lu/ilu/I.LU (Lofty Ones: see footnote 19), and it has been reported that the great God of the Semites was El (the Strong One), from whose name comes the Biblical names Beth-el (the house of God), El-ohim (the Gods), and El-oah (God), and from the same name is derived the name Al-lah (the God).

[76]M. Page 141, line 91 Founded [five] cities in [all] 92, 93 (see footnote 64 and 74), 94, 95, 96, 97, 98, 99, and line 100; and T. Chapter 10.

rule of these cities, you are to give to your sons. Then, you are to build your own city, and call it Nippur, and your temple within your compound, is to be called Ekur," and God Anu now looks at Goddess Ninhursag, and adds, "Goddess Ninhursag, my daughter, you are to stay on Earth and assist God Enlil." Then God Anu looks at God Enlil, and commands, "God Enlil, I also decree that six hundred of our Anunnaki Gods, along with adequate numbers of the Igigi Gods, our regular military, descend immediately to begin work on mining the metals, and on the building of all the support cities, including your city of Nippur, and your temple, Ekur. They shall all be under your command."

Then God Enki suggests, "God Anu, as you have observed on the viewing monitor, it may be possible to perfect Earth creatures, and use them to assist the Igigi Gods in their labor, as we have done elsewhere." God Anu responds, "God Enki, now that your mineral exploration assignment has been accomplished, and with excellent results, you are to remain in the Abzu, and manage the mining operations, under the command of God Enlil. Also, you are to continue your experemental labroratory work on the creatures, to determine if they may be able to perform useful tasks for us." God Anu now gazes at the group, as he announces, "And as for me, I shall return to, and continue to preside over our heavenly planet, Nibiru."

*

From space, we observe Nibiru approaching the Mars area, and passing between Mars and Jupiter, on a plane of orbit that is similar to the 17 degree angle of Pluto, and then we see a close up view of Nibiru, in all its glory.

Then we see the Earth from space, and we zoom in for a close up view from the air, close to the Earth, and we notice the Earth slightly shaking.

From upon the Earth, we observe Nibiru, which is clearly visible against the star-filled sky, radiating and glowing in the northern dawn sky. Its two rings of clouds, within which orbit its four moons, give Nibiru the appearance of a cross in the sky, bathed in a sunset red glow. We look down to the Earth, and we notice moderate Earth tremors with fissures, and volcanic eruptions with lava and thick clouds of smoke, ash, and steam, causing frightened animals to run in all directions, and increasing temperature of the ground and air, and water and steam bubbling up out of the ground, both in the wilderness and in the desert, and surface waters evaporating into a mist, with a rainbow forming in the sunlit distance, and it appears rain is imminent, and mild flooding of fog laden seas and rivers, and mildly blowing winds, causing tree branches to gently sway.

We observe a group of ape-like creatures, huddled together in fear, watching and pointing up at Nibiru in the northern sky at dawn.

From space, we observe Nibiru, after orbiting around the Sun, begin its return into deep space, orbiting away from the Earth.

From upon the Earth, we observe Nibiru growing smaller as it speeds away. Again, we look down to the Earth, and the moderate Earth tremors with fissures and volcanic eruptions and profuse moderate rain are diminishing, and temperatures are decreasing, with a rainbow forming and then slowly fading away in the sunlit distance, and the mild flood waters are receding, and the winds are decreasing, and the animals and relieved ape-like creatures are returning to a calm state of disposition, and the Earth is returning to normal, as the effects of Nibiru subside.

From space, we observe Nibiru speeding out of sight, twinkling as it disappears into the star-filled space of our universe.

*

From the Earth, we see a full Moon in the clear star-filled night sky, and then we look down to the moonlit gently sloping hills of the Abzu.[77]

ABZU
30 Shar Years Later
or
108,000 Earth Years Later
296,800 B.C.

---

[77]M. Page 44, line 29 Within the Apsû, and line 30 Under the heart of the land;and T. page 283 through page 286 (although the evidence is strong for the location of the Abzu/Apsû, which means "watery deep", for there is usually water in deep mine shafts, or "extremely deep hole", to be in southeastern Africa; it is also possible that Ethiopia, at one time called Abyssinia, of which the root word Abzu is very obvious, was a part of the Abzu, and the Abzu included the entire general area, and was the domain of God Enki: see footnote 67).

We look down further and see the entrance to a mine, with tall and thick bushes nearby. The Moon lights up the hillside area, plus there are two glass-like balls of light, one on each side of the entrance, that glow with a soft, firefly-like glow. It is the changing of a shift crew, which change on a staggered schedule throughout the day and night. We observe a group of 20 Igigi Gods, fresh and clean, wearing their blue uniforms. They are being led, marching in orderly formation, carrying their tools, the plastic-like spade and now a plastic-like gold pickax, toward the mine entrance by 2 Anunnaki God officers, crisp and clean, both marching and helmeted, and one wearing a body regulator, wearing their green uniforms. As they are about to enter the mine, we observe another group of 20 Igigi Gods, sweating, dusty and dirty, and noticeably tired, slowly walking out of the entrance, carrying their tools, the plastic-like gold pickax and the plastic-like spade, led by 2 Anunnaki God officers, crisp and clean, both marching and helmeted, and one wearing a body regulator. The Igigi Gods are dragging themselves, in an exhausted slow walk, out of the mine in a somewhat orderly formation.

We scan to the tall and thick bushes near the entrance, and at first, we see only the bushes, but as we zoom in for a close up view, we notice 3 ape-like creatures, 2 males and 1 female, peering through the bushes at the work force, each with a curious look upon their face. They silently watch the changing of the shift crew, composed of the Igigi Gods and the Anunnaki God officers. They are unnoticed, as they silently observe the scene.

Inside the mine, which is also illuminated by identical glowing glass-like balls of light, we hear electrical arcing sounds, as we slowly proceed deeper into the mine. At this point, we do not see the Gods, we only hear the sounds of arcing, and now we begin to hear shoveling sounds. As we go around a corner, we observe the Igigi[78] Gods, sweating, dusty and dirty, and noticeably tired, as they labor very hard in the dusty and dirty mine. They are touching the large rocks with their plastic-like gold pickax; a high frequency device, which generates, stores, and emits electrical energy in the form of a

---

[78]M. Page 43, upper line 1 When the gods (aliens) like men (human beings), 2 Bore the work and suffered the toil, 3 The toil of the gods was great, 4 The work was heavy, the distress was much, 5 The Seven great (highest alien being) Anunnaki (astronauts), 6 Were making the Igigi (lowest alien being) suffer the work, and upper line 7 Anu, their father, was the king.

concentrated laser-like beam, that God Enlil has introduced to the Earth,[79] which is bursting the rocks, like an implosion, into small pieces. The electrical discharge makes the arcing sound as the larger rocks are touched. It is like a forced labor scene, and the pace is quick.

As we scan, we notice one great Anunnaki God commander, who is speaking to 2 Anunnaki God officers, both crisp and clean and helmeted, and one wearing a body regulator. They are standing on a ledge, looking down as they view and supervise the mining. The great Anunnaki God commander speaks in a loud tone to the Anunnaki God officers, as we view the laboring in the background, and orders, "I have appointed both of you as officers, see that they work fast, and do not sit down." One of the Anunnaki God officers eagerly answers, "Yes commander, we shall follow your decree." Then the great Anunnaki God commander turns, and walks toward the exit, and as the 2 Anunnaki God officers watch, their commander walks around the corner and out of view. Both Anunnaki God officers now turn toward the laborers, and one shouts, "Faster, work faster," and then the other shouts, "Over there, faster," as they look down at the laboring Igigi Gods.

Again, we observe the Igigi Gods, as they toil and sweat in the dust and dirt. Some burst the large rocks with the plastic-like gold pickax, while others, using the plastic-like spade, shovel[80] the smaller rocks into the plastic-like baskets, and others take the loaded baskets of rocks, which are carried on their back, and has a plastic-like rope support, which they grasp at their chest,[81] and dump them into a screw-like conveyor,[82] which moves the rocks back into a larger loading container, which is on wide soft wheels.

---

[79]K. Page 52 He (God Enlil) bought the pickax into existence, the "day" came forth (the day arrived when), He introduced labor, decreed the fate, Upon the pickax and basket he directs the "power". Enlil made his pickax exalted, His pickax (handle) of gold (color), whose head is of lapis lazuli (deep blue in color, like a lapis lazuli gem stone), (etc.), Whose tooth (single pick bite/force) is (as powerful as) a one-horned ox ascending a large wall(for practical reasons, I will refer to this sophisticated tool as a plastic-like gold pickax).

[80]M. Page 42, line 20 Toil upon the Igigi, 21 Everywhere they were digging, and line 23 Everywhere they were digging.

[81]K. Page 52, and page 53; and Q. Illustration, Workmen in a Quarry; and U. frontispiece, lower register.

[82]M. Page 44, line 31 Excavating within it, and line 32 Slowly they (the Igigi Gods) raised its head (achieved the objective).

*

It is bright daylight, and from a distance, we see the outside of the oval walled compound of Nippur, the city of God Enlil, which looks somewhat like the city of Eridu, with its colorful central 7 tiered ziggurat, except there are small rectangular buildings outside the wall. All the structures are made of unbaked, sun-dried clay bricks, and except the colorful ziggurat, are coated with cream colored plaster. We observe many sweating, dusty and dirty, and noticeably tired Igigi[83] Gods laboring at a quick pace, supervised by Anunnaki God officers, who are crisp and clean, and some fresh Igigi Gods, led by Anunnaki God officers, also crisp and clean, marching and replacing the tired crews. Some carry the plastic-like spade, or the plastic-like gold pickax, and others carry a plastic-like hod board. All of the Anunnaki God officers are wearing helmets, and some wear a body regulator.

NIPPUR
The Next Day

Inside the oval wall, this base also resembles Eridu. As our eyes scan, we notice the 12 feet high double doors at the main gate, and a small rectangular guard house near it, attached to the wall, like Eridu, and the surrounding 12 feet high oval wall, which has a 7 tiered ziggurat in the center, composed of solid stages, platforms, or tiers, with ramp-like connecting stairs, progressively reduced in height and width, and each coated with a different color plaster, white, black, purple, blue, vermilion, and silver, with Ekur, the gold colored temple of God Enlil, his communication and control center, as the seventh tier, at the top, with antennas atop it, also like Eridu, the only structure inside the wall. The compound inside the wall is complete. We see five craft, which are not being maintained at this time, grouped in an area, near the ziggurat. There is very little activity inside the wall. There are 2 Igigi Gods tending a small sheepfold nearby, composed of a low sun-dried clay brick wall, coated with cream colored plaster, and 6 Igigi Gods working in the courtyard, and pampering the ziggurat, who are sweating, dusty and dirty, and noticeably

---

[83]M. Page 43, upper line 1, 2, 3, 4, and upper line 6 (see footnote 78).

tired. Also, there are 2 Anunnaki God officers, crisp and clean, who are supervising them, and both are wearing helmets and one wears a body regulator. Other Igigi Gods continue serving the highest Gods and performing the lighter and relatively easy domestic work for the highest Gods, as they did at Eridu, and as they do at all the other Earth cities.

Outside the oval wall, we observe 150 Igigi Gods, 120 males and 30 females, who are working very hard, and all are sweating, dusty and dirty, and they are noticeably tired. They are being supervised by 5 Anunnaki God officers, all wearing helmets and some wearing body regulators, who are all crisp and clean, and are ordering them as they perform the work. We notice the glass-like balls of light, positioned around the area, which are used to light the night shift work area. Some Igigi Gods are working with the plastic-like gold pickax and the plastic-like spade, digging irrigation canals[84] that vein through the city, and some carry the plastic-like baskets of clay on their back, away from the canals, to be shaped into bricks, using the plastic-like rope support, which they grasp at their chest, while others carry the sun-dried heavy clay bricks, 12 by 12 by 3 inches, and mortar and cream colored plaster on the plastic-like hod boards, as they build smaller rectangular storage buildings outside, near the wall. Gold, silver, and other precious metal ingots, called zag,[85] which look somewhat like an hourglass in shape, with a lengthways carrying hole in them, are being carried on 2 inch diameter by 6 feet long hollow aluminum-like poles,[86] and then stacked high, crossways, in large quantities near the wall. This labor is similar to the mine labor, like a forced labor scene, and the pace is quick. Also, some Igigi Gods are tending a large flock of sheep and working an area of crops in the large fields.

We notice God Enlil[87] approaching the work area, also outside the oval wall, from the main gate, accompanied by three other of the 7 great Anunnaki Gods, 2 males and 1 female, the commanders of the military, who command all the Anunnaki God officers, who supervise all the subservient Gods for God Enlil: all within the bright sunlight. With them is God Ninurta,[88] a vizier, about 25 years old, the son and chamberlain of God Enlil, whom he is grooming

---

[84]M. Page 43, lower line 5 The Igigi were digging the river (which means digging out the canals, which contained the river water).

[85]T. Page 285, fig. 141.

[86]T. Page 285, fig. 142.

[87]M. Page 43, upper line 8 Their counsellor was the warrior Enlil.

for leadership; and the personal sheriff of God Ninurta, God Ennugi,[89] an Anunnaki God officer, a male, who enforces his decrees. All are wearing helmets and some wear body regulators. God Enlil holds a 3/4 by 3-1/2 by 5-3/4 inches long computer with video. God Enlil is pridefully showing them the construction area.

Near God Enlil, and within his voice range, there is an Anunnaki God officer, a male, giving orders to and directing the clay brick carriers. He shouts to an Igigi God, a male, called God Wêila, "Bring more clay bricks Wêila, quickly. You must work faster," and we observe the sweating face of God Wêila, who looks at the Anunnaki God officer with contempt.[90]

God Enlil, who notices the contemptuous look, gives a despising glance toward God Wêila, and then God Enlil turns and addresses the group, "I have gathered three of you great Anunnaki Gods, the commanders of our military, and my son, God Ninurta, who after my appointment as chamberlain, has become the proud ruler and administrator of his own city, Badtibira,[91] our metal smelting center; and God Ennugi, my sons personal sheriff, who enforces the decrees of God Ninurta," and now looking at the 3 great Anunnaki God commanders, he continues, "you oversee all our operations here on Earth, from the mining in the Abzu, to the building of all our beautiful cities. Under your command, all the Igigi Gods are working with great effort.[92] Keep them working, for we have much more to do here on Earth. Additional support projects are to be built, and new mining sites are to be dug."

Again, we see the Igigi God called Wêila, and upon overhearing the words of God Enlil, the face of God Wêila is filled with anger, but he remains silent.

Then God Enlil demands, "I want maximum production from the Igigi Gods under the command of each of you. They must be forced to work faster. Press them more. Increase the pace. Lengthen their hours." One of the great Anunnaki God commanders responds to God Enlil, "But God Enlil, they have been laboring excessively hard, toiling and suffering[93] for the past forty Shar

---

[88]M. Page 43, upper line 9 Their chamberlain was Ninurta.

[89]M. Page 43, line 10 And their sheriff Ennugi.

[90]M. Page 43, upper line 1, 2, 3, 4, 5, 6, and upper line 7 (see footnote 78).

[91]M. Page 141, line 94 The second, Badtibira, he (God Enlil) gave to the 'nugig' ('nugig' is an epithet for God Nannar, who was given this city after it was first given to God Ninurta); and T. upper page 225; and H. page 71, note 16, and note 17.

[92]M. Page 43, upper line 6 (see footnote 78).

[93]M. Page 43, upper line 1, and upper line 2 (see footnote 78).

years,[94] one hundred and forty-four thousand Earth years, since they arrived on Earth with God Enki, and built the first Earth city, Eridu." God Enlil smugly replies, snapping sharply and angrily, "Well see that they continue. They must be pushed more. I have increased schedules to meet." Then he pauses, looks at the workers, and then looks back at the 3 great Anunnaki God commanders, and orders, "That is all, you may go. You shall spend the night here, and I shall continue with you tomorrow." Then, after all the Gods bow their heads to him, they all walk back toward the main gate in the oval wall, from which they came, and away from the laborers. We scan over the workers, and look up to the Sun, and then look back down to the workers.

We now see time passing through a series of alternating cameo views of the workers toiling, and the Sun slowly setting, and this scene quickly blends into a dark night scene, as the sky begins to fill with twinkling stars, and a full Moon begins to slowly rise.

*

We see the stars twinkling in the night sky, and we look down to the work site, outside the oval wall. The work area is illuminated by the soft, firefly-like glow of the glass-like balls of light placed throughout the area, as well as bathed by the moonlight from the full Moon.

Midnight That Night

The existing quick pace of the laborers has been increased, and their hours have been lengthened. We scan the work area, and again observe the Igigi God called Wêila, who is still working, carrying the clay bricks, and we watch him work for a moment. Then suddenly, he stops working and walks to a group of 10 Igigi Gods who are near him. He stands, with his hands on his hips, and watches them, as they continue to labor and toil. There is an Anunnaki God officer, helmeted, in the distance, who is ordering the laborers as he shouts,

---

[94]M. Page 45, line 10, 11, 12, 13, and line 37 Excessive [work] for 40 years (40 Shar years = 40 x 3,600 = 144,000 Earth years).

"Faster, move faster." He does not see that God Wêila has left his work, as he is shouting and looking in the other direction.

God Wêila addresses the small group of laborers, "We have bore this work, and suffered the toil, our work is heavy, our distress is much."[95] Then, he pauses and slowly looks at the members of the group, and then he announces, "We must revolt!" The group of Igigi Gods look at God Wêila, and then begin to grumble[96] among themselves. Then slowly, one by one, they stop working, as one demands, "We need replacements," and another reflects, "We have labored within the marshes,[97] digging irrigation canals, and mining precious metals in the Abzu,[98] and building cities. We are fed up!" We look at the group, and as we hear more grumbling from them, one declares, "We are tired of work!" Then, beyond the group, others hear the grumbling, stop working, and join the group, adding their complaints. One adds, "I won't labor anymore," and another declares, "We have suffered the work night and day.[99] Let us confront our vizier, God Ninurta, the chamberlain!"[100] Then another adds, "That he may relieve us of our heavy work.[101] I've had enough!" Their din and their numbers quickly swell into a mob scene of all 150 Igigi Gods, with arms, hods, spades, and pickaxes; all swinging and being pushed up and down above their heads in the night air.

From a short distance away, we see the mob outside the oval wall, with the faint outline of the ziggurat in the background. Then the leader of the revolt, God Wêila, suggests,[102] "My brothers,[103] God Enlil, the chamberlain of old time,[104] the counsellor of the Gods, the hero,

---

[95]M. Page 43, upper line 1, 2, 3, and upper line 4 (toil = burden: see footnote 78).

[96]M. Page 45, line 39 They [were complaining], backbiting, and line 40 Grumbling in the excavation.

[97]M. Page 45, line 35 Excavating within the great marsh (the delta area), and line 36 [They] counted [the years] of the toil.

[98]M. Page 45, line 33 Excavating within all the mountains, and line 34 [They counted the years] of the toil.

[99]M. Page 45, line 38 [And] they suffered the work night and day.

[100]M. Page 45, line 41 Let us confront our vizier, the chamberlain (the Igigi Gods were seeking him as a mediator).

[101]M. Page 45, line 42 That he may relieve us of our heavy work.

[102]M. Page 45, line 47 Wê-ila opened his mouth, and line 3 [An Igigi] opened [his mouth].

[103]M. Page 45, line 48, and line 4 [And addressed] the gods, his brothers (brothers = comrades).

[104]M. Page 45, line 49, Enlil, the chamberlain of old time (the older one/previous one), and line 5 The vizier, the chamberlain of old time.

God Enlil;[105] come, let us unnerve him in his dwelling,[106] that he[107] may relieve[108] us of our heavy work!" Someone from the mob shouts, "To the temple of God Enlil, to Ekur," and another yells from the mob, "God Wêila, together, let us break the yoke of oppression," and God Wêila adds, "Now, proclaim war. Let us mingle hostilities and battle."[109] Then someone in the mob throws a small packet towards God Wêila, which lands near his feet, and he picks up the packet, and takes a small cylinder, 1/4 inch in diameter by 2 inches long, out of the packet. He holds it to a plastic-like spade, and presses one end, and an intense laser-like beam of light emits from the other end, and the beam ignites in a flame upon the spade. He sets fire to the spade which is thrust at him, and many of the others touch their plastic-like spades and plastic-like hods to his, and set fire to their own spades and hods in this manner,[110] and pass it on to the others, but the plastic-like gold pickaxes are not set on fire. All of the Igigi Gods follow God Wêila, as they scurry and flow toward the closed main gate in the oval wall, which envelops the compound. They are waving their arms, and waving and pushing their tools up and down, some flaming,[111] as they repeatedly chant, "Relieve us, relieve us," as they rush to the outside of the main gate,[112] with the temple, Ekur, at the top of the ziggurat inside. Then, they begin to surround the oval wall. Soon, the oval wall enclosing the ziggurat is almost surrounded by the protesters and their din, with God Wêila leading the mob, quickly approaching the main gate.

From inside the oval wall, we notice the five craft shimmering in the moonlight. It is very quiet inside the wall, but the mob sounds are quickly beginning to grow, and now become discernible above the previous normal sounds of the work crews. Then we observe an Anunnaki God officer, a male,

---

[105]M. Page 45, line 43 Enlil, and line 45 [Enlil], the counsellor of the gods, the hero, and page 47, line 57, and line 59.

[106]M. Page 45, line 44, and line 46, and pages 47, line 58, and line 60.

[107]M. Page 47, line 6 Replacements Enlil will appoint.

[108]M. Page 47, line 7 Verily he will appoint another.

[109]M. Page 47, line 61, and line 62 (war = revolt, and battle = conflict).

[110]M. Page 47, line 63 The gods heeded his words, 64 They set fire to their tools, 65 Fire to their spades they put, and line 66 And flame to their hods (no mention is made to setting fire to the pickax; probably because of its construction, it simply would not burn; or because it was an extremely powerful device that was too dangerous or explosive to set on fire; or because this was the special tool introduced to Earth by God Enlil, it represented him, and would be too great an affront to set on fire: see footnote 79).

[111]M. Page 47, line 68 They held them (up) as they went.

[112]M. Page 47, line 69 To the gate of the shrine of the hero Enlil.

called Kalkal,[113] who is helmeted, the gate keeper of Ekur, running toward the closed main gate from the small guard house near it. He is disturbed by the growing din of the mob. He reaches the closed gate, and is totally within its shadow just as the mob arrives, and to prevent their entry, he slides a large aluminum-like bolt across the gate, and thus locks it. There are two sides comprising the main gate doors, each composed of solid one foot square panels. The 12 feet high double doors are of aluminum-like metal, and we cannot see through them, except when the central one foot square viewing panel of each door is open. Then God Kalkal unbolts and partially opens one of the viewing panels, and looks out and watches[114] the mob, with the moonlight now upon his face.

Again we see the protesters, with God Wêila directly in front of the gate, as the chant and complaints come from the mob, "Relieve us, relieve us."

Then, we again see God Kalkal inside the gate, as he shuts the viewing panel and bolts it, and now presses his ear to the gate, frightened, as he listens to the chanting mob. He is out of sight, as the gate is shut, now including both central viewing panels.

Again we notice the five craft shimmering in the moonlight, and we zoom toward one of them; the personal craft of God Enlil.

Inside the craft, we observe God Enlil sleeping in his sleep chamber, wearing red tights. The sleep chamber is futuristic looking, with built-in bed and furnishings, somewhat like plastic. From the sleep chamber of God Enlil, located deep within the craft, we cannot hear the protests of the mob.

Meanwhile, inside the main gate, God Kalkal runs from the gate, toward the craft of God Enlil, up the ramp, and through the air flow door, a curtain of moving opaque air.

Inside the craft, we follow him running down a short hallway, through an automatic opening sliding door, and into the sleep chamber of an Anunnaki God officer, a male, called Nusku, who is sleeping in green tights. God Nusku is the personal servant of God Enlil. God Kalkal awakens God Nusku,[115] as he declares, "Wake up God Nusku, there is a disturbance outside." God Nusku wakes up, and together they listen to the barely audible protests of the mob,[116] as God Nusku hastily begins to dress.

113M. Page 47, line 74 Kalkal observed it and was disturbed.

114M. Page 47, line 75 He slid the bolt and watched them.

115M. Page 47, line 76 Kalkal roused [Nusku].

116M. Page 47, line 77 And they listened to the noise (evil conduct) of revolt.

Then we observe the mob, and hear their din, "Relieve us, relieve us."

God Nusku, now fully dressed, with helmet, grabs a double triad[117] laser-like wand, a multipurpose weapon, and attaches it at his waist, and together with God Kalkal, they run back through the short hallway, and out the craft through the opaque air flow door.

Outside the craft, we see them bathed in moonlight, as they run down the ramp, and toward the ziggurat front ramp-like stairs, for a vantage, from which to view the disturbance. They quickly climb the stairs, to the second tier, and from the stage-like tier, they both turn and look down at the mob outside the oval wall, and can now plainly hear the complaints, "Relieve us, relieve us. We have toiled too long. We are tired." They hear the din and see the raised tools, some of which are flaming, waving and moving up and down. The protesters now ring the outside of the oval wall, as they now have completely surrounded[118] it, but they do not attempt to enter the compound. Then we see a close-up view of God Nusku and God Kalkal, as God Nusku reacts, "It is half-way through the watch,[119] midnight. I must wake God Enlil in his sleep chamber," and they both turn, and we observe God Nusku with God Kalkal following, quickly running back down the ziggurat stairs, from the stage-like second tier, and then finally running down the last few steps of the ziggurat, and toward the craft of God Enlil, up the ramp, and through the opaque air flow door.

Inside the craft, we follow them running down a long hallway, to the door of God Enlil, which automatically slides open. God Kalkal waits in the hallway, as God Nusku quietly enters the sleep chamber of God Enlil. Within the sleep chamber, the protests of the mob are not audible,[118] and God Enlil is soundly sleeping. God Nusku gently awakens God Enlil,[120] and as God Nusku bows, he announces, "My Lord, God Enlil, your temple is surrounded,[121] battle has come right up to your gate."[122] As God Enlil is absorbing this fact, he has a

---

[117]E. Cover, fig. 6, and fig. 7.

[118]M. Page 47, line 71, and line 73 Ekur was surrounded, but Enlil did not know.

[119]M. Page 47, line 70, and line 72 It was night, half-way through the watch.

[120]M. Page 47, line 78 Nusku roused [his] lord.

[121]M. Page 47, line 80 My lord, [your] temple is surrounded, and line 82 Enlil, your temple is surrounded.

[122]M. Page 47, line 81, and line 83 Battle (conflict) has come right up to your gate (gate = door).

[123]M. Page 47, line 79 He got [him] out of his bed.

puzzled look on his face. God Enlil gets out of bed[123] and hurriedly begins to dress, with the assistance of God Nusku. He does not wear his helmet.

Again, we observe the mob in revolt outside the wall, chanting, "Relieve us, relieve us."

God Enlil and God Nusku run out of the sleep chamber, where they are joined by God Kalkal, and the three run back through the long hallway, and out the craft through the opaque air flow door.

Outside the craft, we see the three of them, bathed in moonlight, as they run down the ramp, and toward the ziggurat front ramp-like stairs. They quickly climb the stairs, again to the second tier, and from the stage-like tier, all three, with God Enlil in the middle, turn and look down at the chanting mob below. The three can now plainly hear their chant, "Relieve us, relieve us," and see their waving arms, and their tools, some of which are flaming, as they wave and push them up and down in the night air. We notice the face of God Enlil, which is surprised and troubled, as he turns to God Kalkal, and orders, "God Kalkal, see that 'all' the gates are barred,"[124] and God Kalkal bows, and immediately turns and quickly runs back down the ziggurat stairs. God Enlil, now looking excited and tearful, turns and directs God Nusku, "God Nusku, take your weapons and stand before me,"[125] and God Nusku removes the double triad laser-like wand from his waist, and holds it in a raised position, waist high, and stands directly in front of God Enlil.[126] Then God Nusku calmly reflects, "My Lord, sons are your sons, why do you fear your own sons,[127] your subordinates? They are all under your supreme command." God Nusku now pauses for a moment, as he looks at God Enlil, and then suggests, "Send that God Anu be fetched down, and that God Enki be brought to your

---

124M. Page 49, line 87 Nusku, bar your gate, and line 89 Nusku barred his gate (apparently God Nusku had overall authority over the gate, and probably the compound of Ekur, and God Enlil instructed God Nusku, who then probably delegated the task to God Kalkal, the actual gate keeper, who then physically performed the barring/securing of the gate).

125M. Page 49, line 85 Enlil opened his mouth, 86 And addressed the vizier Nusku, and line 88 Take your weapons and stand before me.

126M. Page 49, line 90 (God Nusku) Took his weapons and stood before Enlil; and E. cover, fig. 6, and fig. 7.

127M. Page 49, line 91, 92, 93 My lord, sons are your sons, 94 Why do you fear your own sons, 95 Enlil, sons are your sons, and line 96 (sons = creation, skillfully fashioned surrogate beings, whom the highest Gods had brought into existence to serve them: see footnote 49b).

presence.[128] Call an assembly of the Gods, and resolve the issue." Then, after a brief pause, God Enlil, in a relieved tone, responds, "Yes! An assembly! Send that God Anu be fetched down from Nibiru, and that God Enki be brought from the Abzu to my presence; also, the three great Anunnaki Gods, who are here tonight. We shall meet in the assembly craft, immediately." Then God Nusku bows, and replaces the wand at his waist, and both quickly run back down the ziggurat stairs, from the stage-like second tier.

Again, we see the chanting mob, "Relieve us, relieve us."

We observe God Enlil and God Nusku running down the last few steps of the ziggurat. Then God Enlil runs toward his personal craft,[129] and God Nusku runs toward another craft, which is one of the five shimmering in the courtyard, and we watch God Nusku run up the ramp, and through the opaque air flow door.

Inside the craft, we see God Nusku seated in the computer filled communications craft, in front of a computer which is alive with flashing display lights. He is touching the heat sensitive numbers on the console, as he recites to himself, "God Enki, Abzu. You are immediately summoned to Nippur." He quickly concludes calling God Enki, and then he rises and quickly seats himself in front of another computer console, touches a number, and the computer comes alive with flashing display lights. In the background, there is a transmutation chamber, which appears as a cylinder of dim light. God Nusku touches a sequence of numbers on the console, as he recites to himself, "God Anu, Nibiru. You are immediately summoned to planet Earth, to Nippur," and he quickly concludes fetching God Anu. He now sits and watches the cylinder as it begins to glow, as we look at an increasingly bright beam of light, which is like a particle beam. God Anu quickly materializes[130] within the beam. Then, as the beam diminishes, God Anu walks out of the cylinder of light, and they exchange a bow of the head. God Anu is helmetless.

As God Anu and God Nusku are walking away from the communications craft in the courtyard, in which there are now seven craft, they notice another craft rapidly approaching, with its brilliant glowing firefly-like light emitting

---

128M. Page 49, line 97 Send that Anu be fetched (beamed) down, and line 98 And that Enki be brought (by craft) to your presence.

129M. Page 47, line 84 Enlil withdrew to his dwelling (craft).

130M. Page 49, line 99 He (God Enlil) sent (God Nusku) and Anu was fetched (beamed) down.

from the underside, light up the courtyard area, hover, descend and prepare to land, as the glowing light begins to diminish, and the 4 pods like cattle hoofs extend 6 feet down, and the craft begins to land. They wait for the craft to land, and watch it do so, as God Anu, with a puzzled look upon his face, listens to the chanting, while in the background, the craft lands, making a total of eight craft in the courtyard, and the glowing light further diminishes, and then completely expires, and after a moment, a side of the craft extends down, forming a 4 feet wide ramp and doorway. Then, with the moonlight illuminating the craft, God Enki walks through the opaque air flow door and down the ramp,[131] also with a puzzled look, as his ear catches the din of the mob. He wears a helmet. God Anu and God Nusku are waiting near the ramp, and they and God Enki exchange a bow of the head greeting, and God Nusku announces, "God Enlil has called an assembly." God Anu asks, "What is the purpose, and what is that noise?" God Nusku answers both questions simultaneously, "To resolve the hostilities of the Igigi Gods revolt." Then the three begin to hurriedly walk toward the assembly craft, which is nearby, and as they move toward the craft, we notice a God and Goddess going up the ramp and into the craft through the opaque air flow door. Both are wearing helmets and one wears a body regulator.

Inside the assembly craft, we scan and see a smaller version of the assembly on Nibiru, with self luminous silvery-white metallic colored walls, and complete with a smaller dome viewing monitor in the center, which shows the chanting mob. We notice the 3 great Anunnaki Gods,[132] 2 males and 1 female, standing in back of the special, slightly raised section for the four main Gods, four of the 7 great Anunnaki Gods, and 12 Gods and 7 Goddesses, some of which are seated. Their communication helmets and body regulators are on a ledge that forms the back of the white pew-like molded circular seats. God Enlil is nervously standing in front of the special, slightly raised section. The God and Goddess who have just entered, quickly remove their helmets and body regulator, and place them on the ledge, in back of their seats, and join the Gods, who are quietly conversing, pointing, and making comments, as they watch the mob of Igigi Gods in revolt on the viewing monitor. We hear the low din of their conversations. Then, through the air flow door, a curtain of

---

[131]M. Page 49, line 100 Enki was brought (by craft) also to his presence.
[132]M. Page 49, line 103 With the great Anunnaki present.

moving opaque air, God Nusku ushers God Anu and God Enki into the assembly craft. God Nusku remains standing next to the air flow door. As they enter, the conversations quickly fade to silence, as all the assembly members now take their seats. God Enki removes his communication helmet, which he places on the ledge, in back of his seat. God Anu[133] is in the center, with God Enlil to his right, and God Enki[134] to his left. Goddess Ninhursag has not yet arrived, and her place next to God Enki is vacant, as are several other seats.

God Anu rises in silence, and all the Gods bow their heads to him, and then he motions to his right, to God Enlil, and then God Anu sits down. Then God Enlil rises,[135] and all the Gods bow their heads to him, all except God Anu. He first looks at God Anu, and then he scans the assembly, looking for assurance, and then God Enlil sternly addresses the assembly,[136] "In this assembly of the Gods, I, God Enlil, supreme commander and ruler of the Earth, have brought you to my presence." He pauses, and then resumes in a wavering and inquiring tone, with guilt, as he points to the mob on the viewing monitor, "Is it against me, personally, or as administrator of the Earth, that it is being done?[137] Must I engage in hostilities with my sons?[138] What do my very own eyes see?[139] That battle has come right up to my gate!"[140] God Anu rises as God Enlil speaks, and God Enlil sits down. Then God Anu makes an arm motion to God Nusku, who stands next to the air flow door, and God Anu suggests to God Enlil,[141] "Let God Nusku go out and ascertain the reason why the Igigi Gods have surrounded your temple.[142] To your sons, give a command from us,"[143] and God Anu slowly sits down. Again God Enlil rises,

---

[133]M. Page 49, line 101 Anu, king (ruler) of heaven (space), was present.

[134]M. Page 49, line 102 King (ruler) of the Apsû (Abzu), Enki, was in attendance.

[135]M. Page 49, line 104 Enlil arose [and] they bowed.

[136]M. Page 49, line 105 Enlil opened his mouth, and line 106 And addressed the great [gods].

[137]M. Page 49, line 107 Is it against me that it (revolt) is being done?

[138]M. Page 49, line 108 Must I engage in hostilities with my sons?

[139]M. Page 49, line 109 What did my very own eyes see?

[140]M. Page 49, line 110 That battle (conflict) has come right up to my gate!

[141]M. Page 49, line 111 Anu opened his mouth, and line 112, and page 55, upper line 8 Anu opened his mouth to speak, addressing Enlil.

[142]M. Page 49, line 113 The reason why the Igigi have surrounded your gate, and line 115 Let Nusku go out and [ascertain].

[143]M. Page 49, line 116 A command from us, and line 117 To [your] sons give.

and then instructs God Nusku, "God Nusku, take your weapons, and open the gate.[144] Ask from all of us,[145] who is the instigator of battle,[146] who is the provoker of hostilities,[147] who declared war and brings to me battle?[148] In this assembly, bring his name to God Enlil."[149] As God Enlil slowly sits down, God Nusku bows, and then runs through the opaque air flow door and out of the assembly craft,[150] taking the double triad laser-like wand in his hand, in a ready position.

God Nusku is inside and next to the main gate, and slightly within its shadow. The din of the chanting protesters is very loud, as he unbolts and opens the one foot square viewing panel, at the center of one of the doors, and looks out. As he peers out the small opening, with the moonlight again upon his face, he notices God Wêila in front of the door, and in front of the mob of Igigi Gods.

Outside the gate, we observe the face of God Nusku, as he peers out of the door panel opening, and asks, as he shouts above the din of the chanting mob, "God Wêila, God Wêila, listen," to get the attention of God Wêila. God Wêila looks at the door, and upon seeing God Nusku in the door panel opening, turns and holds up his arms, and asks the protesters, "Quiet, quiet, silence," and they slowly quiet down, as the request for silence spreads around the wall among the protesters.

Meanwhile, inside the assembly craft, we see the quieting mob on the viewing monitor.

Outside the gate, the mob is now somewhat quiet, and we again observe the face of God Nusku in the door panel opening, as he asks,[151] "God Enlil and all the Gods would like to know,[152] who is the instigator of battle, who is the

---

[144]M. Page 51, line 118, 119, 120, and line 121, and page 55, upper line 9 Nusku, open your gate, [take] your weapons with you.

[145]M. Page 51, line 122, 123, 124, 125, 126, and line 127, and page 55, upper line 10 In the assembly of the great gods, bow down, stand up, 11 Speak to them and repeat to them our words, 12 Anu [your father] (your creator) has sent me (God Nusku), and upper line 13 Also your counsellor, [the warrior Enlil] (has sent me).

[146]M. Page 51, line 128 Who is [the instigator of] battle?

[147]M. Page 51, line 129 Who is [the provoker of] hostilities?

[148]M. Page 51, line 130 Who [declared] war (revolt), and line 131 [And brings to me battle] (conflict).

[149]M. Page 51, line 132 [In this assembly], and line 133 [Bring his name] to Enlil.

[150]M. Page 51, line 134 [Nusku went to (out of) the assembly of] all the gods.

[151]M. Page 51, line 135 To the provokers he explained.

[152]M. Page 51, line 136, 137, 138, and line 139.

provoker of hostilities?"[153] Then he pauses, and waits for a reply. God Wêila looks around at the protesters, and then answers, with all the Igigi Gods, in unison and in a forceful tone, "Every single one of us Gods has declared war."[154]

Inside the assembly craft, we notice the astonished reaction of the assembly members to this statement, as they watch on the viewing monitor.

Again outside the gate, God Wêila declares, "Excess toil."[155]

Then, again inside the assembly craft, we see God Wêila speaking on the viewing monitor, "We have ended our toil in the excavations."[156]

Then outside the gate, God Wêila continues, "Our work was heavy, our distress much,"[157] and the mob again begins to loudly chant, "Relieve us, relieve us."

Meanwhile, inside the gate, as God Nusku hears the mob resume the chanting, he slams the door panel shut in disgust, and he replaces the wand at his waist, and runs back through the moonlit courtyard toward the assembly craft.[158]

Inside the assembly craft, we observe God Nusku enter, and now all the assembly members are engaged in heated discussions, and we hear their argumentative words, as some disagree with God Wêila, and react, "Punish them," while others agree with God Wêila, and respond, "Relieve them." As God Nusku approaches the seated God Enlil, the assembly becomes quiet. God Nusku bows, and reports to God Enlil, "Every single one of the Gods has spoken in favor of war with you, God Enlil,"[159] and as God Enlil slowly rises, his eyes begin to tear,[160] and God Nusku walks away to his place next to the air flow door. Then God Enlil turns to his left, toward God Anu, gathers his words, and emotionally appeals,[161] "Noble one, while you are here on Earth, use your authority, your power, while the Anunnaki Gods are present before you in this assembly. Summon one God, the defiant one," and then he looks

---

[153] M. Page 51, line 140 Who is [the instigator of] battle?, and line 141 Who is [the provoker of] hostilities?

[154] M. Page 51, line 146 Every single [one of us gods has declared] war (revolt).

[155] M. Page 51, line 149 [Excessive] toil [has killed (excessively burdened) us].

[156] M. Page 51, line 147 We have ended our toil in the [excavation].

[157] M. Page 51, line 150 [Our] work was heavy, [the distress much].

[158] M. Page 53, line 153 Nusku took [his weapons with him], and line 154 He (God Nusku) went (left the gate), he returned (to the assembly).

[159] M. Page 53, line 164 [Now, every] single one of us gods, and line 165 Has spoken in favor of war (revolt) with (against) Enlil.

[160] M. Page 53, line 166 When Enlil heard that speech, and line 167 His tears flowed.

[161] M. Page 53, line 168 Enlil gathered his words, and line 169 And addressed the warrior Anu.

at the viewing monitor, which shows God Wêila standing in front of the chanting mob. God Enlil now points to him, as he continues, "The leader, God Wêila, and take him with you to our heavenly planet, Nibiru, and there have him put to death,"[162] and we see a close up view of God Enlil as he concludes his emotion filled request, and then he sits down.

Again, God Anu rises and he slowly scans the assembly, and then he answers, "What are we accusing them of? Their work was heavy! Their distress was much! Every day and night they toil, their lamentation is heavy. There is much work to do in order to complete the assigned task here on Earth,"[163] and God Anu pauses. Some of the Gods look at each other with surprised expressions of disbelief, while remaining silent, and then God Anu continues, as he looks to his left, and gazes at the vacant place of Goddess Ninhursag, "Summon Goddess Ninhursag, our chief doctor, midwife of the Gods, wise Mämi,[164] immediately." God Nusku, who is standing next to the air flow door, bows and rushes out through the opaque air flow door, as God Anu continues, "While Goddess Ninhursag, the birth-Goddess, is present on the Earth, let her create offspring, Lullû, primitive man. Let man bear the toil of the Gods!"[165] As he concludes, he remains standing, and observes Goddess

---

[162]M. Page 53, line 170 Noble one, with you to heaven (Nibiru), 171 Carry (use) your authority, take (use) your power, 172 While the Anunnaki (assembly) are present before you, and line 173 Summon one god (the instigator) and have him done to death (kill him), and page 55, upper line 3 You are noble, 4 Take your power, 5, and upper line 7 Summon one and do [him to death].

[163]M. Page 53, line 174, 175, 176, 177, 178 [Every day] and night they toiled, 179 [The lamentation was] heavy, [we could] hear the noise (complaints on Nibiru, either from reports, or by monitor, possibly viewing monitor), 180 There is much work to do, and line 181 [We must complete the assigned] tasks, and page 55, line 1, and line 2, and lower line 3, 4, 5 Every day and night they toiled, and lower line 6 The lamentation was heavy, we could hear the noise (evil conduct).

[164]M. Page 57, line 192 They summoned and asked the goddess, and line 193 The midwife of the gods, wise Mämi (an epithet for Goddess Ninhursag); and E. Page 67, line 7

[165]M. Page 55, lower line 8, and upper line 6, and lower line 9 Let her create Lullû-[man] (primitive man/fool/imbecile), and page 152, note Gii.9; and T. page 302 lulu amelu ("primitive worker") and awilum ("laborer"), and page 307 amelu–a serf, and page 311, and page 183 LU = sheep/shepherded; and G. page 15 LU = amelu, "man", and page 39, note 2 a-PI (= wi or we)-lu (could these syllables possibly mean: wi or we from God Wê-ila/we-e-i-la, mixed with lu from Lullû, primitive man, in order to create mankind, which is part God and part primitive man; and is the correct Hellenic word for this admixture, in this context, δαίμων/daemon, which is any of the secondary divinities, ranking between the Gods and mankind); and H. page 90, note 131 (demon); and D. lower page 94 (Lilith); and M. page 55, lower line 10 Let him bear the yoke of toil, and page 57, line 11 Let him bear the yoke of distress, 12 [Let man carry the] toil of the gods, 1, 2, 3, 4, 5 Let him bear the yoke of toil, 6 Let him bear the yoke of distress, 189, 190, and line 191; and E. page 67, line 4 toil.

Ninhursag coming through the opaque air flow door, being led into the assembly craft by God Nusku. She is wearing a communication helmet, which she removes after entering, and God Nusku leads her to her seat, bows to God Anu, and then returns to stand next to the air flow door. She proceeds to place her helmet on the ledge, in back of her seat, and then turns and remains standing, as she bows to God Anu, and then she and all the other assembly members exchange a bow. Then looking at God Anu, she notes, "I have been listening to the words of the assembly," and she sits down. God Anu, looking at her, commands,[166] "You are the birth-Goddess, creatress of mankind, create Lullû, primitive man, and then mankind, man, that he may bear the yoke of oppression. Let him bear the yoke assigned by God Enlil. Let mankind, man, carry the toil of the Gods!"[167] Then God Anu sits down, and one of the great Anunnaki Gods, a male, one of the group of three sitting behind him, responds, "Yes,"[168] and another of the group of great Anunnaki Gods, a female, reacts, "Yes," and several of the other assembly members add, "Yes, yes."

Then, after this brief emotional display by some of the assembly members, Goddess Ninhursag rises and looks at God Enki, then she scans the assembly, and then addresses the Gods,[169] as she declares, "It is not possible for me to make things, skill lies with God Enki, our chief scientist. Since he can cleanse everything, let him give me the clay, so that I can make it,"[170] and she sits down.

---

[166]M. Page 57, line 192 They summoned and asked the goddess; and E. page 67, line 6.

[167]M. Page 57, line 194 You are the birth-goddess, creatress of mankind (implies that she was already known as the birth-Goddess, and possibly had previous experience creating other "mankind" on other planets), 195 Create Lullû that he may bear the yoke, 196 Let him bear the yoke assigned by Enlil, and line 197 Let man carry the toil of the gods; and E. page 67, line 8, 9, 10 of Enlil, and line 11 toil; and J. page 97, line 28, and page 99, line 29.

[168]M. Page 59, line 218 In the assembly answered 'Yes', and line 219 The great Anunnaki, who administer destinies.

[169]M. Page 57, line 198, and line 199; and E. page 67, line 12, and line 13.

[170]M. Page 57, line 200 'It is not possible for me to make (initiate the essence of animate/living) things, line 201 Skill (ability to prepare the initial essence) lies with Enki (by using the word skill, and from the context, it can be assumed this procedure was previously known), 202 Since he can cleanse (perfect the initial essence of) everything (animate and inanimate), and line 203 Let him give me the clay (perfected initial essence) so that I can make (bring to fruition) it (mankind); and E. page 67, line 14, 15, 16, and line 17 Clay give me (it may seem obvious that use of the word clay is a human metaphorical description of an agent, and not a vessel, which would contain the perfected initial essence. But if we examine this more closely, it may be a reference to the perfected initial essence itself, for the components of clay, its ingredients, are a perfected form of mud, mud with all the proper ingredients to become clay, or essence with all the proper ingredients to become

God Enki rises, looks at Goddess Ninhursag, and responds,[171] "Your compliment is excessive, but enjoyable." He now turns toward the assembly, as he continues, "As you know, I have been moderately experimenting on Earth creatures since I arrived. I shall immediately expand my research, and proceed quickly to fashion the creatures," and he remains standing, as God Anu rises and instructs, "O my son, from your knowledge, work what is wise. Fashion servants for the Gods.[172] Bind upon it the image of the Gods!" Then God Anu sits down, and God Enki continues, "On the first, seventh, and fifteenth day of the month, I shall make a purifying bath. Let one God be selected, and sample genes taken, so that all the Gods may be cleansed in a dipping, represented simultaneously."[173] God Enki pauses, and looks around the assembly, and then he points to the viewing monitor, which shows God Wêila in front of the chanting mob, and God Enki announces, "Select God Wêila, he has personality!"[174] Again, he pauses, and we observe the Gods, as they look at each other with surprised looks, and again God Enki continues, "From his flesh and blood, let Goddess Ninhursag mix clay, as a preserving agent, that God and man may be thoroughly mixed in the clay. So that we may hear the drum, their heartbeat, for the rest of time, let there be a spirit from the Gods flesh. Let it proclaim living man as its sign, so that this be not forgotten, let there be a spirit,"[175] and God Enki sits down, and again we observe the

---

mankind; and if that were correct, clay could be a metaphoric term, an epithet, used by the highest Gods to describe the perfected essence of a complicated process of genetic engineering, and thus very simply and in an understandable human term, inform mankind of this skill); and T. lower page 319 (clay and egg could share the same root word, and therefore be synonymous).

[171]M. Page 57, line 204, and line 205; and E. Page 67, line 18, and line 19.

[172]K. Page 70 O my (God Anu, his father, although the text implies his mother) son, rise from thy bed, from thy knowledge, work what is wise, Fashion servants of (for) the gods, may they produce their (the Gods) bread.

[173]M. Page 57, line 206, and line 207, and page 59, line 208 Let one god (alien) be slaughtered (be the donor), and line 209 So that all the gods (aliens) may be cleansed (represented) in a dipping (simultaneously, implies that the genetic material of the highest Gods was also within the Igigi: see footnote 49b); and E. page 67, line 20, 21, 22, and line 23 And let the gods be cleansed.

[174]M. Page 59, line 223 Wê-ila, who had personality, and page 153, note 223 "that something special was passed on to man from the slain god" (slain = donor).

[175]M. Page 59, line 210 From his flesh and blood, 211 Let Nintu (an epithet for Goddess Ninhursag) mix clay, 212 That god and man, 213 May be thoroughly mixed in the clay, 214 So that we (the Gods) may hear (monitor) the drum (the heartbeat of mankind) for the rest of time (until mankind or the Earth expires), 215 Let there be a spirit from the god's flesh, 216 Let it proclaim living (man) as its sign, and line 217 So that this be not forgotten, let there be a spirit; and E. page 67, line 24, 25, 26, and line 27 Thoroughly united in the clay.

assembly members, as another of the group of great Anunnaki Gods, a male, reacts, "Yes,"[176] and another assembly member adds, "Yes, yes."

Then, after this emotional display, God Anu[177] rises and commands, "I decree that my son, God Enki,[177] our chief scientist, and my daughter, Goddess Ninhursag,[177] our chief doctor, combine their efforts to create man, and bind upon man, our image."[178] Then God Enlil[177] quickly rises and commands, "This, I also decree. Let us clasp hands together." All the Gods rise and clasp hands, and then God Anu instructs, "Let the task begin," and God Enlil demands, "Your oath to us."

Together, all the Gods declare, "The decree of God Anu and God Enlil is binding."

*

It is bright daylight in southeastern Africa, the Abzu, and we see three craft close together in a triangle, and linked together by clear glass-like passageways; the biological laboratory craft of God Enki, the medical craft of Goddess Ninhursag, and a support craft. They are in a clearing, with the forest of tall and thick bushes and trees in the background.

ABZU
Many Months Later

We scan to the back of the craft, where there is a circular laser-like fenced area approximately 200 feet in diameter, with a water pond in one section of the pent up area, near the perimeter. In this pent up area, there are a variety of mutates, which make a diversity of animal sounds, and we see some close up views of them. There is a brown dog with a fish tail. A dog with a fourfold width body. A dog with the head of a pig. A reddish-brown bull with the head of a lion, with long shaggy hair on its head. A brown horse with the head of a

---

[176]M. Page 59, line 218 In the assembly answered 'Yes', and line 219 The great Anunnaki, who administer destinies.

[177]E. Page 71, line 26 Did Anu, Enlil, 27 Ea (Enki), and Ninmaḫ (Ninhursag), and line 28 The great gods, decree (for mankind. Ninmah is an epithet for Goddess Ninhursag).

[178]T. Chapter 11; and V. Genesis 1:26 And God (God Anu) said, Let us make man in our image, after our likeness (similar to this).

dog, with long straight hair on its head. A goat with the head of a dog, and front legs of a bull. A horse with one lion head, and one goat head.[179] Our eyes dwell on the mutates for a length of time, and we observe them from many different angles.

Inside the biological laboratory craft of God Enki, our eyes slowly scan, and we notice the area with glass-like flasks and beakers containing bubbling chemicals of different colors, while the computers are alive with flashing display lights. Then we scan to an area of different size cages of clear plastic-like material, some with small mutates in them, and some empty. We now scan to God Enki, the chief scientist, and Goddess Ninhursag, the chief doctor, who are seated at a laboratory table. There are 10 assistant scientists, 6 Gods and 4 Goddesses, at work in the laboratory. God Enki reports to Goddess Ninhursag, "None of the mutate creatures appears to be suitable, using the cell fusion process and only the genes of the various creatures. But I shall continue these experiments, before I introduce the genes from God Wêila," and then Goddess Ninhursag reflects, "There are an abundance of creatures here in the Abzu, many other species to select from."

After a brief but thoughtful pause, God Enki declares, "The creature from whom I shall fashion servants for the Gods, it exists;[180] the creature exists," as he points to two large cages with solid rear and side walls, and a clear plastic-like door, opposite the laboratory from the other cages, and we see a close up view of the two cages.

In one cage there is a male ape-like creature, and in the other cage there is a female ape-like creature, appearing as they had previously, both with brown skin and brown eyes, and with their approximately 1-1/2 inches long thick straight, but shaggy looking unkempt black-brown body hair, and with long straight black-brown hair on their heads, which is also unkempt. They are at

---

[179]T. Upper page 310 (incidentally, it is reported there was a contest for the protection (naming) of a city in ancient Hellas, and God Enki/Poseidon created (genetically engineered) the horse, and Goddess Ninhursag/Athena created (genetically engineered) the olive (olive tree). The Gods considered the olive was of greater benefit, and named the city Athens after her. If the account is correct, it is possible that the actual creation of the horse and olive took place during this experimentation period; but the competition, presentation of their former creations and naming of the city, obviously took place at a much later time).

[180]K. Page 70 O my (God Enki) mother (although his mother, Goddess Nammu, was not directly involved in the creation of mankind, she may have inspired him to work with Ninmah, an epithet for Goddess Ninhursag, his half-sister), the creature whose name thou hast uttered, it exists (as I further explain under 'Evolution' on page 356 in my epilogue).

the rear of their cages, and are silent, trying to hide from their captors, with expressions of fear upon their faces, and they cannot be seen from the laboratory table.

Then Goddess Ninhursag responds, "These must be the more advanced and highly evolved creatures. I have heard of them," as they rise and walk to the cages for a closer look, and after a long examining look, she continues, "They somewhat resemble us," and God Enki informs, "These creatures were just brought to me from the mine area. They have been evading capture for a long time. They may possibly arbitrarily evolve into mankind, but that shall take hundreds of thousands of Earth years to determine. I have not yet attempted to mutate one of these. I shall now proceed with them, using the cell fusion process. Since the results have not been favorable with the other creatures, I shall now experiment with this more highly evolved creature. They are not as strong as the quadrepedes,[181] but perhaps a combination could be adaptable," and they both look at the caged ape-like creatures.

We observe the male and female ape-like creatures, as they peek at God Enki and Goddess Ninhursag, from half-turned, fearful and suspicious faces.

We see time passing through a series of alternating cameo views of normal clouds passing through the sky, and sunrise to sunset sequences.

*

It is bright daylight, and we see the outside of the biological laboratory craft of God Enki, as it appeared before, one of the three craft close together in a triangle, and linked together by the clear glass-like passageways.

Many Months Later

We scan to the back of the craft, to the circular laser-like fenced area. There are new varieties of mutates, which make snorting sounds, and we see some close up views of them. There is a male ape-like creature with two wings, and long black-brown shaggy hair on its head. Another looks like the

---

[181]M. Page 141, line 50 Animals of all sizes, the quadrupeds (implies that bipeds, ape, primitive man, and early mankind included, were also considered by the Gods, the shepherders, as animals, the shepherded), were placed as a fitting ornament of the plains.

first, except it has two faces. Another looks like the first, except it has four faces. A male ape-like creatures body, but with two heads; one black-brown shaggy haired male ape-like creature, and one black-brown straight haired female ape-like creature. A female ape-like creature has the legs and horns of a goat. A male ape-like creature has a brown horse's legs for arms and legs, but stands upright. A brown horse, but from the shoulders of the horse protrudes a male ape-like creature from the waist up. A reddish-brown bull standing on its rear legs, with the head of a male ape-like creature. A horse standing on all four legs, with the head of a female ape-like creature, but with a fish tail. Some of the bull and horse combinations stand upright, on their rear legs, while others stand on all four legs. Some of the mutates are caressing.[182] (Thus, through a procedure like cell fusion and implantation, various mutates are being created.)

Within the bright sunlight, and from a distance, we observe a group of what appears to be marching Igigi Gods, who are near the area of the mines, and we zoom in for a close up view. There is one Anunnaki God officer, crisp and clean, helmeted and wearing a body regulator, and 15 Igigi Gods, also crisp and clean, and some are leading the way, and others are following 2 mutates, both males, who the Igigi Gods are forcing to do their heavy work, under the supervision of the Anunnaki God officer. The sweating, dusty and dirty snorting mutates are noticeably tired, but passive, for they are carrying large rocks above their heads, and are being prodded as they are being marched to a dumping area near the mines, for the rocks have just been taken from inside a mine. The mutates are both similar. From the waist down, they have a reddish-brown bull's body and rear legs. From the waist up, they have the brown body and arms, and shaggy black-brown body hair of a male ape-like creature, but with more muscular arms, and the head of a male ape-like creature, with long shaggy black-brown hair, including a beard and the horns of a goat. They stand upright on the rear legs of the bull.[183]

---

[182]T. Lower page 309, and page 310, and page 310, fig. 150, and page 39, fig. 12; and J. page XLV, note 1 (in Hellenic); and E. page 77 (in English); and B. page 68 (cylinder seal), and page 116 (cylinder seal); and K. lower page 60, and lower plate XIV; and U. page 40, and page 42; and Y. after page 10, fig. 11, and after page 88, fig. 20; and N. page 81, fig. 44 (upper register), and page 97, fig. 60, and fig. 61 (the bull/ape-like creature, in reverse order, would be similar to the Minotaur of the Minoans, and the horse/ape-like creature would be similar to the Centaur of the Hellenes).

[183]T. Page 67, fig. 27, and page 169, fig. 92, and page 310, fig. 150.

We see time passing through a series of alternating cameo views of normal clouds passing through the sky, and sunrise to sunset sequences.

*

Inside the biological laboratory craft of God Enki, our eyes scan, and we observe the small mutates in some of the different size cages.

Many Months Later

There are 10 assistant scientists, 6 Gods and 4 Goddesses, working among the glass-like flasks and beakers, which still contain bubbling chemicals of different colors, and the computers, which still have flashing display lights. God Enki and Goddess Ninhursag are both standing, looking at a test tube-like vessel with clear fluid in it, that has been singled out from among the others, and God Enki asks, "Goddess Ninhursag, do you know that your nickname among the Anunnaki Gods, is now being commonly used by the Igigi Gods? They also call you Mämi," and Goddess Ninhursag answers, "Yes God Enki, I know. It is amusing. It was brought to my attention yesterday."

Then God Enki reports, "My experiments with the creatures, using only their genes, have been most interesting, but have not completely achieved the desired results. Now that I have acquired this sample, I shall combine the genes from a creature, with the genes from a God," and he picks up the singled out test tube-like vessel, and holds it up in front of Goddess Ninhursag, and he announces, "From God Wêila, the Igigi God who has personality,[184] I have

---

[184]M. Page 59, line 223 Wê-ila, who had personality (good qualities; because he had the vigor and courage to lead the Igigi revolt. If the Igigi were not a sterile being, genetically suspended, possibly an altered higher God, or possibly a hybrid, or possibly a hybrid of the highest Gods themselves, as I put forth in footnote 49b, then instead of creating a new replacement being, mankind, the Igigi alone would have been generatively able to produce their own offspring, and thus replacements, and they would have been allowed to do so, for that would have been the easiest solution; but because they were incapable, that was not possible for them, possibly because they were either altered higher Gods, or a hybrid, and as we all know, hybrids cannot reproduce. Therefore, the creation scenario was necessary, thus confirming my opinion that the being from which mankind was created, the Igigi, was a sterile being. But who were the Igigi created from? Could i-la or ila from God Wê-ila be

recently extracted[185] an inactive sample of generative fluid, a dormant sperm sample, and have activated it in this purifying bath.[186] It is now ready. From his flesh and blood,[187] his genetic material, I have selected a spirit;[188] a special gene, that shall proclaim living man as its sign,[189] so that this shall not be forgotten by man, upon the Earth.[190] I shall now prepare to fertilize an egg from the female creature," and he gestures toward the caged female ape-like creature, as he continues, "with the activated sperm from God Wêila," and through the clear door we see the female ape-like creature in its cage, next to the cage of the male ape-like creature, as God Enki reflects, "an identical procedure was previously successful upon a similar creature, on another planet." Goddess Ninhursag suggests, "We must thank God Wêila for his voluntary participation in this endeavor."

Then God Enki turns to one of his assistant scientists, a Goddess, and instructs, "Prepare the female creature. Let us now take an egg sample from it." The assistant scientist takes a 1/2 inch in diameter by 10 inches long metallic cylinder, and prepares to open the cage of the female ape-like creature, who now withdraws toward the rear of the cage. As the assistant scientist opens the clear cage door and moves toward the female ape-like creature, she touches the creature with the cylinder, and we hear an air release sound. The female ape-like creature is almost instantly sedated, and slowly sinks to the floor of the cage.

Inside the medical craft, we see a hospital setting with 14 rectangular metallic beds, all in a row, and 3 rectangular metallic operating tables, separate and in front of the 14 beds. As we scan the scene, we observe the anesthetized

---

a variant of i-lu or ilu/Lofty Ones/Gods, thus designating the being as a variant being? The clay tablet clues are vague, but in my opinion, from the weight and fabric of the story, it was most certainly from the Anunnaki. As in footnote 173, if all the Anunnaki Gods/fertile aliens are represented by the one Igigi God/sterile alien sample, then the Igigi sample also carried nearly all of the Anunnaki genetic stamp/material, therefore the Igigi and the Anunnaki were genetically closely linked).

185 M. Page 59, line 224 They (the Gods, through Enki and his assistants) slaughtered (obtained generative material) in their assembly.

186 M. Page 59, line 222 He made a purifying bath (as I further explain under 'Sterility' on page 359 in my epilogue).

187 M. Page 59, line 225 From his (the Igigi God, Wêila) flesh and blood.

188 M. Page 59, line 228 From the flesh of the god (alien) [there was] a spirit.

189 M. Page 59, line 229 It (the gene) proclaimed living (man) as its sign.

190 M. Page 59, line 230 And so that this was not forgotten [there was] a spirit.

female ape-like creature, lying on one of the 3 operating tables. The creature is covered, up to its shoulders, with a white plastic-like surgical sheet. There is an intense beam of bright light coming from a small ball, which is above each table and above each bed. There are 4 medical assistants, 2 Gods and 2 Goddesses, led by Goddess Ninhursag, who are performing an egg extraction procedure, as God Enki closely watches. All are dressed in white surgical smocks over their red robes. After a few moments, the medical assistants remove an egg sample from the female ape-like creature, and place it in a small, shallow and clear dish with clear fluid in it, and we see a close up view of the dish. Then Goddess Ninhursag takes the dish and presents it to God Enki, and he declares, "This I shall fashion into Lullû, primitive man," and we see another close up view of the dish.

Inside the biological laboratory craft of God Enki, we see God Enki seated at a circular computer-like microscope, which has a one foot diameter viewing screen. Goddess Ninhursag is standing behind him with his 10 assistant scientists, 6 Gods and 4 Goddesses, as they all quietly watch the screen. We observe the screen and see a clear fluid on the screen. Then God Enki begins programming the heat sensitive console numbers, and the display lights begin flashing. Again we observe the screen, and see a thin clear tube containing an egg, moving toward the center from the left. The egg is deposited onto the center of the screen, within the clear fluid, and the thin clear tube moves back off the screen. Next, we see another thin clear tube containing a sperm, moving toward the center from the right, which pierces the egg and ejects the sperm into the egg. This thin clear tube then moves back off the screen. We now see the fertilized egg, within the clear fluid, and Goddess Ninhursag exclaims, "God Enki, you have fashioned it." God Enki removes the small clear dish from the circular computer-like microscope, and shows it to Goddess Ninhursag, and God Enki informs, "Goddess Ninhursag, I shall now proceed, for it must be purified on the first, seventh, and fifteenth day of the month, in the purifying bath.[191] Through this biological process, the genes shall establish themselves. You may then mix

[191] M. Page 59, line 221, and line 222 He made a purifying bath.

this with the clay,[192] which is the preserving agent, the cytoplasm, so that we may hear the drum, their heartbeat, for the rest of time.[193a] You shall then implant the fertilized egg, with a nip of preserving clay, within the creature from which the egg was removed, and thereby bring it to term. Thus, I shall hybrid a creature, and a God." Then Goddess Ninhursag responds, "God Enki, this I shall do,"[193b] and again, we see a close up view of the dish.

*

It is bright daylight, and we see the outside of the three craft in the Abzu, as they appeared before, close together in a triangle, and linked together by the clear glass-like passageways, with the tall and thick bushes and trees in the background.

---

[192]M. Page 59, line 226 Nintu (and Ninti are epithets for Goddess Ninhursag, meaning Lady Who Gives Life, and could be interpreted to mean Lady of the Rib) mixed clay (see footnote 170).

[193a]M. Page 59, line 227 For the rest [of time they heard the drum] (as I further explain within 'Influence' on page 314 in my epilogue).

[193b]M. Page 59, line 231 After she had mixed the clay, 232 She summoned the Anunnaki (the astronauts), the great (highest) gods (aliens), 233 The Igigi, the great (highest) gods (aliens), and line 234 Spat upon the clay (both groups of aliens gave their approval and blessings to the procedure, but additionally, in the presence of the Anunnaki, the Igigi spat upon the clay, which means the Igigi wanted the results of the procedure to be successful and to emulate them. This is very strong evidence that Wêila was an Igigi, a sterile worker, and not an Anunnaki, however, if the Igigi were also considered as Anunnaki, astronauts, as these lines seem to indicate, which is debatable, the Igigi alone blessed and then spat upon the clay, which I believe is doubtful. Also, the highest and most supreme beings probably would not consider their lower and subservient beings as astronauts, equal in status, but we human beings, observing both of them as Gods/aliens, probably would, and that may be the source of confusion regarding this issue, therefore, viewing them as separate and distinct groups, I do not consider the Igigi as Anunnaki, but I do consider them as Gods/aliens. Furthermore, in line 233, the Igigi are referred to as great gods, which means highest gods, as are the Anunnaki, which is just not correct, for they are the lowest gods, and elsewhere are just called gods or god (alien). I find no other such reference to the Igigi, and because of lack of reinforcement, I must conclude that this reference to great could be a scribal error, or an intentional overstating of their status by an empathic scribe, embellishing the Igigi, elevating them to the Anunnaki status, because it was the Anunnaki, through the Igigi, who made mankind possible: from low/trivial/humble beginnings, come high/important/proud endings: see footnote 184).

ABZU
.56 Shar Years Later
or
2,000 Earth Years Later
294,800 B.C.

We scan to the back of the craft, to the circular laser-like fenced area. There are 8 hybrid creatures, 4 males and 4 females, looking more like mankind now, and we see some close up views of them. They are advanced Homo erectus, with a brain size of approximately 1,000 cubic centimeters. All stand upright on their rear legs, except when drinking or eating, and all make gasp-like sounds. Their bodies have less hair, but are still somewhat shaggy, with about 1/2 inch long body hair, but thinner than before. Their heads are covered with long quite shaggy, almost tight-curly black hair, except one, which has white hair. Most of the hybrids' eyes are still brown, but several are now gray. They are naked, and 4 have white[194] skin, and 4 have brown[194] skin.

---

[194]J. Page 123, and page 125, line 9 [They did] Nin-igi-azag (an epithet for God Enki) [fashion] two small (infant) creatures [. . . . .], 10 [Among] all the beasts he made [their form] glorious, 12 [. . . . .] . . one white (probably pinkish-beige) [and one black (probably dark brown) . . . . .], and line 13 [. . . . .] . . one white (probably pinkish-beige) and one black (probably dark brown) [. . . . .] (I interpret white as similar to the pinkish-beige skin color of Caucasians today, and black as similar to the dark brown skin color of Africans today; which I loosely call white and brown. This seems to imply that white and black creatures/prototype mankind were fashioned/created at the same time from the ape/chimpanzee-like creature, and made glorious or more desirable/useful than any of the other beasts/animals; and perhaps later the white creatures were more highly refined, and therefore their characteristics would be more removed/advanced from the prototype, as they are, while the black creatures were not as highly refined, and therefore their characteristics would be less removed/advanced from the prototype, as they are. Or could the black creatures have been fashioned first, and then the white and/or other types of creatures at a later time, and at different intervals, possibly in the order of black, red, yellow, and white. The black creatures could have been experimental and/or not successful/desirable, for whatever the reason, which would explain and justify the creation of other more removed/advanced creatures/races from the prototype creature/race, although the creation of the red and yellow races are not specifically mentioned in the clay tablets, and also explain the wide diversity of skin color and physical attributes we see today: a selective genetic type of hybridization. And of course it is also possible that the creatures/races were intentionally created different, by the same or different creators, at the same time or at different intervals, at or placed in different locations; white to Europe, yellow to Asia, red to Atlantis, and black remaining in Africa, because our creators knew some day their creatures/human beings, as our programmed intelligence progressed, would be intelligent enough to notice the differences, and the realization of being different would

Some are drinking water from the water pond, lapping it up with their tongue, with their head close to the water, like an animal. Others are eating grass with their mouth, like a sheep, not using their hands. Also, there are gazelles and sheep in the pen, which are also eating and drinking alongside the hybrids. Some hybrids are petting the animals and caressing them. The hybrids look somewhat more like mankind, but retain animal characteristics, and still prefer to associate with the animals.[195]

Then we notice God Enki and Goddess Ninhursag standing outside, near the biological laboratory craft ramp, looking at the hybrids, as he informs her, "Goddess Ninhursag, look at my plants now. Look at Lullû, primitive man, with the animals. At this stage of hybridism, they are still very much like the animals, and prefer their company, eating and drinking as they do. They are even mating with the animals, but as hybrids, they are sterile, and cannot produce offspring." Then God Enki and Goddess Ninhursag turn, and walk up the ramp toward the opaque air flow door.

Inside the biological laboratory craft of God Enki, we see the usual activity; the 10 assistant scientists, 6 Gods and 4 Goddesses, among the glass-like flasks and beakers bubbling chemicals of different colors, and the computers with flashing display lights, as it appeared before. We observe God Enki and Goddess Ninhursag walk through the opaque air flow door and into the craft, as he continues, "They are very close to what I want. Lullû, primitive man, has intelligence equal to the animals. I want mankind's intelligence to be higher than the animals, but lower than the Igigi Gods." Goddess Ninhursag asks, "How shall you achieve that?" God Enki answers, "To complete the process of hybridization, I shall use the gene selector computer to alter their genes, and select the most desirable traits from Lullû, primitive man. Thus, from Lullû, primitive man, I shall create mankind, perfected man," as he points to a computer, which has a one foot square screen, and shows the DNA ladder[196]

---

then stimulate our thoughts as to why the differences, for perhaps the differences are a subtle way to inform us of our genetic creation/manipulation: a type of calling card).

[195]T. Lower page 305, page 306, and page 307, fig. 149; and K. page xvi, and lower page 72 Like mankind when first created (implies successive procedures), They (mankind) knew not the eating of bread, Knew not the dressing of garments, Ate plants with their mouth like sheep, Drank water from the ditch; and F. page 19, 20, 21, and page 22; and Y. after page 10, fig. 11.

[196]M. Page 169, note 49, and sub note 2 mostly designating manufactured objects, seems to mean some kind of lattice-work (the words manufactured objects could be a human metaphorical description of genetic engineering, and the words lattice-work could be a metaphorical description of the DNA ladder of chromosomes).

with 7 genes opposite 7 other genes; a total of 14 genes. At this moment, the computer is off and has no flashing display lights, and we see a close up view of the gene selector computer.

Then God Enki declares, "A new perfected egg, taken from the primitive female hybrid, containing selected, refined traits, taken from the primitive male hybrid, shall soon be ready to process. After processing, using the gene selector computer, a pure genetic mixture shall be completed. I shall bind upon it the image of the Gods.[197] Tests have already shown it to be favorable. In order to protect this pure genetic seed of mankind, you shall implant it, with a nip of preserving clay, within my wife, Goddess Ninki,[198] who shall attentively bring the fertilized egg to term. Thus, mankind shall be created, perfected, many hundreds of thousands of Earth years before their possible arbitrary evolutionary time." (Thus, Homo sapiens, pre-Neanderthal-like mankind, early mankind, is to be created through a procedure like genetic transplant and implantation, by the Gods interceding[199] with genetic engineering in the arbitrary evolutionary progression of the existing ape-like creatures, Homo erectus.)

Goddess Ninhursag asks, "What shall be mankind's fate?" God Enki

---

[197]K. Page 70 Bind upon it the image of the gods.

[198]T. Lower page 313, and page 314 (this could be Goddess Ninhursag).

[199]N. Cover, and page 91, ILL. 52 (the bull (horns) shown here and elsewhere could possibly symbolize the connection between the Gods (extraterrestrials) and mankind, depicting the planet of the Gods (flying/orbiting near and stamping/perturbing the Earth), or the craft of the Gods (sonic boom/loud sounds like a bull), or the "high technological" headdress of the Gods (horns/dual antenna), or the fashioning process (early bull mutates) employed by the Gods to create mankind. This could possibly also depict the Gods originating from or associated with the constellation of Taurus, the bull), see V. Job 38:31 Canst thou bind the sweet influences of Plē̂ī´-ă-dē̇s, or loose the bands of Ō-rī´-o̱n (the Pleiades is in the constellation Taurus; and D. page 70, and page 71); and V. Ezekiel 1:10 (etc.), and they four (the craft legs) had the face (an emblem) of an ox (bull) on the left side (as is explained in A. page 57, and page 58: traveling or coming (flying planet and Gods, see T. page 219, fig. 112; and B. page 128, and page 129) to Earth (winged bull, see E. fig. 13; and B. frontispiece, and page 45; and N. page 196, ILL. 156, and page 248, ILL. 207, and ILL. 208), and employing fertility (bull symbol) techniques to mutate animals, including bulls, and then fashion mankind); and X. middle page 175 When Venus had a coma, the horns of its crescent must have been extended by the illuminated portions of the coma. It thus had two long appendages that looked like a bull's head (the horns could represent the planet Nibiru, itself, not Venus, when it looked like a bull's head as it also went through its phases. Therefore, in any of the aforementioned, the symbolic worshipping of the bull (horns) may have been actually the worshipping of the planet Nibiru, and/or simultaneously the worshipping of the Gods; as I further explain under 'Prophecies' on page 376 in my epilogue).

walks to the gene selector computer, and touches several heat sensitive numbers on the console, and the computer comes alive with flashing display lights, and God Enki answers, "I have decreed mankind's fate,[200] and have already entered their program on the gene selector computer." Now God Enki looks directly at Goddess Ninhursag, and he asserts, "The instructions of God Enlil, are to fashion mankind without our trait, without the ability to reproduce, like the Igigi Gods; and with lives of predetermined years, only thirty to thirty-five thousand Earth years, approximately ten of our Shar years, and not like the immortal Igigi Gods and ourselves. To achieve this, I[201] shall withhold from mankind our qualities, intellect and the capability to reproduce, genetic information from the tree of knowledge computer; and withhold from mankind the gift of immortality, genetic information from the tree of life computer,"[202] and we see close up views of each as he concludes pointing to the two computers, which are off. Each computer has a 6 inch square viewing screen. Then Goddess Ninhursag affirms, "We shall follow the decree of God Enlil," as they now look at the flashing display lights on the gene selector computer.

We see time passing through a series of alternating cameo views of normal clouds passing through the sky, and sunrise to sunset sequences.

*

It is bright daylight, and we see the outside of the biological laboratory craft of God Enki, as it appeared before, one of the three craft close together in a triangle, and linked together by the clear glass-like passageways.

---

[200]K. Page 70 O my (God Enki) mother (although his mother, Goddess Nammu, was not directly involved in decreeing the fate of mankind, she may have inspired him to work with Ninmah, an epithet for Goddess Ninhursag, his half-sister), decree thou its (mankind's) fate.

[201]J. Page 135, line 20 He formed mankind (God Enki, who in this sentence is metaphorically inferred to as Marduk); and V. Genesis 2:7 And the LORD God (God Enki) formed man (mankind, similar to this).

[202]V. Genesis 2:9 the tree (essence) of life (immortality) also in the midst of the garden (base), and the tree (essence) of knowledge (intellect and reproduction) of good and evil (traits, similar to this. Although I refer to and depict both as computers, because of our present state of the art and our own current common genetic engineering experiences, they may be a far more advanced or totally different concept than computers, for certainly the descriptive term "tree" is a human metaphorical characterization of a highly technologically advanced device or procedure).

Many Months Later

We scan to the side of the craft, and near the ramp we observe a large group of Gods. There are 25 Igigi Gods,[203] 20 males and 5 females, one of which is God Wêila, the leader of the revolt, all without helmets or body regulators; 6 Anunnaki God[203] officers, 4 males and 2 females, all helmeted and some wearing body regulators; 3 great Anunnaki Gods,[203] 2 males and 1 female, their commanders, all helmeted and some wearing body regulators; 10 assistant scientists, 6 Gods and 4 Goddesses, all without helmets or body regulators; 4 medical assistants, 2 Gods and 2 Goddesses, all wearing white surgical smocks over their red robes, and all without helmets or body regulators; and 5 Gods and 5 Goddesses, all without helmets or body regulators: most of whom, except the Igigi Gods, are also considered astronauts, Anunnaki Gods. God Enki and Goddess Ninhursag are standing in front of the group, at the foot of the ramp, both are without helmets or body regulators. All the Gods are expectantly looking up at the opaque air flow door.

Goddess Ninhursag turns and addresses the group,[204] "I have assembled you today, in order that you may be the first to welcome a new creature,[205] mankind, to the Earth. Since giving birth to him six Earth months ago, Goddess Ninki, the wife of God Enki, has cared for him as if he were her own son, the son of a God," and Goddess Ninhursag now turns and looks up at the opaque air flow door, and announces, "Goddess Ninki, present mankind to the Gods."

Then Goddess Ninki steps through the opaque air flow door, and stands at the top of the ramp, within the bright sunlight. She is holding a naked 6 month old in appearance, male baby;[206] first pre-Neanderthal-like mankind, early mankind, Homo sapien, who at maturity will have a brain size of approximately 1,300 cubic centimeters. She holds him up for their approval. The baby has white skin, tight-curly black hair, brown eyes, and is normal in every way. Goddess Ninki caresses the baby, as the group cheers, "Welcome mankind, welcome mankind." Then an Igigi God, a male, exclaims, "You have relieved us," as he beams with joy, and another Igigi God, a female, shouts, "Mankind is welcome," as she also beams with joy.

---

[203]M. Page 59, line 232, and line 233 (see footnote 193b).

[204]M. Page 59, line 235, and line 236.

[205]M. Page 59, line 229 It (the results) proclaimed (revealed) living (man, newly created mankind) as its sign.

[206]T. Page 315, fig. 153 (this could be Goddess Ninhursag).

Again, Goddess Ninhursag turns and addresses the joyful group of Gods, and particularly the Igigi Gods, as she continues, "You commanded me a task, I have completed it. I have removed your heavy work, I have imposed your toil on man. You raised a cry for mankind, I have loosed the yoke, I have established your freedom."[207] Then 6 Igigi Gods, including God Wêila, and the 3 great Anunnaki God commanders, and 3 Anunnaki God officers, all run to Goddess Ninhursag, and kiss her feet,[208] and God Wêila recalls to Goddess Ninhursag, "Formerly, we used to call you Mämi. Now, let your name be Mistress-of-All-the-Gods,"[209] as he looks at her and thankfully smiles.

Then God Enki declares, "His name shall be Adapa, our model for man," and we see a close up view of Adapa, as he is being caressed by Goddess Ninki, as God Enki continues, "and mankind's nickname shall be, the black headed people.[210] And as for me, Adapa shall be regarded as my newly created plant,"[211] and God Enki chuckles at his own facetious remark. Then the group also chuckles at the remark regarding plant. After a brief pause, after the chuckling subsides, God Enki continues, "Among all the forms, I have made his glorious,"[212] and again we see Goddess Ninki caressing Adapa, as God Enki continues, "Adapa is genetically clean. He is perfect." God Enki

---

[207]M. Page 59, line 237, 240, and line 241, and page 61, line 242, and line 243.

[208]M. Page 61, line 244 They (Anunnaki and Igigi) heard this speech of hers, and line 245 They (the two distinct groups) ran together and kissed her feet, (saying).

[209]M. Page 61, line 246 Formerly we used to call you Mämi, and line 247 Now let your name be Mistress-of-All-the-Gods (Bēlet-kāla-ilī).

[210]M. Page 141, line 47, and line 48 (after the Gods) Had created the black-headed people, and page 169, note 48; and J. page 98, note 7, and page 226, note 2; and T. page 321, and page 322; and K. upper page 42 O Utu (Sun God), shepherd of the land (the Sun leads the land, or crosses the land), father (guardian, not creator) of the black-headed people (mankind), lower page 52, and lower page 61; and L. lower page 285 through middle page 286 (since the color black is an absence or lack of color, could the identification black-headed people, an epithet for mankind because of black hair, represent and symbolize an absence or lack of completeness of function for human beings; and since the Gods hair color is described as white, as reported in footnote 21, a combination of all colors, could that represent and symbolize a totality and completeness of function for the supreme beings).

[211]M. Page 117, line 7, and line 11, and page 119, line 5, 12, and line 19, and page 121, line 35 (etc.) my plants (my creation = mankind); and K. middle page 62 After the water (essence) of creation (mankind) he (God Enki) had decreed (developed/programmed), (etc.), Like plant and herb (God Enki) had clothed the land (covered the Earth with mankind. It should be obvious that the reference to mankind as plants is more than facetious, for all the Gods considered God Enki, as he considered himself, a kind of horticulturist of mankind. Through this perspective, we can better understand and appreciate the term "seed of mankind").

[212]J. Page 125, line 10 [Among] all the beasts he (God Enki) made [their form] glorious.

and Goddess Ninki have grown to adore Adapa, and they have become the first advocates of mankind, since God Enki is the true creator of mankind, and both of them are beaming with pride and joy.

Then God Enki turns to one of the great Anunnaki God commanders, and instructs, "Report this to God Enlil, he shall be pleased. Our tests indicate that while his animal characteristics have been suppressed, his animal intelligence level has been elevated, and is now higher than the animals, but lower than the Gods.[213] He cannot reproduce, he is a sterile hybrid,[213] and he is not immortal, his days are numbered. Being genetically programmed,[214] I have established his functions and guidelines, and given him a maximum physical capacity, and a minimum mental capacity. This shall assure the Gods that our hybrid workers shall not develop their intellect, and they shall always be dominated and controlled by the Gods. They shall be ideally suited for menial work, and they shall remain helpless and clinging, not able to sustain themselves, dependent upon the Gods; responsive, passive and obedient to our commands. Mankind shall serve all the Gods. I have not given him a serious, earnest and intellectual disposition, characteristic of the Gods. And tell God Enlil, that I have fixed upon man, the image of the Gods,[214] as God Anu and God Enlil had decreed."

God Enki now turns to Goddess Ninhursag, and directs, "Come Goddess Ninhursag, to the house of destiny,[215] the creation chamber of the Gods," and they both begin to walk toward the medical craft, which is one of the other two, of the three triangular linked craft.

*

We observe the inside of the medical craft, which contains the area called the creation chamber, with the 14 rectangular metallic beds, all in a row, and

---

[213]T. Page 329 through page 331; and V. Psalm 8:5 For thou has made him (mankind) a little lower than the angels (a little lower than the Anunnaki Gods, and most certainly also deliberately lower than the Igigi Gods, similar to this; as I further explain under 'Sterility' on page 359 in my epilogue).

[214]Q. Middle page 204; and V. Genesis 1:27 So God (God Anu and/or God Enlil through God Enki) created man (mankind) in his (God Anu and/or God Enlil through God Wêila) own image, in the image of God (God Anu and/or God Enlil through God Wêila) created he (God Anu and/or God Enlil through God Enki) him (mankind, similar to this).

[215]M. Page 61, line 249 They entered the house of destiny, and line 250 Did prince Ea (God Enki) and the wise Mämi (Goddess Ninhursag).

the 3 rectangular metallic operating tables, separate and in front of the 14 beds.

Several Hours Later

It appears as it did before, with an intense beam of bright light coming from a small ball, which is above each table and above each bed, except now, as our eyes scan, we notice a birth-Goddess[216] lying on each of the 14 beds, each covered, up to her shoulders, with a white plastic-like surgical sheet.

As an automatic sliding door opens for them, God Enki and Goddess Ninhursag now enter the chamber,[215] both dressed in white surgical smocks over their red robes, and also entering, as an automatic sliding door opens for them, but coming from another chamber, are the 4 medical assistants, 2 Gods and 2 Goddesses, also dressed in white surgical smocks over their red robes. They all walk toward the first of the 14 beds. God Enki holds a glass-like flask, which contains a catalytic agent, and Goddess Ninhursag holds a gold tray, which contains fertilized eggs to create males and females, embedded within the clay preserving agent. We observe God Enki holding the glass-like flask, and then we see a close up view of the flask. It is one-third full, with a deep red wine-like color chemical. There is a glass-like rod in the flask, extending 4 inches out of the mouth. We observe Goddess Ninhursag holding the gold tray, and then we see a close up view of the tray.

There are 14 empty, small clear glass-like dishes upon it. Next to the dishes, on two separate clear glass-like plates, like butter plates, one on the right and one on the left of the tray, and under two clear glass-like covers over the two plates, there are two bars of pink clay, one on each plate. The bars are each 1/8 inch thick by 6 inches long by 1 inch wide. They are scored in 1/8 inch square segments. Also on the tray, there are several gold instruments. There is a small spade-like medical instrument, used for cutting and opening flesh, and a curved tweezers,[217] and a 1/4 inch in diameter by 10 inches long

---

216M. Page 61, line 251 With the birth-goddesses assembled, and page 63, line 8 The wise and learned (Goddess Ninhursag), and line 9 Twice seven (14) birth-goddesses had assembled (the clay tablets are silent as to the identity of the birth-Goddesses, or living incubators, who probably were not the highest ranking Goddesses; but probably were Anunnaki/astronaut females, female medical assistants within the medical group under the control of Goddess Ninhursag; and the remote but interesting possibility exists that the birth-Goddesses were Igigi females, for the Igigi were the lowest ranking Gods, and performed all the menial work).

217T. Page 41, fig. 14 (cylinder seal); and B. page 149 (cylinder seal).

rod, with a concave end. Then Goddess Ninhursag uncovers the clay bars,[218] takes the spade instrument, and begins to section 1/8 inch cubes, 7 of them from each of the bars, and places one in each dish, 7 to the right, and 7 to the left. As she does this, God Enki places a drop of the red catalytic chemical on each cube of clay,[219] by touching the cube with the end of the glass-like rod. After he is finished, God Enki declares, "It is ready. You may now proceed."[220]

Then Goddess Ninhursag notes, "From the seed of mankind,[221] Adapa, and prepared for me by you, God Enki,[222] I have nipped off fourteen pieces of clay.[223] These, I shall implant within the fourteen birth-Goddesses assembled here. Seven on the right,[224] to produce males,[225] and seven on the left,[226] to produce females.[227] They shall count the months,[228] and bring it to

---

[218]M. Page 61, line 4 [She] put [her hand out] to her clay; and K. page 70 Mix the heart of the clay that is over the abyss (Abzu, this took place in the Abzu).

[219]M. Page 61, line 252 He trod the clay in her presence; and K. page 70 The good and princely fashioners (chemicals) will thicken the clay.

[220]M. Page 61, line 254 Ea (God Enki, the chief scientist), seated before her (Goddess Ninhursag, the chief doctor), was prompting her, 1 [Prince] Ea spoke, and line 2 Seated he was prompting her; and K. page 70 Ninmah (the earth-mother goddess) (an epithet for Goddess Ninhursag) will work above thee, Ninti (goddess of birth) (an epithet for Goddess Ninhursag, the chief doctor, the primary birth-Goddess) will stand by thee at thy fashioning (during this stage of the procedure, they were working together).

[221]J. Page 135, line 21 The goddess Aruru (an epithet for Goddess Ninhursag) together with him (God Enki, who in this sentence is metaphorically inferred to as Marduk) created the seed of mankind (initial mankind).

[222]M. Page 61, line 253 She kept reciting the incantation, 255 After she had finished her incantation, and line 3 Prompting her she recited the incantation.

[223]M. Page 61, line 256 She nipped off fourteen pieces of clay, and line 5; and C. page 143 a thing nipped out of clay; and V. Genesis 2:7 And the LORD God (God Enki) formed man (mankind) of the dust (clay) of the ground (similar to this); and W. page 37, Volume II We created man of an extraction of clay, (etc.); thereafter We produced him as another creature (similar to this: see footnote 170).

[224]M. Page 61, line 257 Seven she put on the right, and line 5.

[225]M. Page 63, line 9 Seven produced males.

[226]M. Page 61, line 258 Seven on the left, and line 6 (the phrases, Seven she put on the right, and, Seven she put on the left, are very enigmatic, and could be a reference to a procedure of implanting seven fertilized eggs on the right side of the uterus, programmed to produce males, and implanting seven fertilized eggs on the left side of the uterus, programmed to produce females; or it could be a reference to a procedure whereby seven eggs and/or sperms from the right ovary and/or testicle were harvested, fertilized, programmed, and implanted to produce males, and seven eggs and/or sperms from the left ovary and/or testicle were harvested, programmed, united, and implanted to produce females).

[227]M. Page 63, line 10 [Seven] produced females, 11 The birth-goddess, creatress of destiny, 12 They (the birth-Goddesses) completed them in pairs, and line 13 They completed them in pairs in her presence; and V. Genesis 1:27 male and female created he (God Enki) them (mankind, similar to this).

[228]M. Page 63, line 277 The birth-goddesses were assembled, and line 278 And Nintu (an epithet for Goddess Ninhursag) [sat] counting the months.

term. When the tenth month arrives,[229] I shall cover my head, and perform the midwifery.[230] The umbilical cord cutter I shall use,"[231] and she holds her 4 inch diameter gold symbol pendant up from around her neck, as she declares, "for when the birth-Goddesses give birth,[232] I shall say, I have created, my hands have made it."[233] Then she looks at the gold tray, and again, we see a close up view of the tray.

After a brief pause, Goddess Ninhursag adds, "Because mankind shall be a sterile hybrid, unable to reproduce, I shall continue this process,[234] with the assistance of the birth-Goddesses, and create additional numbers of mankind, as many as are needed." As we carefully observe, they prepare to implant the first birth-Goddess with the nip of clay containing a fertilized egg, which is placed on the concave end of the gold rod, by using the gold curved tweezers. (Thus, a procedure like artificial insemination and implantation is used to produce additional numbers of sterile mankind by the birth-Goddesses.)[235]

---

[229]M. Page 63, line 280 [At the] destined [moment] the tenth month was summoned, 281 The tenth month arrived, and line 282 And the elapse of the period opened the womb.

[230]M. Page 63, line 283 With a beaming, joyful face, and line 284 And covered head she performed the midwifery.

[231]M. Page 61, line 260 She used the umbilical cord cutter, and line 7 For (the removal of) hair, she used the cutter of the umbilical cord (incidentally, the spiral or double spiral, similar to the symbol for Goddess Ninhursag; as I illustrate on page 308 in my comments, and similar to the Hellenic letter Ω, is also thought to be an ancient symbol for Atlantis; the home (away from Nibiru) of Poseidon/God Enki, where, with the consorting efforts of a 'mortal woman' named Cleito, the only daughter of an 'earth-born primeval' father whose name was Evenor (Eve?) and mother was Leucippe, 'begat and brought up five pairs of male children' (five sets of twins, obviously genetically engineered/selected), 'the eldest, who was king, he named Atlas, and from him the whole island (Atlantis) and the ocean received the name of Atlantic,' according to Plato).

[232]M. Page 63, line 286 She girded her loins as she pronounced the blessing.

[233]M. Page 63, line 289 I have created, my hands have made (brought to fruition) it (mankind); and K. page 70 Ninmah (an epithet for Goddess Ninhursag) will bind upon it (mankind) the image of the gods, known as mankind (it should be obvious that the epithets Ninmaḫ and Mämi for Goddess Ninhursag, director of the birth-Goddesses/surrogate mothers, because of her duties, could also be acknowledged as creatress/mother of mankind/humans, and could appropriately be the root words for our present endearing reference to our own mothers as ma(mah) and mama from Ninmaḫ, and mom and mommy from Mämi. Additionally, the Gods could be considered our genetic parents, and our actual earthling mothers and fathers our biological parents; and this realization we should all acknowledge and distinguish: see footnote 164 and 209).

[234]M. Page 63, line 16 That Bēlet-ilī, the wise Mämi, may be honoured, and page 65, line 295, 296, and line 297 (praises for Nintu and birth-Goddesses).

[235]T. Chapter 12, and page 314, fig. 151, and fig. 152 (the birth-Goddesses are surrogate mothers for the invitro or externally fertilized eggs, what we loosely call test tube babies, which are to be the artificially fertilized children of mankind; carried, incubated, and born by the birth-Goddesses: see footnote 216).

*

It is bright daylight, and from a distance, we see the oval walled compound of Nippur, the city of God Enlil, with its colorful central 7 tiered ziggurat, and temple with antennas atop it, as the seventh tier, as it appeared before. There is someone working outside in the fields, canals, and on the building of the smaller rectangular storage buildings, still the only structures outside the oval wall, but we cannot distinguish them at this distance.

NIPPUR
2.50 Shar Years Later
or
9,000 Earth Years Later
285,800 B.C.

We scan and see a close up view of the area outside the oval wall. We observe laboring workers, who are pre-Neanderthal-like mankind, early mankind, 70 in number, 35 males and 35 females, all naked.[236] They are all sweating, dusty and dirty, and noticeably tired, with white pigment skin, brown eyes, and with long shaggy hair of tight-curls, like the Gods, except it is long on both males and females, and it is black in color. Mankind is silly, nonsensical, frivolous and hilarious. They only blurt crude sounds, which have no resemblance to articulated words, and they are all very passive, working and laboring very hard. It is like a forced labor scene, and the pace is quick. They are using the tools of the Igigi Gods,[237] the plastic-like gold pickax and the plastic-like spade, and digging the irrigation canals and carrying the plastic-like baskets of clay on their back, away from the canals, to be shaped into bricks, using the plastic-like rope support, which they grasp at their chest;

---

[236]T. Page 96, fig. 45, and page 103, fig. 49, and page 326, fig. 154, and fig. 155, and page 275, fig. 134 (early naked mankind, steering and poling a boat for a God); and N. page 39, fig. 18.

[237]K. Lower page 52 He (God Enlil) placed it (his special pickax) as a gift in their (mankind's) hands, (etc.), They (the Anunnaki Gods take from the Igigi Gods and) give the pickax to the black-headed people (mankind) to hold (to use), and upper page 53 (wielded by mankind) The pickax and the basket build cities (mankind labors with the advanced technology tools of the Gods: see footnote 79 and 110).

building the smaller rectangular storage buildings, working in the fields, tending the sheep, and serving all the Gods, including the Anunnaki Gods, and the Igigi Gods. They carry the sun-dried heavy clay bricks, and carry the mortar and cream colored plaster on the plastic-like hods. They stack the metal ingots, zag, and carry them on the aluminum-like poles. Mankind is performing all the labor the Igigi Gods provided previously.[238a]

As we continue to scan the area, we observe more sweating, dusty and dirty, and noticeably tired mankind. The females are working alongside the males, at every level of labor. They are all between 15 and 23 years of age in appearance, and there are no small children, children from the birth-Goddesses, among the workers. The sterile babies, born by the implanted birth-Goddesses, are raised in a special nursery area, and they are cared for like the sheep,[238b] until they are old enough to work; until they are 15 years of age in appearance. Some males and females are caressing the sheep, but there is no caressing among male and female mankind. There are 4 Igigi Gods, all without helmets or body regulators, as well as 6 Anunnaki God officers, all helmeted and some with body regulators. The Anunnaki [239] Gods are crisp and clean, and are supervising the work, now with the assistance of the crisp and clean Igigi Gods.

Then we notice God Enlil and God Enki walking outside the wall, surveying the work, and we see a close up view of them, as God Enlil professes, "God Enki, the black-headed people, your plants, have replaced the Igigi Gods labor everywhere, and they are very well pleased, for they toil no more. They are free from work, and they complain no more. You created mankind with the proper amount of intellect; higher than the animals, but lower than the Igigi Gods. Mankind's mental capacity is perfect for performing menial work, for they shall never rise above their servant status.[240] You sent me the

---

[238a]V. Genesis 2:8 And the LORD God (God Enlil) planted a garden (established cultivation at Nippur, the city/base of God Enlil) eastward (east of the Abzu, or Africa) in Ē´-dĕn (Sumer); and there he put the man (mankind) whom he had (decreed to be) formed, and 2:15 And the LORD God (God Enlil) took the man (mankind), and put him into the garden (base) of Ē´-dĕn to dress it and to keep it (to care for it, similar to these).

[238b](One of the definitions in Webster's dictionary of sheep is "a person who is silly, stupid, meek, timid, defenseless, etc.," which would be an appropriate description of early mankind: see footnote 165 and 181).

[239]K. Upper page 60 The Anunnaki, the great gods, In thy (mankind's) midst (in Sumer) have taken up their dwelling place (live in Sumer), In thy (mankind's) large groves they (the Anunnaki Gods) consume (their) food.

[240]K. Lower page 52 Upon his (God Enlil's decreed creation) black-headed people he looked steadfastly (approvingly).

white[241] skin color trait of mankind, and kept the brown[241] skin color trait to work the mines in the Abzu," and God Enki responds, "Yes God Enlil, each type has been given specific traits to make them more adaptable to different kinds of work and conditions. Do they cause you trouble?" God Enlil replies, "No, they do not cause trouble. They are passive and very obedient,[242] but they are also very frivolous and silly, hilarious and noisy. Without intelligence, they are but fools," and they pause to look at them, and they both smile at mankind, being amused by their behavior.

Again we observe mankind, silly and frivolous, working very hard. As God Enlil and God Enki conclude smiling at the workers, God Enlil cautions, "God Enki, give attention. I must warn you. Do not extend to mankind intellect and the capability to reproduce, genetic information from the tree of knowledge computer; or the gift of immortality, genetic information from the tree of life computer. For I, as ruler of the planet Earth, do not want your creation, mankind, to achieve our level, the level of the Anunnaki Gods," and God Enki assures, "Do not concern yourself, God Enlil. They are sterile, and thus, they cannot reproduce, for I have programmed sterility and death into their genes. After thirty to thirty-five thousand Earth years, approximately ten of our Shar years, their body cells shall die, and as each death occurs, mankind shall have to be replaced by the birth-Goddesses."

Then we notice an Igigi God, a male, as he shouts orders to several mankind males and females who are working, "Faster, you must work faster," and they cower, tremble and obey, and then another Igigi God, a female, shouts at mankind, "Over there, faster," and they also cower, tremble and obey.

*

We see a full Moon in the clear star-filled night sky, as seen from upon the Earth, and then we look down to the moonlit gently sloping hills of the Abzu.

---

[241]J. Page 123, and page 125 (see footnote 194).

[242]K. Lower page 59 O Sumer, great land, of the lands of the universe, (Sumer is) Filled with steadfast brightness, the people from sunrise to sunset (are) obedient to the divine decrees, and page 60 O house (land) of Sumer, may thy stables be many, may thy cows multiply, May thy sheepfolds be many, may thy sheep be myriad, May thy decrees stand, May thy steadfast people lift hand (pray) to heaven, May the Anunnaki decree the fates in thy midst (in Sumer).

ABZU
That Night

We look down further and see the entrance to a mine, somewhat as it appeared before, with the tall and thick bushes nearby, and the Moon lighting up the hillside area, plus the two glass-like balls of light, one on each side of the entrance, that glow with a soft, firefly-like glow. It is the changing of a shift crew. We observe a group of mankind, 20 in number, 10 males and 10 females, all naked, and all with brown pigment skin, brown eyes, and long shaggy black hair of tight-curls. Mankind is silly, nonsensical, frivolous and hilarious. They also only blurt crude sounds, which have no resemblance to articulated words, and they are all very passive, and all are fresh and clean. They are being led, marching in orderly formation, carrying the tools of the Igigi Gods, the plastic-like gold pickax and the plastic-like spade, toward the mine entrance by 4 Igigi Gods, all without helmets or body regulators, and 4 Anunnaki God officers, all helmeted and some with body regulators. All the Anunnaki Gods and the Igigi Gods are crisp and clean. As they are about to enter the mine, we observe another group of mankind, 20 in number, 10 males and 10 females, all naked, sweating, dusty and dirty, and noticeably tired, slowly walking out of the entrance, being led by 4 Igigi Gods, all without helmets or body regulators, and 4 Anunnaki God officers, all helmeted and some with body regulators. They are dragging themselves, in an exhausted slow walk, out of the mine, in a somewhat orderly formation, carrying the tools of the Igigi Gods, the plastic-like gold pickax and the plastic-like spade, while all the Anunnaki Gods and the Igigi Gods are crisp and clean.

We scan to the tall and thick bushes near the entrance, and at first, we see only the bushes, but as we zoom in for a close up view, we notice 2 ape-like creatures, 1 male and 1 female, peering through the bushes at the new work force, mankind, being led by the Gods. They are unnoticed, as they silently watch the changing of the shift crew, each with a curious look upon their face. Then, they look at each other in puzzled disbelief.

Inside the mine, which is also illuminated by identical glowing glass-like balls of light, we hear the electrical arcing sounds, as we slowly proceed deeper into the mine. At this point, we do not see mankind, we only hear the sounds of arcing, and now we begin to hear shoveling sounds. As we go around a corner, we observe mankind, sweating, dusty and dirty, and noticeably tired, as they passively labor very hard in the dusty and dirty mine. They are touching the large rocks with the plastic-like gold pickax, the high frequency, energy concentrating laser-like device, and bursting the rocks, like an implosion, into smaller pieces, as the Igigi Gods had done. They toil and sweat in the dust and dirt. It is also like a forced labor scene, and the pace is quick, with some using the plastic-like spade, shoveling the smaller rocks into the plastic-like baskets, and others taking the loaded baskets of rocks on their back, while holding the plastic-like rope support, which they grasp at their chest, and dumping them into the screw-like conveyor, which then moves the rocks back into the larger loading container, which is on wide soft wheels, as we saw before. Mankind is performing every task that had previously been the responsibility of the Igigi Gods. There is no caressing among male and female mankind. They are also between 15 and 23 years of age in appearance, 70 in number, 35 males and 35 females.

As we scan the scene, we notice the Anunnaki God officers, who are crisp and clean, standing on ledges, looking down and supervising the work, now with the assistance of the crisp and clean Igigi Gods. One Igigi God, a male, shouts down at mankind, "Work harder, faster," and they cower, tremble and obey, and then another Igigi God, a female, shouts down at mankind, "Over there, keep working," and they also cower, tremble and obey.

*

It is bright daylight, in the Abzu, and we see the outside of the three craft; the biological laboratory craft, the medical craft, and the support craft, close together in a triangle, and linked together by the clear glass-like passageways, in the clearing, as they appeared before, with the tall and thick bushes and trees in the background.

Several Days Later

We scan to the back of the craft, to the circular laser-like fenced area, where we observe white and brown skin color naked children, children from the birth-Goddesses, up to 15 years of age in appearance, young mankind, and we see some close up views of them. The children are noisy, hilarious and silly. All have a passive disposition, and also only blurt crude sounds, which have no resemblance to articulated words. They are being raised here, 210 in total number, 105 males and 105 females; 14 of every age level, 7 males and 7 females, up to age 15 in appearance. They all have black, medium length shaggy hair, with tight-curls, and brown eyes. They are groomed and clean. Also, there are gazelles and sheep in the pen, and some children tend the sheep, while others are caring for a small area of crops. Some children eat grass, like the gazelles and sheep, using only their mouth and not using their hands, while others lap water at the water pond, like the animals, and alongside the animals. Some children caress the animals. There are 4 adults, naked mankind, 2 with white pigment skin, 1 male and 1 female, and 2 with brown pigment skin, 1 male and 1 female, within the fenced area with the children, and they are watching over the children. Their hair is like the children's, as are their eyes, and they also are groomed and clean.

This is the breeding and special nursery area for young mankind, born by the implanted birth-Goddesses. We notice a sheep stray to the edge of the pen, with the tall and thick bushes outside of it in this area. The white adult mankind male goes to retrieve it, and as he reaches the laser-like fence near the sheep, he looks into the bushes, and sees a male ape-like creature staring back at him from within the bushes. They both silently stare for a moment, each with a curious and puzzled look upon their face, and then they both run away from each other, frightened at what they have just seen.

From a short distance, we observe God Enki and Goddess Ninki, his wife, seated beneath a tree and within its shade, away from the laser-like fenced area. With them is Adapa, who is also seated, and who is now 25 years of age in appearance, but actually he is 9,000 years old, for 360 of Adapa's years equal 1 of our years today, and his programmed longevity is to be approximately 30 to 35 thousand years, or 100 of our years today. He is naked, and broadly smiling, with black, tight-curly hair on his head, which has been cut to a medium length, like the tight-curly white hair of God Enki. His eyes are brown, and he is clean, pampered, and well taken care of, as he is the personal favorite of God Enki and Goddess Ninki.

We zoom in for a closer view, and notice Goddess Ninki teaching smiling Adapa how to count on his fingers, as she raises another of his fingers with each number she pronounces. God Enki watches with pride and pleasure, as

Goddess Ninki encourages Adapa, "One," and he answers slowly and crudely, "Ooonn," and she asks, "Two," and he answers, "Ttttuu," and she asks, "Three," and he answers, "Tthhhee." Goddess Ninki, pleased with progress, then caresses Adapa. God Enki reflects, "Look Goddess Ninki, look at the breeding nursery for young mankind, born by the birth-Goddesses. Even though we groom them, mankind still eats grass with the gazelles and sheep, and drinks with the game at the drinking place."[243]

We zoom to the pen, and again see the children eating grass and drinking water next to the animals, in the same manner as the animals.

Then, after a brief moment, we zoom back, and God Enki continues, "Their heart delights with the animals at the water,[244] and mankind even mates with the animals. My creation is closer to the animals, than to the Gods. It is a pity, Goddess Ninki, that my plants, mankind, being hybrid and sterile, cannot procreate," and Goddess Ninki responds, "Yes God Enki, it is regrettable that God Enlil has decreed it to be so for the black-headed people. For we both have become increasingly attached to mankind, and particularly to Adapa." God Enki declares, "Yes Goddess Ninki, it would be wonderful if mankind were higher in intellect, and could enjoy the pleasures of love, and be able to reproduce, to know their own kind, and to bear their own offspring, like us, instead of being brought forth by our birth-Goddesses." Goddess Ninki cautions, "But God Enlil has warned you against giving this knowledge to mankind," and God Enki replies, "Yes he has, regrettably. He has decreed against this."

Then, after a pause, we see a mischievous smile, and a sparkle in the eyes of God Enki, as he resolves, "But, if I can change mankind's status, indirectly, and without God Enlil immediately becoming aware, I shall not be breaking his decree, mankind shall. For mankind is not bound by the decrees of God Enlil. After all, God Enlil may accept it! For when we eventually leave this planet, Earth, it would be a fitting remembrance to the Gods if my creation could continue on their own, without the support and guidance, and dependence upon the Gods," and again we zoom to the pen, and observe young mankind.

---

[243]T. Page 307, fig. 149; and K. page xvi, and page 73; and F. page 19, line 39, and line 40, and page 21, line 3 (Who) eats grass with the gazelles, and line 4 Drinks with the game at the drinking-place.

[244]F. Page 19, line 41, and page 21, line 5 (Whose) heart delights with the animals at the water.

*

Inside the biological laboratory craft of God Enki, we scan over the glass-like flasks and beakers, still bubbling chemicals of different colors, and the computers, all of which are off, except one, the tree of knowledge computer, which has a 6 inch square viewing screen.

Several Months Later

We notice God Enki and Adapa, alone in the laboratory. God Enki is preparing and looking at a test tube-like vessel containing a sample of blood, which he has just taken from Adapa. He turns to the tree of knowledge computer, and touches the heat sensitive console numbers in sequence, and the computer responds with flashing display lights, as he programs it. Then God Enki declares, "To all mankind, through you Adapa, I shall bestow the gift of knowledge, genetic information from the tree of knowledge computer, which shall give you a higher intellect and reproductive capabilities; so you shall have wisdom, and shall be able to procreate, and thus, mankind shall become and call each other man and woman, and mankind shall multiply upon the face of the Earth, until the end of time. For I have grown very close to mankind, and especially you Adapa, and after the Gods leave the Earth, I should not want mankind to cease, but to continue progressing upon the face of the Earth. Then mankind shall be truly independent, to do as you please, and free from domination by the Gods. Mankind shall be free." Then, we see alternating views of Adapa and the computer, with Adapa smiling at the flashing display lights.

We see time passing through a series of alternating cameo views of normal clouds passing through the sky, and sunrise to sunset sequences. This blends into additional cameo views of pre-Neanderthal-like mankind, early mankind, progressing in culture and appearance, and their transition to Cro-Magnon mankind, recent mankind, through: naked mankind, male and female,[245] in the garden of God Enlil at Sumer, only caressing the animals, and with

---

[245]V. Genesis 2:25 And they were both naked, the man and his wife, and were not ashamed (for they were generatively inactive, similar to this).

no sexual relations between mankind, and no children conceived or born by mankind; God Enki altering and activating the genes of Adapa,[246] and Adapa and a female, Eve, discovering each other, freely caressing, and making love, conceiving, and Eve bearing a child effortlessly,[247] like the animals, also others discovering each other, freely caressing, and making love, conceiving, and the females bearing children effortlessly,[247] and copulation everywhere among mankind;[248] mankind increasingly becoming less frivolous and realizing they are man and woman, now that they have reproductive capabilities,[249] and they cover their naked bodies with aprons made of fig leaves;[250] God Enlil

---

[246]K. Page x; and T. page 295, fig. 147 (as I further explain under 'Sterility' on page 359 in my epilogue).

[247]M. Page 63, line 14 Since Mämi conceived the regulations for the human race, and page 105, line 5 Regulations for the human race Mämi conceived, and page 63, line 15,17, 18 And when the pregnant woman gives birth, 19 Let the mother of the babe sever herself, and line 20 The man to [the young lady], and page 105, line 6 The male to the young lady, and page 63, line 21, and page 105, line 4, and page 63, line 272 The young lady to the young man gives her breasts, and page 105, line 7 To (to could be scribal error) the young lady to the young man gives her breasts, and page 63, line 273 The young man to the young lady gives his beard, and page 105, upper line 9 The young man to the young [lady gives his beard, and page 63, line 274 The young lady kisses the cheek of the young man, and page 105, line 8 The young lady kisses the cheek of the young man, and page 63, line 275 Let the young lady take the young man to the open air shrine and street, and page 105, upper line 10 Let the young [lady] take the young man to the open air shrine and street, and page 63, line 276 Becoming one, the wife and her husband, 290, 291, and line 292, and page 65, line 294 (the brick herein described, and as on page 61, line 259, and line 6, and page 63, line 15, and line 288, and page 153, philological note 259, has an enigmatic function, which seems to be associated with conception and child bearing, but its precise function is elusive), 299 When evening the bed is laid, 300 Let the wife and her husband lie together, 301 When, to institute marriage, 302, 303, 304, and line 305 Conceiving at the destined moment (the preceding lines seem to indicate instructions and regulations pertaining to courtship, marriage, conception, and child bearing, and were imposed upon mankind through Goddess Ninhursag; either by Goddess Ninhursag herself, or possibly Goddess Ninhursag and God Enki together, with Goddess Ninhursag implementing the rules under the direction of God Enki, as occurred after the flood, in M. page 103, line 43, 44, 45, 46, 47, 48, 49, 50, 1, 2, 3, 4, 6, 8, 10, and line 11).

[248]M. Page 141, line 49 Animals multiplied everywhere (animals = people, early people were considered by the Gods as animals: see footnote 181); and T. page 327, fig. 156.

[249]T. Page 323 through upper page 332; and V. Genesis 3:7 And the eyes of them both were opened (bestowed with generative ability); and they knew that they were naked (had attained generative activity, similar to this).

[250]W. Lower page 172 through upper page 173, Volume 1 so they took to stitching upon themselves leaves of the Garden (similar to this); and V. Genesis 3:7 and they sewed fig leaves together, and made themselves aprons (to conceal their generative ability by covering the genitals, similar to this).

walking in his garden at Sumer, in the morning, calls Adapa and Eve, and mankind, and they fearfully hide among the trees;[251] God Enlil scolds Adapa and Eve, and mankind, in his garden at Sumer, for learning the secret of reproduction; God Enlil gives coats of animal skins to Adapa and Eve,[252] and mankind, and they wear clothes; mankind serves all the Gods;[253] the behavior of proliferating mankind toward themselves regresses and becomes promiscuous uncontrollable evil conduct,[254] and God Enlil sends mankind out of his garden at Sumer, toward the east, to the Zagros Mountains, as mankind becomes savage-like in behavior; God Enlil places monitors and flaming sword-like lasers that rotate, at the east of Sumer, near the foothills of the Zagros Mountains, to keep savage mankind out;[255] the ground yields thorns and thistles, and mankind eats herbs and bread after sweating and very hard labor in the fields;[256] women sorrow during childbirth with great effort,[257] and many childbirths, and mankind becomes more savage, with much rivalry for mates, and killing for mates, and continuing prolific sexual relations, and a perversity of sexual relations, including polygamy, homosexual relations,[258] sodomy, and even cannibalism; all savage traits are arrested and slowly subside, with assistance from the Gods, except women bearing children with effort; mankind holding babies, raising children, grooming themselves, preparing meals, and eventually using clay cups and clay utensils, all without handles; mankind is domesticated by the Gods, and worships and respects the

---

[251]V. Genesis 3:8 And they (Adam and Eve) heard the voice of the LORD God (God Enlil) walking in the garden (Sumer) in the cool of the day (morning); and Ăd´-ăm and his wife hid themselves from the presence of the LORD God amongst the trees of the garden (similar to this).

[252]V. Genesis 3:21 Unto Ăd´-ăm also and to his wife did the LORD God (God Enlil) make coats of skins, and clothed them (similar to this).

[253]T. Page 329, fig. 157.

[254]M. Page vi, postscript 'noise' in the Epic means or implies 'evil conduct'.

[255]V. Genesis 3:24 So he (God Enlil) drove out the man (mankind), and he placed at the east of the garden of Ē´-dĕn (Sumer) Chĕr´-ū-bims (monitors, possibly the Igigi Gods 'Those Who See and Observe', and/or possibly young Angels/young Anunnaki Gods), and a flaming sword (lasers) which turned every way, to keep the (mankind) way (away) of (from) the tree of life (similar to this).

[256]V. Genesis 3:18 Thorns also and thistles shall it bring forth to thee; and thou shalt eat the herb of the field, and 3:19 In the sweat of thy face shalt thou eat bread (similar to these).

[257]V. Genesis 3:16 Unto the woman (Eve) he (God Enlil) said, I will greatly multiply thy sorrow and thy conception, in sorrow thou shalt bring forth children (similar to this).

[258]V. Genesis 4:7 And unto thee (Cain) shall be his (Abel's) desire (sexual preference), and thou (Cain) shall rule over him (Abel, similar to this).

Gods,[259] and again works for and serves all the Gods; mankind is allowed to re-enter and populate in Sumer by decree of God Enlil, and eventually begin living in clay brick homes, and teaching their children to read and write pictographs on clay tablets. This blends back into time passing through a series of alternating cameo views of normal clouds passing through the sky, and sunrise to sunset sequences.

*

We see a bright blue daylight sky with normal clouds, and then we look down, and from a distance, we observe the outside of the oval walled city of Shuruppak,[260] the city of Goddess Ninhursag, with its colorful 300 foot high central 7 tiered ziggurat,[261] composed of solid stages, platforms, or tiers, with ramp-like connecting stairs, progressively reduced in height and width, and each coated with a different color plaster, white, black, purple, blue, vermilion, and silver, and gold colored temple, which is the communication and control center of the Gods,[262] for their exclusive use only, as the seventh tier, at the top, with antennas atop it. Everything is normal, including the pace. It is like all the other cities, with all the structures made of unbaked, sun-dried clay bricks, and except the colorful ziggurat, are coated with cream colored plaster, then coated with a cream colored wash, which now have 8 feet high single story rectangular residential dwellings, with 7 feet high walled rectangular entry courtyards, sprawling around the 12 feet high oval wall for one mile in all directions from the ziggurat and temple compound, in addition to the smaller rectangular storage buildings. Most of the houses are small, but several are

---

[259]T. Upper page 332 through middle page 334; and V. Genesis 4:26 then began men (mankind) to call upon (invoke) the name of the LORD (similar to this. In this context, initially, this is most certainly a reference to God Enlil (see footnote 75); but it could be any one or all of the four main Gods; and eventually, all of the highest Gods in general; and synonymously, it could be a reference to Nibiru. Thus, religion begins upon Earth).

[260]M. Page 141, line 97 The fifth (city), Šuruppak, he (God Enlil) gave to Sud (an epithet for Goddess Ninhursag, meaning nurse); and F. page 80, line 11 Shurippak, a city which thou knowest; and H. page 75, line 31 its kingship to Shuruppak was carried, 32, and note 31.

[261]T. Page 35, fig. 11, and page 276 through page 279; and B. page 123; and Y. page 141 through page 145; and L. page 137.

[262]F. Page 80, line 13 That city (Shuruppak) was old (long established), and the gods were in its midst (were very involved).

larger in height and overall size, and a few of the larger houses are inside the oval walled compound,[263] and are visible from outside the oval wall. Also, we notice a wide river to the east, close to the city, with clear water in it, and irrigation canals veining through the area.

SHURUPPAK
68.94 Shar Years Later
or
248,200 Earth Years Later
37,600 B.C.

We zoom in and see many cameo views, as we now begin our observations of Cro-Magnon mankind, recent mankind, fundamentally, mankind as we know ourselves today. The wind is normal, being variable, it ranges from almost calm to very light, also there is normal rainfall. The Euphrates[264] River is wide, and slowly flowing, with clear water, and at a full, normal water level of 35 feet. There are tall, lush, thick reeds along and near the river, and because of their abundance, they are flattened and bound with reeds into 3 inch diameter by 5 feet long reed bundles, and are used almost everywhere as a substitute for wood, which is scarce.[265] People are fishing in the river with flawless nets, and there are fish jumping in the water. People are also catching birds with bow and arrow. There are large and small, rectangular shaped, freshly pitch coated reed[266] boats in the river, which are entirely constructed of reed bundles, and are a common and familiar sight. The canals are filled with water, at the normal 10 feet water level, with clear water, and are irrigating the crops. People are digging with wood hoes and wood shovels in the canals,[267] keeping them open and maintaining them. They scoop and shovel mud from under the water, and pull it up over the top of the bank, and some people are using reed baskets, with a rope support, which they grasp at their chest, to haul the mud away on their back,[268] to be fully utilized as clay. Fields and

263 I. Illustration opposite page 77 (similar to this).
264 F. Page 80, line 12 [And which] is situated [on the bank of] the river Euphrates.
265 L. Page 101, and page 103; and N. middle page 20.
266 L. Upper page 104.
267 B. Page 26.
268 Q. Illustration, Workmen in a Quarry; and U. frontispiece, lower register.

orchards have black soil and lush crops of grapes, figs, dates, olives, apples, pears, pomegranates, almonds, walnuts, pistachios, corn, wheat, barley, and other grains, as well as lentils, cucumbers, lettuce, onions, beans, and other vegetables. The fields are weedless, with the adults, and children of various ages, tending the crops.[269] The sheep and goats are fat, and the children are tending, herding, and milking them. Adults are plucking the sheep and collecting the wool, as others spin and weave the wool into cloth, as they do with flax,[270] for making linen.

Streets are clean and well kept, and are teeming with mankind,[271] and sprinkled in are some young and older Gods and Goddesses, most of whom are also considered astronauts, Anunnaki Gods, Anunnaki God officers, and Igigi Gods, with most wearing helmets and some wearing body regulators, except the Igigi Gods. All of the children of mankind, of all ages, are playing in the streets, some with their fat pet dogs, while others play games. The streets are clean and even, and are not paved with clay bricks, but are of smooth compacted sand. The rectangular courtyards, surrounded by the 7 feet high sun-dried clay brick walls, coated with cream colored plaster, and a cream colored wash, are clean, and are paved with baked clay bricks. The outsides of the houses are clean and well kept, and are in excellent condition, and they look new, with fresh cream colored plaster and cream colored wash over the sun-dried clay brick walls, and with freshly clay coated nearly flat[272] reed bundle roofs. Wood front doors are highly polished. Small rectangular reed huts, with reed doors, used as storage buildings, and entirely constructed of reed bundles, are attached to the back of the houses, and are in excellent condition, clean and well kept. Inside, the houses are clean and orderly, and mostly consist of two or three rooms, with pale red sun-dried clay brick floors, and cream color plastered walls. Furniture is made with highly polished wood, and consists of chairs, and rectangular tables and benches, all in excellent condition. Window draperies are like new, clean, and well sewn. U-shaped open top baked clay brick fireplaces, one in each kitchen area, are full of reed bundles, which are the common fuel.

---

[269]M. Page 65, line 328, 329, 330, 331, 332, 333, 334, 335, 336, 337 With (common wood) picks and spades they (mankind) built the shrines, and line 338 They built the big canal banks; and B. page 86, and page 87.

[270]L. Page 104.

[271]K. Page xii Mankind, the teeming multitude.

Some adults, 20%, pray, sitting cross legged, as they bow their heads, after placing offerings of large round loaves of sesame coated bread at a wall niche altar in their homes,[273] or at any of six smaller temples near the base of the ziggurat, which are 10 feet in height, and have a base of 10 feet by 10 feet. They pray to various personal Gods, and implore, "God Anu," "God Enlil," "God Enki," "Goddess Ninhursag," and "GODS OF THE UNIVERSE™,"[274] and also they petition, "accept this offering."

The people's clothes are exceptionally clean, crisp and well sewn, and beautiful, and consist of: white tunic-like knee length skirts for working in; and white plain toga-like full length robes for everyday; and colorful full length robes with edging, for special events, which are only draped over one shoulder. Priests, members of the council, and special individuals clothes consist of colorful full length robes with edging for everyday and special events, like the special events clothes of the people, except their robes cover both shoulders. All adults wear leather wedge-like sandals[275] with straps, imitating the gold sandals of the Gods; and some people wear a flawless wire earring with stone pendants,[276] imitating the clip on electromagnetic body regulator of the Gods; and some people wear an intricately detailed wood or silver bracelet on their wrist,[277] imitating the gold watch-like bracelet instrument of the Gods; and some men wear an impeccable, white turban-like cloth hat,[278] imitating the communication helmet of the Gods; and some men wear a colorful cloth head band; and some women wear a faultless, wide colorful cloth head band ribbon in their hair, and large crescent shaped earrings[279] on special occasions.

Some adults and children, 20%, wear a 4 inch diameter clay, wood, gold,

---

[272]Y. Lower page 158 (except no middle open court, and only single story at this time); and B. upper page 33.

[273]B. Page 109; and O. lower page 41, and ILL. 19 (similar to this); and Y. page 159.

[274]M. Page 143, line 143 An (Anu), Enlil, Enki (and) Ninhursag, and line 144 The gods of the universe.

[275]T. Page 137, fig. 72; and E. fig. 7 right.

[276]Q. Illustration, King Assurbanipal (see my illustration on page 311 in my comments).

[277]Q. Illustration, Winged Figure.

[278]T. Page 35, fig. 9 left; and L. after page 64, Gudea; and N. page 110, ILL. 73, and page 130, ILL. 92, and page 131, ILL. 93, and page 136, ILL. 97 left, and page 137, ILL. 98.

[279]T. Page 45, fig. 18; and B. page 3; and Y. after page 10, fig. 3; and N. page 93, ILL. 54.

or silver symbol pendant,[280] hanging around their neck on a silver chain necklace, imitating and identifying with the four main Gods. These are the highly worshipping people, especially the people that give offerings and pray. All the people in this region have white pigment skin, with full faces and fair features,[281] and look like Caucasian mankind of today, with long legs and broad shoulders. All are well groomed and clean, with combed hair and natural shaped beards. Some adults have tight-curly black hair, and some have wavy black hair, and some have straight black hair, all of medium length on both men and women. The adults all have brown eyes. They converse with each other, and display affection toward, and freely caress each other. They no longer caress the animals. There are visible families, with mothers nursing babies. They use pale red and cream decorated clay utensils,[282] tableware, and vases, and use clay cups for drinking, and reed baskets of various sizes, all without handles, and they use flint knives.[283] They roast birds, ox and lamb on wood skewers, and smoke-dry fish in U-shaped open top baked clay brick fireplaces, one located inside, in each kitchen area, or in the other located outside,[284] in back of their houses. They eat large round loaves of sesame coated bread, which are baked in dome top ovens built into the rear portion of the fireplaces, and drink the clear river water, beer, and red or white wine, and there are full vases and drinking cups with all three everywhere. Mothers open their doors and welcome their older daughters to their homes, and younger daughters are helping mothers prepare meals, and with the household duties.

The people are joyous, partying, singing, dancing, and merrymaking, and some adults and children play musical instruments, like the tambourine, harp, drum, and flute.[285] They are not hilarious and silly, as when mankind was first created, for this trait has subsided, but they do have a good sense of humor. Mankind is now comparatively more intelligent and knowledgeable,

---

[280]T. Page 137, fig. 72, and page 235, fig. 117; and E. fig. 4 (see my illustration on page 308 in my comments).

[281]N. Page 43, ILL. 22.

[282]O. Lower page 64, and page 65, ILL. 40; and N. page 47, ILL. 25 (similar to these).

[283]O. Page 19, ILL. 3, and page 47, ILL. 24, and page 87, ILL. 55; and N. page 31, ILL. 12 (similar to these).

[284]O. Page 105, ILL. 89 (similar to this).

[285]L. Page 100.

and has become independent, disobedient and defiant; mankind has come of age. They are assertive to other people, and to the Gods, including the Anunnaki Gods, and the Igigi Gods, all whom they labor for and serve.[286]

An Anunnaki God officer, a female, helmeted and wearing a body regulator, crisp and clean, shouts to a laboring, slightly dusty and dirty, but not sweating or noticeably tired brick carrying human, "Faster, work faster," and now mankind responds, "I am working fast enough." Then an Igigi God, a male, without helmet or body regulator, crisp and clean, shouts to another laboring, also slightly dusty and dirty, but not sweating or noticeably tired wood hod carrying human, "More, carry more," and again now mankind responds, "I shall not carry more."[287a]

Some people have been appointed as group leaders over other people,[287b] who carry the clay bricks and the wood hods containing mortar and cream colored plaster, and some people are marching, in somewhat orderly formation, carrying their own tools; a wood hoe, a wood spade, a flint pickax, and a wood hod board, as the group leader, a man, leads them under the supervision of the Anunnaki Gods, with the assistance of the Igigi Gods. Mankind carries flint and wood tools instead of the plastic-like tools of the Igigi Gods, primarily because the Gods no longer trust mankind with their sophisticated equipment, and because there are not enough of their plastic-like tools for the great numbers of mankind. Mankind is now at a stage where they are imitating the Gods, in every way, through improvisation. Other men have been appointed as high Priest and King, in many of the cities, holding both titles simultaneously. Many of the people sign their name and status, by using personalized pictograph cylinder seals,[288] which some wear as a pendant around their neck. People are teeming, they are everywhere in the city.

The children are well groomed and clean, like the adults. Some of the 10 year old in appearance and younger children have blue or green eyes, like the Gods, and some have tight-curly blond hair, also like the Gods, and some

---

[286]J. Page 89, line 8 That the service of the gods may be established, and that [their] shrines [may be built] (reason why mankind was created).

[287a]M. Page 65, line 336 (mankind/humans) Were becoming stiff necked (stubborn = disobedient).

[287b]K. Page xi Workers would have no controller, no supervisor (without the organization of the Earth by God Enlil).

[288]B. Page 192, Chapter 15, especially upper page 197.

havc straight blond hair, and somc havc various shadcs of brown hair. All thc other 10 year old in appearance and younger children, and all the children older than 10 years of age in appearance, have brown eyes and tight-curly, wavy or straight black hair, like all the adults. All the age groups have medium length hair. The small children play with toys in the courtyards, or behind the houses, and older children play games. All the children are learning to read and write with pictographs, on 3/4 by 3-1/2 by 5-3/4 inches long clay tablets, a clay version of the small hand held computer with video that some of the Gods carry.[289] Everywhere, mankind is imitating the advanced technology of the Gods, using the available materials.[290]

*

From space, we observe Nibiru approaching the Mars area, and passing between Mars and Jupiter, on a plane of orbit that is similar to the 17 degree angle of Pluto, and then we see a close up view of Nibiru, in all its glory.

Then we see the Earth from space, and we zoom in for a close up view from the air, close to the Earth, and we notice the Earth slightly shaking.

From upon the Earth, we observe Nibiru, which is clearly visible against the star-filled sky, radiating and glowing in the northern dawn sky. Its two rings of clouds, within which orbit its four moons, give Nibiru the appearance of a cross in the sky, bathed in a sunset red glow. We look down to the Earth, and we notice moderate Earth tremors with fissures, and volcanic eruptions with lava and thick clouds of smoke, ash, and steam, causing frightened animals to run ın all dırectıons, and ıncreasıng temperatures ot the ground and aır, and water and steam bubbling up out of the ground, both in the wilderness and in the desert, and surface waters evaporating into a mist, with a rainbow forming in the sunlit distance, and it appears rain is imminent, and mild flooding of fog laden seas and rivers, causing thorough irrigation of the crops in the fields,

---

[289]B. Page 70.

[290]L. Chapter 3, and especially lower page 109 (perhaps the technique of artificial fertilization of plants was taught to mankind by the Gods, probably much like the process the Gods used to create mankind); and Q. Chapter IV, and lower page 305 through page 331 (it is possible that the colorful clothing of mankind was an attempt to emulate the colors of the ziggurat, which would of course also be a symbol of reverence of the Gods, a way of wearing and consequently displaying their reverence: see footnote 66).

and mildly blowing winds, causing tree branches to gently sway.

We observe a group of people, sitting cross legged, bowing their heads and prostrating themselves,[291] with mixed expressions of fear, submission, humility, subordination, and reverence, as some grasp their God Anu/Nibiru symbol necklace pendant, which looks like a cross,[292] and hold it up in acknowledgment, as they look up at Nibiru in the northern sky at dawn. The people respond, as one person acclaims, "The Planet of Crossing,[293] Nibiru," and another professes, "The day of the Lord."[294] Then another announces, "The Bringer of Purification,"[295] and another proclaims, "Exalted One."[296] Then another declares, "The Lord of the Pure Incantation,"[297] and another asserts, "The Creator of the Earth."[298] Then another exclaims, "Lord of the Deep,"[299] and another affirms, "The Planet with Wings."[300]

From space, we observe Nibiru, after orbiting around the Sun, begin its return into deep space, orbiting away from the Earth.

From upon the Earth, we observe Nibiru growing smaller as it speeds away. Again, we look down to the Earth, and the moderate Earth tremors with fissures and volcanic eruptions and profuse moderate rain are diminishing, and temperatures are decreasing, with a rainbow forming and then slowly fading away in the sunlit distance, and the mild flood waters are receding, and

---

[291]J. Page 97, line 24, and page 227, line 24.

[292]L. After page 64, upper center symbol on Stele of Ur-Nammu.

[293]T. Page 218, and page 230, fig. 115, and upper page 361 It is the god Nibiru ["Planet of Crossing"] (which crosses through our solar system and/or appears as a cross in the sky); and J. page 109, line 109 (interpreted); and O. page 65 ILL. 40, i and l, and upper page 78 illustrations, and lower page 97, and page 109, ILL. 95, a right; and X. middle page 188 as a cross; and N. page 111, ILL. 74 copper disk (perhaps), and upper page 114 'sun-disks'.

[294]V. Amos 5:18 Woe unto you that desire the day of the LORD! (Nibiru, similar to this); and T. lower page 362 (see footnote 259).

[295]J. Page 97, line 19 (could imply a purifying/distilling of the waters of the Earth, after they evaporate and mist up, and then condense and fall as rain: cleansed/pure water. Or, this epithet could have a far deeper meaning. Could the genetic code or programming of the life systems on Earth, the plants, animals, and even mankind, be perfected and/or refined, thus purified, through an increased rate of growth or development, as similarly described in X. lower page 192, after their stimulation by Nibiru).

[296]J. Page 95, line 13 (largest/most powerful planet: see footnote 14 and 15).

[297]J. Page 97, line 26.

[298]J. Page 103.

[299]J. Page 77, line 143 (interpreted: wide/elliptical/deep space orbit).

[300]J. Page 75, line 128, and page 77, line 141 (both interpreted); and T. page 219, fig. 112; and B. page 128, and page 129; and L. Chapter 4.

the winds are decreasing, and the animals and relieved people are returning to a calm state of disposition, and the Earth is returning to normal, as the effects of Nibiru subside.

From space, we see Nibiru speeding out of sight, twinkling as it disappears into the star-filled space of our universe.

*

We see a bright daylight scene, with God Enlil and God Enki walking within the bright sunlight in Shuruppak, on a smooth compacted sand road, in back of the last row of houses, next to the crops in the fields. Both are wearing a tall, fez-like casual hat of gold-like metallic material, and a gold-like metallic clip on electromagnetic body regulator on one ear; in addition to their full length, from elbow to ankle, red robe with gold trim at the bottom, neck, and elbow, 5 inch wide gold waist band, gold wedge-like sandals with straps, gold winged sphere symbol of Nibiru on their robe, above the left breast, gold watch-like bracelet instrument, and their 4 inch diameter identifying gold symbol pendant, hanging around their neck on a gold chain necklace.

SHURUPPAK
Several Days Later

The city stretches two miles in the background, and the colorful 7 tiered ziggurat, within the oval walled compound, is nearly fully visible in the center of the city, one mile away. A few paces behind God Enki, walks his personal human servant, the high Priest and King[301] of Shuruppak. His name is Ziusudra.[302] He has brown eyes and black tight-curly medium length hair,

---

[301] Y. Page 128.

[302] M. Page 143, line 145 At that time, the king Ziusudra, the anointed [one]; and U. page 17 through page 19; and G. page 42, number 2; and H. page 75, note 32, and page 76, note 34 (Zi-u-sud-ra/Ziusudra is the Sumerian name for the King; which in Akkadian is Utnapishtim; and in Babylonian is Atra-hasis; and in Hebrew is Noah; and in Hellenic is (Deucalion) Xisuthros or Sisithros; and the Hindu name for Ziusudra is Satyavrata, both Atra-hasis reversed; and in Chinese, it is reported to be Fo-hi. It is interesting to note that an element within the name Ziusudra, Sud, is also the nurse epithet for Goddess Ninhursag; could it be that his name reflects the nurse/doctor/medical aspect of Goddess Ninhursag, or

and a black tight-curly natural beard of medium length. He wears a special full length purple robe with gold edging, which covers both shoulders, and a tall, purple fez-like hat. He also wears a gold 4 inch diameter God Enki symbol pendant, hanging around his neck on a gold chain necklace, and a gold winged sphere symbol of Nibiru on his robe, above the left breast, like the Gods wear, and a gold wire earring with gold pendants on one ear, and a gold bracelet on his wrist, and he carries a gold scepter and a gold ring, as symbols of his authority. Ziusudra appears to be 20 years old. The Gods do not age, and look the same as they appeared previously, for they long ago skillfully mastered and employed the art of genetic engineering, and through its application have been able to alter their genes and suspend their aging process.

Then God Enlil, in a stern tone of voice, admonishes God Enki, "Mankind flourishes God Enki, look at them, here at the city of Shuruppak. They have multiplied[303] rapidly, and are gaining intellect. Their status has improved. God Enki, when I realized you were indirectly involved, I was at first very angry, but now, I am not. You should not have given mankind genetic information from the tree of knowledge computer, a higher intellect, and the ability to reproduce. When you first created mankind, the black-headed people, they were passive and naked, like the animals, and even mated with them. They did not realize they were naked, and were genetically unable to procreate. Then you, God Enki, by altering their genes like a cunning, clever serpent,[304] gave mankind reproductive knowledge, to know each other sexually. But with this knowledge, mankind deviated from your programmed genetic evolution, and returned to their earliest animal behavior, and they became savage, and

---

his identification of being from the city of Shuruppak, which is the city of Sud, Goddess Ninhursag: see footnote 260).

[303]M. Page 67, line 353 [When the land extended] and the peoples multiplied, and page 107, line 1, and page 141, line 49 (animals = people: see footnote 181 and 248).

[304]T. Page 331, and page 333, fig. 158; and V. Genesis 3:1, 3:2, 3:3, 3:4, 3:5, 3:6, 3:7, and 3:13 (similar to these. A part of the Egyptian (Magan) symbol for their God Ptah (in Hellenic ἑπτά = seven/seventh: first and highest numerically of a group of seven rulers) was the entwined serpent/snake; and the Sumerians/Babylonians considered God Enki their wise, clever, and cunning God, and one of his symbols was the entwined serpent/snake; and even our medical profession today has adopted as their symbol the entwined serpent/snake, which could be a subconscious vestige of days long past. In the Bible, the Hebrews state that the serpent (symbol for God Enki/Ptah) beguiled Eve with forbidden fruit. Therefore, I believe God Enki, Ptah, and the Hebrew serpent who benevolently encouraged Eve, are one and the same God/supreme being, who was also known as Kronos/Poseidon, and Neptune).

they proliferated, and they began to kill each other, competing for mates as their sexual awareness grew. They also began to practice sodomy, even cannibalism. Their evil conduct toward each other was very poor; it was unacceptable."

"But those early widespread savage animal traits have diminished, and have been almost entirely eliminated through our domestication and civilization program. And now, through progressively increasing wisdom, attributed to your genetic programming, they are reaching new levels in achievement and development. But I have recently been informed, this new intellectual gain shall soon adversely affect their body processes, which had been retarded through your genetic programming, and they shall experience a rapid advance in their aging process, causing their long lives to be shortened,[305] returning to the longevity of the creatures from which they were created." Then after a very brief pause, God Enlil accuses, "God Enki, you have broken my decree!" God Enki quickly responds, "But mankind, themselves, learned the secret, God Enlil. Mankind, not I, has broken your decree. Mankind has come of age."

Then, the three stop, with Ziusudra in the background conversing with several workers tending the crops at a normal pace, and God Enlil turns to God Enki, and reflects, "When I first saw they had acquired this capability, but with a depraved savagery, I angrily expelled mankind from this land, because mankind had become as one of us, acquiring knowledge, and lacking only immortality, which I feared mankind might also acquire.[306] But slowly, as their conduct toward each other improved, and they began to respect each other and the Gods, my anger diminished. I allowed mankind to return, for all the Gods had become dependent on their labor, for mankind gives sustenance to all the Gods.[307] But with increasing intellect, mankind is now growing defiant, assertive and disobedient. We are losing our dominance. Our control is being challenged. You changed mankind, God Enki, and the people have multiplied and wandered over the Earth. Look at mankind now!"

God Enlil points to a busy street, which extends toward the central part of

---

[305]C. Lower page 82 for through the perfidy of man the potential span of human life has been reduced.

[306]V. Genesis 3:22 And the LORD God (God Enlil) said (to God Enki), Behold, the man (mankind) is become as one of us, to know good and evil (can reproduce): and now, lest he put forth his hand, and take also of the tree of life, and eat, and live for ever (attain immortality, similar to this).

[307]M. Page 67, line 339 For food for the peoples, for the sustenance of [the gods] (reason why mankind was created: see footnote 286).

the city, away from where they stand, and we zoom in for a close up view, and notice the street is teeming with people; and sprinkled in are some young and older Gods and Goddesses, Anunnaki God officers, and Igigi Gods, most wearing helmets and some wearing body regulators, except the Igigi Gods.

Then after a brief moment, we zoom back, and God Enki defends, "But God Enlil, mankind highly worships us, they dutifully respect us, with reverence. It has taken a long time to reach this level of worship. Especially in time of need, they petition us, as personal Gods, bringing special offerings,[308] and asking for favors. How can you forget their daily worship and special offerings?" God Enki pauses, and then he continues, "All mankind is not defiant, assertive and disobedient. Look at Ziusudra!" God Enki partially turns his head and looks back at Ziusudra, and motions with his arm, and they both fully turn, and look at Ziusudra.

We observe Ziusudra, conversing with the workers tending the crops. Then God Enki continues, "The tenth generation from the seed of Adapa. My servant, whom I elevated to high Priest and King, here at Shuruppak, and who administers the provisions, and lives within the temple wall.[309] He is very obedient and faithful, and for this, I have bestowed kingship upon him. Since the time of Adapa, Goddess Ninki and I have had many loyal servants, and we were very proud of them. Ziusudra, as his father before him, Ubar-Tutu, is a loyal and faithful servant. Daily he stands in attendance at the temple,[310] and prostrates himself with humility in reverence.[311] He is a very pious man."

As they again turn and resume walking, God Enlil recounts, "Because you genetically engineered reproductive capabilities for mankind, and gave our guarded secret to them, our sons, the young Gods, are beginning to take mates for themselves;[312] the daughters of mankind, seeing that they are fair

---

[308]M. Page 143, line 146 He made offerings of sesame bread at the temple of Enki.

[309]M. Page 133, line 6 I am Atra-ḫasīs (Ziusudra), and line 7 I lived in (within) the temple (oval walled compound) of Ea (God Enki), my lord.

[310]M. Page 143, line 148 Every day he stood constantly present at his temple.

[311]M. Page 143, line 147 With humility (and) well chosen words, in reverence held (God Enki).

[312]K. Page 100 In my (the Gods) city my friends (Gods) have taken wives (humans) unto themselves, My neighbors (Gods) have taken wives (humans) unto themselves; and C. page 360 [heavenly beings (Gods/astronauts/supreme beings) did indeed come down to earth and seduce mortal (human) women]; and D. lower page 62 space man (Gods/astronauts/supreme beings) crossbred with the indigenous (human) population (in Hellenic, the mixed offspring of God or Goddess and human being were called a demigod or hero: see footnote 165).

and compatable.[313] This has been occurring for the past one Shar year, three thousand six hundred Earth years." Just as God Enlil concludes, two 10 year old in appearance boys, one with blue eyes and one with green eyes, and both with straight medium length blond hair,[314] run past them very closely. Both God Enlil and God Enki look at the running boys, and we see a close up view of the running boys.

God Enlil then looks at the back of the houses, and indicates with his arm, and they both look in that direction, and we notice a young God freely caressing a 16 year old in appearance human female with straight black hair, necking with her behind a house, close to a reed hut. God Enlil continues, "There is increased intermating between the Gods and mankind.[315] The young Gods and Goddesses, our sons and daughters, are defiling themselves. Previously, the evil conduct of mankind toward mankind was unacceptable, but tolerated, for the Gods were not threatened. But now, this new evil conduct of mankind toward the Gods goes far beyond that, for now the Gods are threatened. We are directly threatened."

God Enlil then looks at the back of another house, and indicates with his arm, and they both look in that direction, and we notice a young Goddess freely caressing a 16 year old in appearance human male with wavy black hair, necking with him behind a house, close to a reed hut. God Enlil cautions, "There interbred offspring; as for the Gods, have degenerated genes, deficient and imperfect, which are mankind's traits; and as for mankind, have advanced genes, superior and excelling, which among other Godly traits, increases their wisdom, and further threatens our existing imperiled dominance. This

---

[313]V. Genesis 6:2 That the sons of God (Gods/astronauts/supreme beings) saw the daughters of men (human beings) that they were fair (attractive and/or skin color was like their own); and they (Gods) took them (for themselves; human) wives of all which they chose (similar to this: see footnote 21. Genesis 6:3 is elaborated upon under '120 Years' on page 361 in my epilogue).

[314]C. Page 351, (i) (a) human children begotten by gods (astronauts, supreme beings); and V. Genesis 6:4 There were giants (Anunnaki) in (upon) the earth in those days; and also after that (during subsequent orbits of Nibiru), when the sons of God (Gods/astronauts/supreme beings) came in unto (mated with) the daughters of men (human women), and they bare (had) children to (by) them (similar to this. It has been suggested that the Hebrew rendering of Anunnaki progressed from Annakum, meaning belonging to or coming from the Anunnaki, to Anakim/Anak, which means giant; therefore astronaut = Anunnaki = giant: see footnote 48).

[315]K. Upper page 101 Uncooked meat he (mankind) eats, During his life he (mankind) has no house (wealth), When he dies he (mankind) lies unburied (all of these I interpret as descriptions of early mankind); and Y. page 120.

displeases me, God Enki, and I warn you, I shall not permit this new evil conduct toward the Gods to continue indefinitely, mankind shall soon force me to limit this wicked, intolerable behavior![316] I have repented that I decreed mankind be created. My heart grieves."[317] God Enki exonerates, "But Ziusudra is pure and clean. His lineage, his ancestors, his family, have not interbred with the Gods. There are many others like him." God Enlil confirms, "Yes, I know," and he pauses for a moment, and again he warns, "Wide understanding you perfected for mankind,[318] wisdom you have given them, to them you have given knowledge, eternal life you have not given them. God Enki, I decree that you do not give mankind genetic information from the tree of life computer, and they also become immortal, as the Gods."

From a short distance away, we see God Enlil and God Enki, walking on the compacted sand road, next to each other, with Ziusudra following them, a few paces behind God Enki: all within the bright sunlight.

We see time passing through a series of alternating cameo views of normal clouds passing through the sky, and sunrise to sunset sequences, and sprinkled throughout the passing time, we see various young Gods and Goddesses caressing various young human females and males; all with various types of black hair and brown eyes, and all between 16 and 19 years of age in appearance.

*

It is bright daylight with normal clouds, and we see the inside of the oval walled city of Nippur, the city of God Enlil, with its colorful 7 tiered ziggurat in the center, with the temple Ekur and the antennas atop it, as the seventh tier. The city now looks like Shuruppak, and has a few larger rectangular houses with rectangular walled courtyards inside the oval wall, except there are no smaller temples near the base of the ziggurat.

---

[316] V. Genesis 6:5 And GOD (God Enlil) saw that the wickedness of man was great in (upon) the earth (similar to this).

[317] V. Genesis 6:6 And it repented the LORD (God Enlil) that he had (given the decree that) made man on the earth, and it grieved him at his heart (similar to this).

[318] F. Page 23, line 22, and page 22, line 29, and line 34.

NIPPUR
.33 Shar Years Later
or
1,200 Earth Years Later[319]
36,400 B.C.

We scan and notice mankind, all with black tight-curly hair and brown eyes, working at a normal pace in the courtyard, pampering the ziggurat, and tending the small sheepfold nearby, composed of the low cream colored wall. They are all slightly dusty and dirty, but not sweating or noticeably tired. All wear God Enlil symbol necklaces. There are 2 Igigi Gods, 1 male and 1 female, crisp and clean, both without helmets or body regulators, and 2 Anunnaki God officers, 1 male and 1 female, crisp and clean, both helmeted and one wearing a body regulator, supervising the work. There are five craft grouped in an area, near the ziggurat, one of which is the assembly craft. We notice 2 Gods and a Goddess, all helmeted and one wearing a body regulator, quickly walking toward the ramp.

Inside the assembly craft, our eyes scan, and it appears as it did before, a smaller version of the assembly on Nibiru, complete with a smaller dome viewing monitor in the center, which shows mankind in the present, at a normal pace, being defiant, assertive and disobedient; and young Gods and Goddesses caressing young humans with black and brown variations of wavy and tight-curly hair. All the main Gods, except God Anu, are present. Three of the 7 great Anunnaki Gods are seated behind the slightly raised section, and are quietly conversing, as are the 19 Gods and 10 Goddesses. God Enlil, God Enki, and Goddess Ninhursag sit at the slightly raised section for the four main Gods, four of the 7 great Anunnaki Gods. All the Gods are watching the viewing monitor, and some are pointing at it in disgust. Their communication helmets and body regulators are on the ledge that forms the back of the white pew-like molded circular seats. The 2 Gods and Goddess rush in through the opaque air flow door, and quickly remove their helmets and body regulator, and sit down.

---

[319] M. Page 67, line 352 Twelve hundred years [had not yet passed], and page 107, line 1.

God Enlil rises, and the conversations quickly fade to silence, and all the Gods bow their heads to him, and God Enlil sternly addresses the assembly,[320] "In this assembly of the Gods, I, God Enlil, supreme commander and ruler of the Earth, have brought you to my presence. The black-headed people, mankind, have multiplied[321] too rapidly. They are too numerous, and no longer passive," and we see the present on the viewing monitor, which shows a street teeming[322] with people, and a few Gods and Goddesses, Anunnaki God officers, and Igigi Gods.

The viewing monitor then shows the homes of mankind, sprawling out from the oval walled compound, and teeming people among the homes, with young blond children, and the original blond 10 year olds in appearance, which have now grown to be 13 years of age in appearance, among the black haired children and adults. Young Gods and Goddesses are freely caressing young humans, all with black and brown variations of wavy and tight-curly hair, in back of the homes, near the reed huts, then God Enlil reports, "They have extended themselves, and are now in many places on the Earth. They are gaining intelligence. They are independent, assertive and disobedient,[323] and I hear[324] them both day and night. The Gods are increasingly mating with mankind. Mankind is seriously threatening the Gods."

We now look at the viewing monitor, which shows a young God and a young human female, rolling around together and freely caressing in the tall grass, away from the houses.

Then we observe the concerned faces of the Gods in the assembly, some of which are angry expressions, then God Enlil continues, as he points to the viewing monitor, "This mating with mankind is decreasing the purity of the Gods' genes, and degenerating them. Mankind is increasingly becoming a threat to the Gods' purity. The Gods are defiling themselves; and mankind, as a result of this interbreeding process, has acquired even more intellect, and additionally endangers our already jeopardized control. This new wicked,

---

[320] M. Page 67, line 357 [And (God Enlil) addressed] the great gods, and page 107, line 4, and line 5.

[321] M. Page 67, line 353 [When the land (domain of mankind) extended] and the peoples multiplied, and page 107, line 1.

[322] M. Page 67, line 354 The [land] was bellowing [like a bull].

[323] M. Page 67, line 355 The gods got disturbed with [their uproar], and page 107, line 2, and line 7 I have got disturbed [with] their noise.

[324] M. Page 67, line 356 [Enlil heard] their noise (certainly he saw their evil conduct as well).

evil conduct[325] and assertiveness of mankind, this disobedience, I shall not further permit. This noise,[326] the assertiveness and disobedience of mankind, is a threat to us, and is too intense for me,[327] and with their uproar I am deprived of sleep.[328] This noise of mankind, their assertiveness and disobedience, you, God Enki," and God Enlil points to his left, across the empty seat of God Anu, and we observe God Enki, calm and collected, "have now created."

He pauses, and scans the assembly for a moment, looking for assurance, and then he declares, "I, God Enlil, having been given the divine decree of supreme power and function of God Anu on Earth, by God Anu, president of the assembly, to rule over the Earth, decree that mankind shall be diminished. I decree a plague,"[329] and we hear a gasp from some of the assembly members, as he elucidates, "let disease, sickness, plague and pestilence blow upon mankind like a tornado."[330]

He pauses, and all the assembly members look at the viewing monitor, which now shows the future effects of the plague on mankind, showing bright daylight with normal clouds, and swirling disease germs within a light and normal wind, and coughing, dizzy, feverish and sick, diseased young and old people, all in the future, and then the viewing monitor returns to the present. Some of the Gods in the assembly gasp, and look at each other in disbelief, while others smile in approval.

Then, God Enlil commands, as he scans the assembly, again looking for assurance, "God Namtara," and we observe God Namtara in the assembly,

---

[325]M. Page vi, postscript 'noise' in the Epic means or implies 'evil conduct' (see footnote 116, 163, and 254).

[326]M. Page vi, postscript (see footnote 325); and H. page 59, note 113 Enlil and ḫubūru, "noise" (noise = evil conduct/clamor for independence).

[327]M. Page 67, line 358 The noise (evil conduct) of mankind [has become too intense for me], and page 107, line 6.

[328]M. Page 67, line 359 [With their uproar] I am deprived of sleep, and page 107, line 3, and line 8 [With] their uproar sleep does not overcome me (deprived of sleep = no peace of mind due to worry/anxiety because of the eroding of absolute authority over mankind, thus diminishing control over mankind. Also, noise and uproar appear to be synonymous, and both could mean evil conduct).

[329]M. Page 67, line 360 Command let there be plague, and page 107, line 9.

[330]M. Page 107, line 11 Let disease, sickness, plague and pestilence, and line 12 Blow upon them like a tornado (diminish mankind as a tornado would).

with an appalled look on his face, "our God in charge of disease control, since all the Gods are immune, you shall release disease germs among the people, and cause sickness and death among mankind, that their numbers be reduced."[331]

Again, he pauses, and all the assembly members look at the viewing monitor, which now shows many microscopic magnifications of various live viruses.

Again, some of the Gods in the assembly gasp, and look at each other in disbelief, while others smile in approval, and then we notice God Enki, sadly looking at the viewing monitor, which has now returned to the present, and again shows the teeming people in the streets.

God Namtara rises, and affirms, "It shall be done immediately."

God Enlil commands, "Let us clasp hands together."[332] All the Gods rise and clasp hands, and then God Enlil demands, "Your oath to me."

Together, all the Gods declare, "The decree of God Enlil is binding."

*

We see a bright blue daylight sky with normal clouds, and then we look down, and from a distance, we observe the outside of the oval walled city of Shuruppak, with its central 7 tiered ziggurat, and temple with antennas atop it, as the seventh tier, and homes sprawling around the wall for one mile in all directions from the ziggurat and temple compound, and the wide river banks, close to the city, and the irrigation canals veining through the area. Ziusudra, the loyal servant of God Enki, is high Priest and King.

SHURUPPAK
Many Months Later

We zoom in and see many cameo views. The wind and rainfall are both

---

[331]M. Page 107, line 10 Let Namtar diminish their noise.
[332]M. Page 107, line 13 They commanded and there was plague.

normal, but within the air there are disease germs swirling[333] over the land. The Euphrates River is wide and slowly flowing with clear water, and at a full, normal water level of 35 feet. There are tall, lush, thick reeds along and near the river. The birds and fish are diseased and look sick, and are thin. Fish are jumping in the water, but no one is net fishing, and birds are flying in the air, which no one is catching with the bow and arrow. The large and small, rectangular shaped, pitch coated reed boats, the larger in the river, and the smaller up on the banks, show signs of neglect. They need new pitch and some have holes, where the reed bindings have broken, and the 3 inch diameter reed bundles hang down, exposing the 6 inch diameter, triangular shaped, reed bundle framework, composed of three, 3 inch diameter reed bundles. The canals are water filled at the normal 10 feet water level with clear water, and are irrigating the crops, but no one is working on the canals to maintain them, and some of the banks show signs of neglect and have collapsed, and narrow the water flow, as some of them are almost filled with mud, and have a shallow water level. The fields and orchards have black soil and lush crops, but no one is tending the crops, and they look neglected, with weeds growing among the crops, while some of the crops are unpicked, and are rotting where they grow and fall. The sheep and goats are all diseased, and some are dead, and others are lying sick on the ground, and all of them look thin, as they shake with fever chills. No one is tending them, and the animals that are able, wander at will, unattended, and no one is milking or plucking them.

The pace is slow, and the streets have very little activity and are almost empty, and show signs of neglect, with weeds growing in many, many places. Some look uneven, with dips in them, and there is litter in the streets. There are no children playing in the streets. Diseased and thin dogs roam the streets, and some lie dead where they fell. The courtyards have many weeds and litter, and are unswept and look neglected. The cream colored plaster and sun-dried clay bricks on the outsides of the homes are beginning to crumble, and the homes look dirty and not well kept, in poor condition, and show signs of neglect, and the clay coated nearly flat reed bundle roofs need repair, as the clay has crumbled off in some spots. The wood front doors are dull from lack of attention, and some small reed huts have holes in the walls, and are in need

[333] M. Page 107, line 14 Namtar diminished their noise, 15 Disease, sickness, plague and pestilence, and line 16 Blew upon them like a tornado.

of repair, as some reed bindings have broken, and the 3 inch diameter reed bundles hang down, exposing the 6 inch diameter, triangular shaped, reed bundle framework. The ziggurat and oval wall are in the same condition of neglect as the outsides of the homes. The insides of the homes are unkempt and neglected, with unswept floors and buckled pale red sun-dried clay bricks in the floors, caused from rain water leaking in through the roofs. The cream colored plaster is crumbling, and we notice dull and broken furniture, and soiled and unraveled window drapes, and there are no reed bundles in the fireplaces. Everywhere, the people are coughing, are dizzy, have chills and fever, headaches and aches, and are sick and dying from various diseases. Some are lying in bed, and others on the ground, close to the houses, or in the streets where they fell. The others are in a dismal, listless, and quiet state. They are weary and distressed from chills and fever, and are perspiring.

Their clothes are worn, soiled and wrinkled, and are coming apart at the seams. Their sandals are worn, and some have broken straps. The pendant earrings and bracelets are in need of repair, and most have been removed, and the men's white turban-like hats and colorful head bands are soiled, and most have also been removed. The women have no colorful head band ribbon in their hair, and wear no large crescent shaped earrings. 20% of the adults and children continue to wear their symbol necklaces of the four main Gods, which are still in a reasonably clean condition, with their original chains also in good condition, and not broken or replaced by a substitute, but their pendant cylinder seals have been abandoned. Their shoulders are broad, and their legs are long, but they are poorly groomed and dirty, with unkempt hair and beards. There is little conversation, and they are nearly too sick to caress, and there is almost no child bearing, and there are nearly no mothers nursing babies. Some clay utensils are broken, and no other animal, ox or cow, or bird is being roasted, or fish smoke-dried, for they all are too sick to be eaten, and the people are too sick to eat them. There are some large round loaves of sesame coated bread on the tables, some of which are moldy. Although there is abundant clear river water, beer, and wine to drink, the people are too sick to consume them, and there are full vases and drinking cups with all three everywhere.

20% of the adults continue to pray to various personal Gods and Goddesses, sitting cross legged and bowing their heads in front of the niche altar in the wall of their homes, after they place large round loaves of sesame coated bread as offerings. They are somber, with no merrymaking or music, with musical instruments being set aside, and in a neglected state. There are no laboring workers visible, and the people are very sparsely seen outside their homes in the city streets. The children are as sick as the adults, and suffer the same symptoms, and they do not play their games, but lie about the homes without vigor, and there are very few infant children. Mothers that are able, open their doors and welcome their older daughters to their homes, and the younger daughters that can, try to help their mothers. The clay school tablets have been abandoned. The conditions are desperate.

*

From a distance, we notice a craft rapidly approaching, glistening in the bright sunlight, with its brilliant glowing firefly-like light emitting from the underside, hover, descend and prepare to land outside the city, away from the homes, in an open area near the river, as the glowing light begins to diminish, and the 4 pods like cattle hoofs extend 6 feet down, and the craft begins to land.

Several Days Later

We observe a sick Ziusudra, who is near the open area comforting several sick people, who have fallen next to the road. As soon as he sees[334] the craft descending, he slowly moves toward the open landing area. His clothing is worn, soiled and wrinkled, but he is wearing his gold God Enki symbol necklace, gold winged sphere symbol of Nibiru, and gold bracelet, and he looks weak and tired from disease. He is sick and is perspiring from fever and chills. Ziusudra now appears to be 23 years old.

Then the craft lands, and the glowing light further diminishes, and then completely expires, and after a moment, a side of the craft extends down, forming a 4 feet wide ramp and doorway. Then, God Enki walks through the

[334]M. Page 67, line 364, and line 365, and page 107, line 17 The discerning one, the man Atra-ḫasīs (Ziusudra), and line 18 Kept an open ear [to his lord], Ea (God Enki).

opaque air flow door and down the ramp of the craft. He is alone, and is wearing his full length, from elbow to ankle, red robe with gold trim, wide gold waist band, gold wedge-like sandals with straps, gold winged sphere symbol of Nibiru, gold watch-like bracelet instrument, and his gold God Enki symbol necklace, and he is without a helmet or body regulator.

At almost the same time, Ziusudra arrives to greet him, and he bows down, and prostrates himself, and then he stands,[335] and they exchange greetings, and then warmly embrace.

After their long embrace, Ziusudra pleads,[336] "God Enki, my personal God, the God who elevated me to high Priest and King of the city of Shuruppak, the God who created us. I petition you, my Lord, how long shall God Enlil impose disease upon us, forever?[337] His anger consumes the land.[338] The people are dying. They suffer sickness and headaches, aches and dizziness, chills and fever. Childbearing has almost stopped. Mankind groans.[339] Even our animals are affected. God Enki, since you created us,[340] I pray, will you remove this plague, this pestilence from our midst?[341] I have prepared offerings for you, God Enki, for it is your duty, as my personal God, to care for us in time of need." Then God Enki advises,[342] "My loyal, faithful and pure servant, Ziusudra. Gather your elders,[343] and counsel[344] the city elders. I can not directly interfere with the task of God Namtara, who follows the decree of God Enlil."

He pauses, and then we again see a mischievous smile, and a sparkle in the eyes of God Enki, as he resolves, "But, I can give you a plan for relief. Of this plan, you shall advise the elders." God Enki now whispers into the ear of Ziusudra,[345] but we cannot hear his words.

---

[335]M. Page 123, upper line 3 [Atra-ḫasīs] (Ziusudra) bowed down, he prostrated himself, he stood up

[336]M. Page 67, line 368, and line 369, and page 107, line 21, and line 22.

[337]M. Page 69, line 371 Will they (the assembly) impose disease on us [for ever]?

[338]M. Page 107, line 24, and line 26 [The disease] from the gods is consuming the land (mankind).

[339]M. Page 107, line 23, and line 25 Ea (God Enki), lord, the human race is groaning.

[340]M. Page 109, line 27 Since you created us.

[341]M. Page 109, line 28 [Will you] remove the disease, sickness, plague and pestilence?

[342]M. Page 69, line 372, and line 373, and page 109, line 29.

[343]M. Page 69, line 374 The elders gather (hold a meeting at your house).

[344]M. Page 69, line 375 And counsel (advise them) in the house (at the meeting).

[345]M. Page 69, line 385 Atra-ḫasīs received the command.

It is bright daylight, and the wind is normal and the air is filled with swirling disease germs, and we see the outside of the large house of Ziusudra, and the gate leading to the courtyard, which is within the oval wall at Shuruppak, near the neglected ziggurat. The house is also neglected.

The Next Day

The city elders, 15 males and 5 females, all wearing their symbol necklaces, are gathering at the house of Ziusudra for a council meeting.[346] Some are being brought by ox drawn four wheel cart,[347] and others are slowly coming on foot. All are sick and coughing, with fever and chills, and perspiration. They are going through the gate and courtyard, and into the house.

Inside the house, there is a large nicely furnished but neglected central room, imitating the assembly of the Gods, with wood benches arranged in a circle for the elders, and a large wood chair for Ziusudra. Ziusudra, wearing his gold God Enki symbol necklace, and the other adornments that he had previously worn, is welcoming the group, and they quickly sit down, and then he sits down, and the group, which had been sparingly conversing, quiets and looks at him, and Ziusudra declares,[348] "I have gathered you in council, to propose a plan for relieving us of this plague. We should advise all our people, to cease praying and cease offering, to our various personal Gods and Goddesses.[349] Command, and let the heralds proclaim. Make a loud noise, an announcement in the land, so all can hear."[350]

He pauses, and scans the council for a moment, and we observe the faces of the elders, as they wait for words of relief, then Ziusudra commands, "We should all jointly direct our full attention to God Namtara,[351] and only God

346M. Page 69, line 386 And gathered the elders to his gate.

347Y. After page 50, fig. 14.

348M. Page 69, line 387, and line 388.

349M. Page 69, line 393 [Do not reverence] your gods, and line 394 [Do not] pray to your [goddesses], and page 109, line 31.

350M. Page 69, line 391 [Command] that heralds proclaim, and line 392 And make a loud [noise] in the land, and page 109, line 30.

351M. Page 69, line 395 [But seek] the door of [Namtara].

Namtara." The elders look at each other, with puzzled expressions on their faces, then Ziusudra elucidates, "We shall all pray our obedience to God Namtara. We shall build a temple for him,[352] and observe his rites,[353] and give him our undivided devotions, and ask for his blessings. We shall take offerings of baked loaves of sesame meal to his temple door. He should be pleased.[354] If this plan works, God Namtara shall be put to shame by the gifts, and become embarrassed, and he shall lift his hand,[355] and stop the plague, and sickness shall leave us." One of the elders exclaims, "Let us try the plan!" Another elder shouts, "Let the plan begin immediately!"

*

It is bright daylight, and we see a new temple,[356] near the neglected others, now a total of seven smaller temples, next to the base of the neglected ziggurat at Shuruppak.

Several Weeks Later

There are people gathered here, sitting cross legged, bowing their heads, praying to God Namtara.[357] Large round loaves of sesame coated bread are piled in front of the temple door.[358] The worshippers chant, "God Namtara, lift your hand, you are highly worshipped by mankind," and they keep repeating the chant, and we observe them chanting, as their voices fade away.

We see time passing through a series of alternating cameo views of normal clouds passing through the sky, and sunrise to sunset sequences, and sprinkled throughout the passing time, we see improving conditions, and rapidly

---

[352]M. Page 69, line 401 They built a temple for Namtara in the city.

[353]M. Page 109, line 32 observe his rites.

[354]M. Page 69e 396 [And bring a baked (loaf)] in front of it, and line 397 The offering of sesame-meal may be pleasing to him, and page 109, line 33, and line 34.

[355]M. Page 69, line 398, and page 109, line 35, and line 36 Then he will be put to shame by the gift and will lift his hand (the plague).

[356]M. Page 69, line 400 The elders hearkened to [his] words, and line 401 They built a temple for Namtara in the city.

[357]M. Page 69, line 405 They did [not] reverence their gods, and line 406 They did [not] pray to [their goddesses], and page 71, line 407 But they sought [the door] of Namtara.

[358]M. Page 71, line 408 And [brought] a baked (loaf) in front of [it].

decreasing disease germs that are swirling in the air, within the normal wind, steadily diminish and disappear.[359] The sickness slowly abates and the people again become healthy at the wide, slowly flowing Euphrates River with clear water, with jumping fish which look healthy, and healthy looking birds in the air, with lush and thick reeds along and near the river. The pace is again normal, and people are again catching fish with a net, as the fish jump in the water, and people are catching birds with bow and arrow, as they soar in the air. People are repairing the holes and the broken bindings, and applying a fresh coating of pitch to the large, rectangular shaped, reed boats in the river, and to the small, rectangular shaped, reed boats on the banks. People are working on the canals, making them as wide and as deep as they were previously. People are tending the crops, removing the weeds, and picking the fruit. The sheep and goats regain their healthy look, and children are tending, herding, and milking the flocks, while adults pluck them.

The streets are again filled with people, and the people are maintaining them by smoothing the dips in the streets, and removing the weeds and litter from them. There are children of all ages playing in the streets, some with their healthy looking dogs. The courtyards are being cleaned and swept. The outsides of the homes are being repaired, new sun-dried clay bricks and cream colored plaster are being installed, and a fresh coating of cream colored wash is being applied, and the nearly flat reed bundle roofs are being coated with fresh clay after they are repaired. The wood front doors are being polished, and again have a glossy finish. The holes and broken bindings in the small reed huts are being repaired. The ziggurat and oval wall are in the same condition of repair as the outsides of the homes. The insides of the homes are being cleaned and swept, and the broken pale red sun-dried clay bricks in the floors are being replaced, and the walls are being repaired with the cream colored plaster. Furniture is being repaired and polished, and drapes are resewn and cleaned, and there are reed bundles in the fireplaces. Everywhere, the people begin to look healthy. The beds are empty and the people again have vigor and energy, and they are again conversing. No dead people are visible, as they have been recently buried. Their clothes are again crisp and clean, and their sandals are repaired, and the straps are replaced. Their pendant earrings

---

[359]M. Page 71, line 409 The offering of sesame-meal was pleasing to him, and line 410 [He was put to shame] by the gift and lifted his hand (the plague).

and bracelets are being repaired and worn, and the men's white turban-like hats and colorful head bands are being cleaned and worn. The women again wear a wide and colorful head band ribbon in their hair, and large crescent shaped earrings.

More symbol necklaces of the four main Gods are being worn, all with original chains, now by 40% of the adults and children, and pendant cylinder seals are again being worn. The people again groom their hair and natural beards. People are again freely caressing, and many mothers are nursing babies. In the kitchen, the broken clay utensils are all being replaced, and the people roast birds, ox and lamb, and smoke-dry fish, in the inside or outside fireplaces. All mothers again open their doors and welcome their older daughters to their homes, and younger daughters are again helping mothers. Freshly baked large round loaves of sesame coated bread are on the tables, and people again drink the clear river water, beer, and wine, and there are full vases and drinking cups with all three everywhere. More adults, now 40%, are praying, sitting cross legged and bowing their heads in front of the niche altar, in the wall of their homes, after they place large round loaves of sesame coated bread as offerings. The people are cleaning and repairing their musical instruments, and are beginning to again play them. The people are again joyous, as partying, singing, dancing, and merrymaking again begin. Children of all age are reading and writing pictographs on clay tablets. People are again laboring and working, and they again begin to show their independence, and become defiant and disobedient, as their assertiveness to the Gods, as well as to other people, resumes. Mankind has now full recovered[360] from the plague, and has returned[361] to normal behavior.

*

It is bright daylight, still with normal clouds, and we see the inside of the oval walled city of Nippur, the city of God Enlil, as it appeared before, with its colorful 7 tiered ziggurat in the center, with the temple Ekur and the antennas atop it, as the seventh tier.

---

[360]M. Page 71, line 412 [Plague] left them.
[361]M. Page 71, line 413 To normal they returned.

NIPPUR
.33 Shar Years Later
or
1,200 Earth Years Later[362]
35,200 B.C.

We scan and notice fully recovered mankind, all with black tight-curly hair and brown eyes, working at a normal pace in the courtyard, pampering the ziggurat, and tending the normal looking sheep in the small sheepfold nearby. Again, they are all slightly dusty and dirty, but not sweating or noticeably tired. All wear God Enlil symbol necklaces. There are 2 Igigi Gods, 1 male and 1 female, crisp and clean, both without helmets or body regulators, and 2 Anunnaki God officers, 1 male and 1 female, crisp and clean, both helmeted and one wearing a body regulator, supervising the work. There are six craft grouped in an area, near the ziggurat, one of which is the assembly craft. We notice 2 Gods and 2 Goddesses, all helmeted and two wearing a body regulator, quickly walking toward the ramp.

Inside the assembly craft, our eyes scan, and it appears as it did before, a smaller version of the assembly on Nibiru, complete with a smaller dome viewing monitor in the center, which shows mankind in the present, at a normal pace, fully recovered from the plague, and returning to normal behavior of being defiant, assertive and disobedient; and young Gods and Goddesses caressing young humans with black and brown variations of wavy and tight-curly hair. All the main Gods, except God Anu, are present. Three of the 7 great Anunnaki Gods are seated behind the slightly raised section, and are quietly conversing, as are the 17 Gods and 8 Goddesses. God Enlil, God Enki, and Goddess Ninhursag sit at the slightly raised section for the four main Gods, four of the 7 great Anunnaki Gods. All the Gods are watching the viewing monitor, and again some are pointing at it in disgust. Their communication helmets and body regulators are on the ledge that forms the back of the white pew-like molded circular seats. The 2 Gods and 2 Goddesses

[362]M. Page 71, line 416, and page 73, line 1 Twelve hundred years had not yet passed (as I further explain under 'Longevity' on page 353 in my epilogue).

rush in through the opaque air flow door, and quickly remove their helmets and body regulators, and sit down.

God Enlil rises, and the conversations quickly fade to silence, and all the Gods bow their heads to him, and again, God Enlil sternly addresses the assembly,[363] "In this assembly of the Gods, I, God Enlil, supreme commander and ruler of the Earth, have brought you to my presence. The black-headed people, mankind, have somehow recovered from the plague, and further extended themselves over the land. Do not favor them.[364] They are not diminished![365] Again they have multiplied,"[366] and we see the present on the viewing monitor, which shows a street teeming[367] with people, and a few Gods and Goddesses, Anunnaki God officers, and Igigi Gods.

The viewing monitor then again shows the homes of mankind, sprawling out from the oval walled compound, and teeming people among the homes, with young blond children, and the former blond 13 year olds in appearance, which have now grown to be 16 years of age in appearance, among the black haired children and adults. Young Gods and Goddesses are again freely caressing young humans, now some with straight blond hair, but most with black and brown variations of wavy and tight-curly hair, all taking place in back of the homes, near the reed huts, then God Enlil reports, "They have become more numerous than before![368] They are again increasingly threatening the Gods, and degenerating the Gods. Formerly the young, but now the older Gods as well, are mating with mankind!"

We now look at the viewing monitor, which shows 35 to 45 year old in appearance Gods and Goddesses, freely caressing young humans. Some are leaning against trees, and others are lying near the river bank, close to the tall reeds, away from the houses.

Then we observe the concerned faces of the Gods in the assembly, some of which are angry expressions, then God Enlil continues, "Their independence, defiance and disobedience,[369] again cause us to have less dominance over

---

[363]M. Page 73, line 6, and page 109, line 37.

[364]M. Page 109, line 38 Do not favor them.

[365]M. Page 109, line 39 The peoples are not diminished, (etc.).

[366]M. Page 73, line 2 When the land extended and the peoples multiplied.

[367]M. Page 73, line 3 The land was bellowing like a bull.

[368]M. Page 109, line 39 The peoples are not diminished, but have become more numerous than before!

[369]M. Page 73, line 4 The god (God Enlil) got disturbed with their uproar, and page 109, line 40 I have got disturbed [with] their noise (evil conduct).

them. Mankind has returned to their former behavior, and they are again asserting themselves, and again threatening the Gods. This noise, the wicked, evil conduct[370] and assertiveness and disobedience of mankind, is again a threat to us, and is again too intense for me,[371] and I hear[372] them constantly, and because of it, I am again deprived of sleep."[373] He pauses, turns to his left, and glancingly looks suspiciously at God Enki, and we observe God Enki, calm and collected, exchange the glancing look of God Enlil.

Now God Enlil scans the assembly for a moment, looking for assurance, then he declares, "With the supreme power granted me by God Anu, I, God Enlil, decree that mankind must be suppressed,[374] there must be no rejoicing among them!"[375] He pauses, and momentarily scans the assembly, again looking for assurance, and we observe the faces of the Gods in the assembly, as they expectantly wait for the decree of God Enlil, and then he elucidates, "A famine! I decree a famine,[376] to cut off food supplies from the people.[377] Let there by a scarcity of fruit and vegetables to satisfy their hunger, and let their stomachs become empty."[378] Some of the Gods in the assembly gasp, and look at each other in disbelief, while others smile in approval, then God Enlil commands, "God Adad," and we observe God Adad in the assembly, with an aghast look on his face, "our God in charge of rain distribution, withhold your rain, make it scarce,[379] and release it as snow, at the far north and south axis of the Earth, where there are no people."

Again he pauses, and all the assembly members look at the viewing monitor, which now shows the future effects of the diverting of rain, and the resulting famine on mankind; showing the dim polar regions, completely covered with low dark clouds, being blanketed by heavy snow, which is rapidly

---

[370]M. Page vi, postscript (see footnote 325 and 326).

[371]M. Page 73, line 7 The noise of mankind has become too intense for me.

[372]M. Page 73, line 5 Enlil heard their noise.

[373]M. Page 73, line 8 With their uproar I am deprived of sleep, and page 109, line 41 [With] their uproar sleep does not overcome me.

[374]M. Page 73, line 21 [Mankind] must be suppressed.

[375]M. Page 73, line 20 There must be no rejoicing among them.

[376]M. Page 109, line 50 Let pestilence be laid on the peoples.

[377]M. Page 73, line 9, and page 109, line 42 Cut off food supplies from the peoples.

[378]M. Page 73, line 10 Let there be a scarcity of plant-life to satisfy their hunger, and page 109, line 43 Let plant life be in short supply in their stomachs.

[379]M. Page 73, line 11 Adad (above, the storm God, in charge of rain, lightning, and thunder) should withhold his rain, and page 109, line 44.

accumulating to a great depth; upon the flat ice sheet areas at the north axis, and upon the frozen mountainous areas at the south axis. (Thus, the most recent major ice age is about to begin.)

God Enlil continues, and we see the balance of his command as he speaks the words, in the form of cameo views of the future, on the viewing monitor, "Hold the rain, so the annual spring flood shall not irrigate the fields.[380] Let the wind blow and parch the ground, and the clouds thicken, but not release a downpour.[381] Let the black fields become white, and produce salt.[382] Let no vegetables shoot up, no cereals grow.[383] Let the field diminish their yields,[384] that the womb of mankind may be constricted, and give birth to no child. May there not be births,[385] for there must be no rejoicing among them.[386] The renewed threat of mankind to the Gods is intolerable. Mankind must be suppressed."[387]

Again, we observe some of the Gods in the assembly gasp, and look at each other in disbelief, while others smile in approval, and then we notice God Enki, sadly looking at the viewing monitor, which has now returned to the present, and again shows the teeming people in the streets.

God Adad rises, and affirms, "I shall follow your decree immediately."

God Enlil commands, "Let us clasp hands together." All the Gods rise and clasp hands, and then God Enlil demands, "Your oath to me."

Together, all the Gods declare, "The decree of God Enlil is binding."

*

We scan an obscure daylight sky, now with only a few very thinly scattered

---

[380] M. Page 73, line 12 And below, the (annual river) flood should not come up from the abyss (water was under the control of God Enki), and page 109, line 45.

[381] M. Page 73, line 14 Let the wind blow and parch the ground, and line 16 Let the clouds thicken but not release a downpour.

[382] M. Page 73, line 19, and page 109, line 47 Let Nisaba (probably Goddess of crops) turn aside her breast, Let the black fields become white, and line 48 Let the broad plain produce salt.

[383] M. Page 109, line 49 Let the earth's womb rebel, Let no vegetables shoot up, no cereals grow.

[384] M. Page 73, line 18, and page 109, line 46 Let the fields diminish their yields.

[385] M. Page 73, line 22 May there not be births, and page 109, line 51 That the womb may be constricted and give birth to no child.

[386] M. Page 73, line 20 There must be no rejoicing among them.

[387] M. Page 73, line 21 [Mankind] must be suppressed.

dark clouds, and then we look down, and from a distance and through dust and sand filled air, we observe the outside of the oval walled city of Shuruppak, with its central 7 tiered ziggurat, and temple with antennas atop it, as the seventh tier, and homes sprawling around the wall for one mile in all directions from the ziggurat and temple compound, and the wide river banks, close to the city, and the irrigation canals veining through the area. Ziusudra, the loyal servant of God Enki, is high Priest and King.

SHURUPPAK
Many Months Later

We zoom in and see many cameo views. The wind is normal, being variable, it ranges from almost calm to very light, as it lightly blows dust and sand everywhere, for there is no rain.[388] The Euphrates River is narrow and scarcely flowing, and the water is muddy, and it has dropped until it is only 25 feet deep.[389] There are few thin fish jumping in the water, and many people, at a slow pace, are fruitlessly net fishing. People are also trying to catch the very few remaining thin birds with bow and arrow, but they cannot be caught. The tall and thick reeds along and near the river bank are drooping. The large and small, rectangular shaped, pitch coated reed boats are all on the dried river banks, and they show signs of neglect. They need new pitch, and many have holes where many reed bindings have broken, and many 3 inch diameter reed bundles hang down, exposing the 6 inch diameter, triangular shaped, reed bundle framework. Very little water trickles through the parched beds of the irrigation canals, and there is very little greenery along the canals, because the water level in the river is now at a point equal to the canal tie-ins. The water is not reaching the crops away from the canals. No one is working on the canals to maintain them, and some of the banks show signs of neglect and have turned to sand, which has collapsed or blown into the nearly dry canals. There are no crops away from the canals, as the fields are dry,[390] and are

---

[388] M. Page 111, line 54 Adad above made his rain scarce.

[389] M. Page 111, line 55 Below, (the river) was blocked up and did not raise the (annual) flood from the Abyss.

[390] M. Page 111, line 56 The fields diminished their yields, and line 57 Nisaba turned aside her breast.

beginning to turn white from the salt,[391] but near the canals, grains and vegetables are sparsely growing,[392] and yield very small amounts. Dust and sand are lightly blowing everywhere. The orchard trees, the roots of which go deeper into the ground than the grains and vegetables, are bearing scrawny fruit, but no one is tending the orchards, however small the fruit may be, and most of them have been thoroughly picked by the hungry people. The sheep and goats are very thin, and there are not many to be seen, as they are now guarded, and kept in the homes for protection, to be rationed and eaten as food, as they now give no milk, and most have been thoroughly plucked.

The streets have very little activity, and are almost empty and neglected, and show signs of blowing sand that has formed small mounds everywhere. There is litter, but no weeds in the streets, and there are no children playing in them. No dogs are seen in the streets, as they are also protected, like the sheep and goats, and kept in the homes, to be rationed and eaten as food, even if they are very thin. The courtyards have small mounds of sand, blown in by the light wind, as well as litter, but no weeds, and are unswept and look neglected. The cream colored plaster and sun-dried clay bricks on the outsides of the homes are beginning to crumble, and the homes look dirty and not well kept, and are in poor condition, and show signs of neglect, with small mounds of sand accumulating at the corners and in the doorways, and the clay coated nearly flat reed bundle roofs need repair, as the clay has crumbled off in spots. The wood front doors are dull from lack of attention, and many small reed huts have large holes in them, and are in need of repair, as many reed bindings have broken, and many 3 inch diameter reed bundles hang down, exposing the 6 inch diameter, triangular shaped, reed bundle framework, with sand accumulating around them. The ziggurat and oval wall are in the same condition of neglect as the outsides of the homes, with small mounds of sand also accumulating around them. The insides of the homes are unkempt and neglected, with unswept floors and small piles of wind blown sand beneath many small holes in the roofs. The dust and sand, being lightly blown by the normal and variable winds, are everywhere. The cream colored plaster is crumbling, and we notice dull and broken furniture, and soiled and unraveled

---

[391]M. Page 111, line 57 The black fields became white, and line 58 The broad plain produced salt.

[392]M. Page 111, line 58 Earth's womb rebelled, and line 59 No vegetables shot up, no cereals grew.

drapes, and there are no reed bundles in the fireplaces. The people look somewhat emaciated and ravaged, thin and weak from the starvation and thirst caused by the famine, which was brought on by the drought. Everywhere, the people are in various stages of moderate suffering, with some lying in bed, too weak to walk, and others lying on the ground, close to the houses, or in the streets where they fell; dead, or in a near death state. The others are in a dismal, listless and quiet state. They are weary and distressed from lack of nourishment.[393]

Their clothes are worn, soiled and wrinkled, and are coming apart at the seams. Their sandals are worn, and some have broken straps. The pendant earrings and bracelets are in need of repair, and most have been removed, and the men's white turban-like hats and colorful head bands are soiled, and most have also been removed. The women have no colorful head band ribbon in their hair, and wear no large crescent shaped earrings. 40% of the adults and children continue to wear their symbol necklaces of the four main Gods, most of which are dirty, and a few have unraveled pieces of string as chain substitutes, but their pendant cylinder seals have been abandoned. Their shoulders are slightly pulled in, and they are slightly hunched over as they slowly walk. They are poorly groomed and dirty, with unkempt hair and natural beards. We notice two men arguing and weakly pulling each other, fighting over a meager carrot that one of them has just unearthed next to a nearly dry canal, and weak and starving people,[394] picking small vegetables[395] from the same area. There is little conversation, and they are nearly too weak to caress, and there is almost no child bearing,[396] and there are nearly no mothers nursing babies. Some clay utensils are broken, and no bird is being roasted, or fish smoke-dried, as they have mostly all been eaten, and no other animal, ox or cow, is being roasted, as most of the thin animals have already been eaten, and the few remaining are also being rationed. There are a few small round loaves of sesame coated bread remaining, due to the lack of adequate grain, and these are being rationed by cutting them into very small pieces. The people drink muddy river water, for there is no beer or wine, and there are half-full vases and drinking cups with muddy river water everywhere.

---

[393]M. Page 111, line 60 Pestilence was laid upon the peoples.

[394]M. Page 109, line 52 They cut off food supplies from the peoples.

[395]M. Page 109, line 53 Vegetables were in short supply in their stomachs.

[396]M. Page 111, line 61 So that the womb was constricted and gave birth to no child.

40% of the adults continue to pray to various personal Gods and Goddesses, sitting cross legged and bowing their heads in front of the niche altar in the wall of their homes, after they place some of the very small pieces of sesame coated bread as offerings. They are somber, with no merrymaking or music, as the musical instruments have been set aside, and are now in a neglected and dusty state. There are no laboring workers visible, and the people are sparsely seen outside their homes in the city streets. The children are in the same condition as the adults, and they do not play their games, but lie about the homes without vigor, and there are very few infant children. Very few mothers open their doors and welcome their older daughters to their homes, and the younger daughters that can, try to help their mothers. The clay school tablets have been abandoned. All mankind is hungry and thirsty. The conditions again are desperate.

*

From a distance, in daylight, and although obscured through the dust and sand filled air, again we notice a craft rapidly approaching, with its brilliant glowing firefly-like light emitting from the underside, hover, descend and prepare to land outside the city, away from the homes, in an open area near the river, as the glowing light begins to diminish, and the 4 pods like cattle hoofs extend 6 feet down, and the craft begins to land, all as the normal and variable wind lightly blows the dust and sand.

Several Days Later

We observe a famished Ziusudra, who is near the open area, and who is again comforting several starving people, who have fallen next to the road. As soon as he sees the craft descending, he slowly moves toward the open landing area. His clothing is worn, soiled and wrinkled, but he is wearing his gold God Enki symbol necklace, gold winged sphere symbol of Nibiru, and gold bracelet, and he looks thin and weak from starvation. Ziusudra now appears to be 26 years old.

Then the craft lands, and the glowing light further diminishes, and then completely expires, and after a moment, a side of the craft extends down, forming a 4 feet wide ramp and doorway. Then, God Enki walks through the opaque air flow door and down the ramp of the craft. He is alone, and is wearing the same red robe and adornments that he had previously worn, and he is without a helmet or body regulator.

At almost the same time, Ziusudra arrives to greet him, and he bows down, and prostrates himself, and then he stands, and they exchange greetings, and then warmly embrace.[397a]

After their long embrace, again Ziusudra pleads,[397b] "God Enki, my personal God, the God who elevated me to high Priest and King of the city of Shuruppak, the God who created us. I petition you, my Lord, how long shall God Enlil impose famine upon us? How long are we without rain to go?[398] Rain and irrigation have left.[399] Our supplies are disappearing,[400] the people are weak, dying from starvation. They cannot continue their work in the fields.[401] God Enki, my personal God, help mankind! Help us![402] God Enki, intervene for mankind![403] I beg of you, end this drought, so the famine shall abate. I have prepared offerings for you, God Enki, for it is your duty, as my personal God, to care for us in time of need." Then God Enki sympathetically advises, "My loyal, faithful and pure servant, Ziusudra. This famine, that God Enlil imposed in the great assembly,[404] I can not end. I can not directly intervene with the task of God Adad, who acts by the decree[405] of God Enlil."

He pauses, and then we again see a mischievous smile, and a sparkle in the eyes of God Enki, as he resolves, "But, as last time with the pestilence, I can give you a plan for relief. Gather your city elders,[406] and counsel[407] them. This is what you shall advise them of." God Enki now whispers into the ear of Ziusudra, but we cannot hear his words.

397a M. Page 123, upper line 3 (see footnote 335).
397b M. Page 75, line 4 Atra-ḫasīs appealed to his lord.
398 M. Page 75, line 3 How long are we without rain to go?
399 M. Page 75, line 2 Rain and irrigation have left.
400 M. Page 75, line 5 Their supplies began to disappear.
401 M. Page 75, line 6 In the fields, diminished is their work.
402 M. Page 75, line 7 Help mankind, help us.
403 M. Page 75, upper line 8 Enki, intervene for mankind I shouted.
404 M. Page 75, upper line 11 The famine, Enlil imposed in the great assembly.
405 M. Page 75, upper line 12 I can not intervene by oath.
406 M. Page 75, upper line 13 Gather your [elders] (hold a meeting at your house).
407 M. Page 75, upper line 14 [And] counsel (advise them) [in the house] (at the meeting).

*

It is daylight, with dust and sand obscuring our vision, as the normal and variable wind lightly blows the dust and sand, and we see the outside of the large house of Ziusudra, and the gate leading to the courtyard, which is within the oval wall at Shuruppak, near the neglected ziggurat. The house is again neglected.

The Next Day

The city elders, 15 males and 5 females, all wearing their symbol necklaces, are again gathering for a council meeting. Some are being brought by hand drawn four wheel cart, and others are slowly coming on foot. They are thin, weak and frail, worn and a little hunched over as they slowly walk. They are starving, with hollow cheeks, and dry thirsty looking parched lips. They are going through the gate and courtyard, and into the house. Small mounds of sand are piled everywhere, as the light and variable blowing wind shifts and adds to the mounds.

Inside the house, we again see the large nicely furnished but neglected central room, imitating the assembly of the Gods, with wood benches arranged in a circle for the elders, and a large wood chair for Ziusudra. The dust and sand are everywhere, with small piles of wind blown sand accumulating beneath the many small holes in the roof. Ziusudra, wearing his gold God Enki symbol necklace, and the other adornments that he had previously worn, is welcoming the group, and they quickly sit down, and then he sits down, and the group, which had been sparingly conversing, quiets and looks at him, and Ziusudra declares, "I have again gathered you in council, to propose a plan for terminating the drought and famine. As we have done before, we should advise all our people, to cease praying and cease offering, to our various personal Gods and Goddesses.[408] Command, and let the heralds proclaim. Make a loud noise, an announcement in the land, so all can hear."[409]

---

[408]M. Page 75, lower line 9 Do not reverence your gods, and lower line 10 Do not pray to your [goddesses].

[409]M. Page 75, line 15 [Command that] heralds [proclaim], and lower line 8 [And make a loud] noise in the land.

He pauses, and scans the council for a moment, and we observe the faces of the elders, as they wait for words of relief, then Ziusudra commands, "To God Adad,[410] and only God Adad, we should all concentrate our obedience and offerings!" The elders look at each other, with puzzled looks on their faces, then Ziusudra elucidates, "Ask for the blessings of God Adad, and bestow upon him our full devotions. As we did for God Namtara, we shall do for God Adad. We shall build a temple for him,[411] and take offerings of baked loaves of sesame meal to his temple door. These gifts should be pleasing to him.[412] He shall be put to shame, and he shall lift his hand,[413] and cause rain to fall."

Again he pauses, and scans the council, and again we observe the faces of the elders, who are now trying to smile. Then Ziusudra softly reveals, "But rain in a different form," and again he pauses, and scans the council. Again the elders' faces look puzzled, then Ziusudra continues, "He shall rain down a mist in the morning, and rain down a dew in the night, so that the fields shall furtively bear[414] crops, and the famine shall end. For by giving us a mist and dew, and not rain, God Adad shall not be breaking the decree of God Enlil." One of the elders probes, "Watering the Earth, without the knowledge of God Enlil?" Another elder inquires, "How can this trick work?" Another elder concedes, "Surely, God Enlil knows of our ways?" Then Ziusudra instructs, "Do as I command, and we shall prevail."

*

Through the normal and variable wind, which lightly blows the dust and sand, we see a new temple,[415] near the neglected others, now a total of eight smaller temples, next to the base of the neglected ziggurat at Shuruppak.

Several Weeks Later

---

[410]M. Page 75, lower line 11 But seek [the door of] Adad.

[411]M. Page 75, line 20 They built a temple for Adad in the city.

[412]M. Page 75, lower line 12 And bring a baked (loaf) [in front of it], and lower line 13 [The offering of sesame-meal] may be pleasing to him.

[413]M. Page 75, lower line 14 Then he will be put to shame [by the] gift and will lift his hand (the drought).

[414]M. Page 75, line 16 He may rain down a mist in the morning, 17 And may furtively rain down a dew in the night, and line 19 So that the fields will furtively bear grain.

[415]M. Page 75, line 20 They built a temple for Adad in the city.

There are people gathered here, sitting cross legged, bowing their heads, praying to God Adad.[416] The few remaining very small round loaves of sesame coated bread are piled in front of the temple door,[417] but not nearly as much as before, for the shortage of grain continues. The worshippers chant, "God Adad, lift your hand, you are highly worshipped by mankind," and they keep repeating the chant, and we observe them chanting, as their voices fade away.

We see time passing through a series of alternating cameo views of only a few very thinly scattered dark clouds passing through the obscure sky, and obscure sunrise to sunset sequences, and sprinkled throughout the passing time, we see this plan also working.[418] Within the lightly blowing dust and sand, we notice a mist in the morning,[419] increasing in the air, and collecting on the pomegranates, as well as on the other crops, and stimulating their growth. Then we notice a dew in the night,[420] collecting on the pomegranates, and the other crops, and also stimulating their growth. Both in the morning and at night, we notice mist and dew, as drops of water forming on the crops, and then dropping to the ground and running together to form little trickles of water, and the many little trickles of water running together to form larger trickles of water, all of which are watering the soil and the crops. The dust and sand that had been lightly blowing everywhere, slowly stop, as the ground is now moist enough to hold the soil from the normal and variable light wind, and the air slowly clears, and a bright blue sky prevails, but still with only a few very thinly scattered dark clouds. Some of the crops come to life and begin to sprout,[421] but the trickles of water are insufficient, and do not reach all of the crops, and as a result, the drought and famine conditions are very slow to improve, and are not fully relieved.

The Euphrates River is still narrow, and scarcely flowing, and the water is still muddy and at the 25 feet deep level. There are a few almost normal looking fish jumping in the water, and many people, at an almost normal pace, are net fishing for the few remaining fish, with poor catch results, and arguing and

---

[416]M. Page 75, line 23 They did not reverence their gods, 24 They did [not] pray to their goddesses, and line 25 But they [sought] the door [of Adad].

[417]M. Page 75, line 26 [And brought] a baked (loaf) in front of it.

[418]M. Page 77, line 27 The offering of sesame-meal was pleasing to him, and line 28 He was put to shame by the gift and lifted his hand (the drought).

[419]M. Page 77, line 30 In the morning he rained down a mist.

[420]M. Page 77, line 31 And furtively rained down a dew in the night.

[421]M. Page 77, line 33 [The fields] furtively bore grain.

fighting over the few fish and almost normal looking few remaining birds that are pursued, and finally caught. The tall and thick reeds, which are drooping along and near the river bank, begin to straighten and reach upwards. The large and small, rectangular shaped, reed boats are all on the dried river banks, but they are being repaired. Reed bundles and bindings are being replaced, and fresh pitch is being applied to the reeds. Very little water trickles through the parched beds of the irrigation canals, and there is still very little greenery along the canals, because the water level in the river is still at a point equal to the canal tie-ins, and the water is not reaching some of the crops, away from the canals. People are working on the canals, removing the collapsed and blown in sand from them, and trying to dig them deeper than the river water level. Some of the fields are again turning black from the mist and dew, as some of the crops away from the canals begin to sprout. The orchard trees, all of which were bearing scrawny fruit, because of their deep roots, are now bearing normal fruit, as they are among the first crops to show signs of recovery, and people are again tending the orchards. The sheep and goats are almost back to their normal full look, and they are no longer very thin, and they have become more numerous. They are now kept in the courtyards or in back of the homes, and not in the homes, but are still in a guarded and protected state, and the children are again tending, herding, and milking them, and the adults are again plucking them.

The streets again have activity, as the last mounds of sand are being removed, and litter cleared. We begin to see people in the streets, and children of all ages resume playing games with vigor in the streets. The dogs, now almost back to their normal look, are kept with the sheep and goats, and are also still being somewhat protected. The courtyards have been swept clean, and there are no mounds of sand in them. The cream colored plaster and sun-dried clay bricks on the outsides of the homes are under repair, and are being freshly coated with a cream colored wash, and all the sand has been removed from the doorways, and the nearly flat reed bundle roofs are being repaired, and fresh clay is being applied. People are polishing the wood front doors, again giving them a glossy finish. The small reed huts are being repaired, and the reed bundles and bindings are being replaced, and sand is being swept from around them. The ziggurat and oval wall are in the same condition of repair as the outsides of the homes. The insides of the homes are again clean and well kept. All the dust and sand have been removed, and the floors have been swept. The crumbling cream colored plaster is being repaired, and the furniture is being repaired and polished. The soiled and unraveled drapes are being resewn and cleaned, and there are reed bundles in the fireplaces. The people again look healthy, and everywhere they are in various stages of recovery, with people beginning to leave their beds. The people begin to eat meals, but the portions are smaller than normal, which include some of the pomegranates. They are now partially recovered, and we observe a continued very slow recovery, but they have not fully recovered to their previous normal level. They begin to converse and argue with each other, as they are now receiving some nourishment. No dead people are visible, as they have been recently buried. Their clothes are again clean and crisp looking, almost as they had appeared before, and their sandals have been repaired, and the broken straps replaced. Their pendant earrings and bracelets are being repaired and worn, and the men's white turban-like hats and colorful head bands are also being cleaned and worn. Some women begin to wear their wide, brightly colored head band ribbon in their hair, and their large crescent shaped earrings.

Many of the adults and children, now 60%, wear symbol necklaces of the four main Gods, all with repaired chains, and some begin to wear their pendant cylinder seals. Their slightly pulled in shoulders begin to broaden again, and their slightly hunched over slow walk begins to straighten up and quicken to an almost normal pace. Their grooming is improving, with hair and natural beards again well kept. Mankind is again caressing, but less than normal, and with some restraint, and child bearing resumes, with many mothers nursing babies, but not as many as normal. The broken clay utensils are being replaced, and ox and lamb are again being roasted, as well as a few birds, and a few fish are being smoke-dried, but not at the previous normal level. There are larger round loaves of sesame coated bread, due to the increased supply of grain, but not as large as normal. In addition to the muddy river water, some beer and wine is being poured, and there are half-full vases and drinking cups with all three everywhere. Many adults, now 60%, pray to various personal Gods and Goddesses, sitting cross legged and bowing their heads in front of the niche altar, in the wall of their homes, after they place less than normal size round loaves of sesame coated bread as offerings. Some merrymaking and music playing begins, as the instruments are dusted off and repaired, but not on a widespread basis. Laboring workers are again visible, and the people are again seen outside their homes in the city streets. The children of all ages are in the same condition as the adults, slowly recovering. Almost all mothers open their doors and welcome their older daughters to their homes, and the younger daughters again help their mothers, and the clay school tablets are again used for reading and writing pictographs. Mankind is only moderately recovered, as the conditions for mankind are slowly improving,[422] but have not reached a full level of normalization,[423] and have generally declined.

*

It is bright daylight, still with only a few very thinly scattered dark clouds, and we see the inside of the oval walled city of Nippur, the city of God Enlil, as it appeared before, with its colorful 7 tiered ziggurat in the center, with the temple Ekur and the antennas atop it, as the seventh tier.

NIPPUR
.33 Shar Years Later
or
1,200 Earth Years Later
34,000 B.C.

We scan and notice moderately recovered mankind, all with black tight-curly hair and brown eyes, working at an almost normal pace in the courtyard, removing the last small mounds of sand, pampering the ziggurat, and tending the almost normal looking sheep in the small sheepfold nearby. Again, they are all slightly dusty and dirty, but not sweating or noticeably tired. All wear God Enlil symbol necklaces. There are 2 Igigi Gods, 1 male and 1 female, crisp and clean, both without helmets or body regulators, and 2 Anunnaki God officers, 1 male and 1 female, crisp and clean, both helmeted and one wearing a body regulator, supervising the work. There are seven craft grouped in an area, near the ziggurat, one of which is the assembly craft. We notice 3 Gods and 2 Goddesses, all helmeted and three wearing a body regulator, quickly walking toward the ramp.

---

[422]M. Page 77, line 34 Slowly the famine left them.
[423]M. Page 77, line 35 Slowly to their [fields] they returned.

Inside the assembly craft, our eyes scan, and it appears as it did before, a smaller version of the assembly on Nibiru, complete with a smaller dome viewing monitor in the center, which shows mankind in the present, at an almost normal pace, moderately recovered from the drought and famine, and returning to normal behavior of being defiant, assertive and disobedient; and young and older Gods and Goddesses again freely caressing young humans with straight blond hair, and black and brown variations of wavy and tight-curly hair. All the main Gods, except God Anu, are present. Three of the 7 great Anunnaki Gods are seated behind the slightly raised section, and are quietly conversing, as are the 18 Gods and 9 Goddesses. God Enlil, God Enki, and Goddess Ninhursag sit at the slightly raised section for the four main Gods, four of the 7 great Anunnaki Gods. All the Gods are watching the viewing monitor, and again some are pointing at it in disgust. Their communication helmets and body regulators are on the ledge that forms the back of the white pew-like molded circular seats. The 3 Gods and 2 Goddesses rush in through the opaque air flow door, and quickly remove their helmets and body regulators, and sit down.

God Enlil rises, and the conversations quickly fade to silence, and all the Gods bow their heads to him, and again, God Enlil sternly addresses the assembly,[424] "In this assembly of the Gods, I, God Enlil, supreme commander and ruler of the Earth, have brought you to my presence." Now in an angry tone, he reports, "Mankind, somehow, is evading my decree! Their noise, the assertiveness and disobedience of mankind, is again too intense for me, and as a result of it, I am again deprived of sleep![425] Their threat to us continues! Could it be possible that one of the Gods here, is intervening my commands?"

God Enlil now pauses, and very slowly scans the assembly, looking for assurance. He first looks at, and stops at God Namtara, then he stops at God Adad, and finally he stops at God Enki. As God Enlil stops at each of them, all three Gods' faces mirror the same indifferent look, then God Enlil reflects, "I decreed a plague, they survived! I decreed a famine, they appear to be surviving the famine! They are recovering! Mankind continues to threaten the Gods!"

God Enlil then turns around to the ledge behind him, where there is a spherical object concealed under a white plastic-like cloth. He unveils and

---

[424]M. Page 117, line 1.

[425]M. Page 117, line 2 [The noise of mankind] has become [too intense for me], and line 3 With their uproar I [am deprived of sleep].

reveals the object, a large well ripened and full-sized pomegranate,[426] upon a clear glass-like tray, and he takes the pomegranate from the tray. He turns back toward the assembly, and holds it up for all the Gods to see, and we see a close up view of the pomegranate, as he holds it up, then God Enlil exclaims, "Look at this pomegranate! Does this look like the fruit of a famine?"

Again he scans the assembly, looking for assurance, and we notice a slight smile on the face of God Enki, then God Enlil asserts, "I shall not permit the full recovery of mankind to occur!" He pauses, and looks at the viewing monitor, and we see the present on the viewing monitor, which shows a street teeming with people, and a few Gods and Goddesses, Anunnaki God officers, and Igigi Gods.

The viewing monitor then again shows the homes of mankind, sprawling out from the oval walled compound, and teeming people among the homes, with young blond children, and the former blond 16 year olds in appearance, which have now grown to be 19 years of age in appearance, among the black haired children and adults. Young and older Gods and Goddesses are again freely caressing young humans, some with straight blond hair, and now with blue eyes, but most with black and brown variations of wavy and tight-curly hair, all taking place in back of the homes, near the reed huts.

Then we observe the concerned faces of the Gods in the assembly, some of which are angry expressions, as they expectantly wait for the decree of God Enlil, then God Enlil declares, "I shall increase the famine that has been upon mankind, and broaden the magnitude. With my supreme power and rule over the Earth, granted me by God Anu, I, God Enlil, decree that God Adad," and again God Enlil scans the assembly, looking for assurance, and then he looks at God Adad, and we observe God Adad, who now has a concerned look on his face, and then God Enlil commands, "personally, and under the close supervision of God Anu, shall guard the upper regions,[427] and withhold 'all' forms of rain from the people, and cause it to be released at the north and south axis of the Earth."

Again he pauses, and all the Gods look at the viewing monitor, which now shows the future effects of the renewed diverting of rain, and the resulting

---

[426]M. Page 81, line 12 A pomegranate (he brought) before the assembly, and line 13 He (God Enlil) was filled with anger [at the Igigi] (they instigated the situation because they revolted for replacements, and their replacements, mankind, have become a problem).

[427]M. Page 117, line 4 Command that [Anu and Adad] guard [the upper regions].

severe famine on mankind; showing the polar regions, still completely covered with low dark clouds, being blanketed by additional heavy snow, which is still rapidly accumulating to a very great depth; upon the flat ice sheet areas at the north axis, and upon the frozen mountainous areas at the south axis.

Then God Enlil imposes, "I shall personally guard the middle Earth,[428] the surface air currents, the winds, and prevent any form of relief, especially birds, from reaching mankind through the air." We continue to see the future on the viewing monitor, which now shows a bird being blown through dust and sand filled air, and away from the people, by the mild but forceful and constant wind, against its direction of flight, and against its will.

Then God Enlil commands, "God Enki," and we observe God Enki, calm and collected, "shall guard the seas, together with his plants,[429] mankind, and prevent mankind from obtaining fish from the seas. This shall be done through gravity control. By increasing the gravity over the smaller seas, near the people, the water level of the smaller seas and of the rivers near the people shall be reduced. This shall drive the fish away from the people, and prevent any fish from being taken from the smaller seas and the rivers." We continue to see the future on the viewing monitor, which now shows, amidst the constant and mildly blowing dust and sand, the water level of the slightly muddy smaller seas near the people dropping, and then the water level of the muddy rivers dropping and surging toward the lowered slightly muddy smaller seas. The fish are jumping in the surging waters, as they are carried away from the people, and toward the center of the larger unaffected clear seas.

We again observe some of the Gods in the assembly gasp and grimace, and look at each other in disbelief, while others smile in approval, then God Enlil elucidates, "By these means, the drought and famine, the yoke of oppression, shall be renewed, with more intensity, and affect the people more

---

[428]M. Page 117, line 5 That Sin and Nergal guard [the middle earth] (presumably God Enlil is guarding the lower Earth. Sin is a son of God Enlil, and Nergal is a son of God Enki).

[429]M. Page 117, line 6 That the bolt, the bar [of the sea], and line 7 Ea (God Enki) may guard together with [his plants] (his creation = mankind. We still do not fully understand many of the metaphors of the ancients, 'the bolt, the bar of the sea', being one of them. Although we are somewhat able to grasp its abstract meaning by the direction of the surrounding words and sentences, its exact meaning is still elusive, and the interpretation I render is just one of many that could be applied to account for the exercise of control of the waters of the Earth. This concept certainly qualifies to stretch the boundaries of our understanding; as I state on page i in my dedication: see footnote 42).

severely than before." He pauses, and scans the assembly, looking for assurance, and then he again holds up the pomegranate, and exclaims, "There is to be no means of nourishment reaching mankind! Their threat to the Gods shall cease!"

God Enki then looks at God Namtara and God Adad, with a hopeless and sad expression on his face, and we observe the faces of God Namtara and God Adad, as they return a glancing look to God Enki, with concerned expressions on their faces, and then we notice God Enki, sadly looking at the viewing monitor, which has now returned to the present, and again shows the teeming people in the streets.

God Enlil commands, "Let us clasp hands together." All the Gods rise and clasp hands, and then God Enlil demands, "Your oath to me."

Together, all the Gods declare, "The decree of God Enlil is binding."

*

Again we see an obscure daylight sky, now with only several very thinly scattered dark clouds, and then we look down, and from a distance and through dust and sand filled air, we observe the outside of the oval walled city of Shuruppak, with its central 7 tiered ziggurat, and temple with antennas atop it, as the seventh tier, and the homes sprawling around the wall for one mile in all directions from the ziggurat and temple compound, and the wide river banks, close to the city, and the irrigation canals veining through the area. Ziusudra, the loyal servant of God Enki, is high Priest and King.

SHURUPPAK
6.94 Shar Years Later
or
25,000 Earth Years Later
9,000 B.C.

We zoom in and see many cameo views. The constant wind is forceful,[430]

---

[430]M. Page 117, line 9 Sin and Nergal guarded the [middle] earth.

as it mildly blows dust and sand everywhere, for there is no rain, mist, or dew. The Euphrates River is very narrow and stagnant. The banks are parched, and the water is a muddy greenish color, and has now dropped until it is only 3 feet deep.[431] There are no fish jumping in the water, and no people are net fishing, as the fish have all been carried out to the center of the larger unaffected clear sea.[432] We see no birds flying in the air, as they have either died or been eaten, or have been blown away from the people. There are no tall and thick reeds along and near the river banks, only very small reeds are growing at the edge of the stagnant river, and they are scrawny and look dry, and are drooping. There are skeletal remains of the 6 inch diameter, triangular shaped, reed bundle framework of the large and small, rectangular shaped, pitch coated reed boats, scattered along the upper parched river banks. Very few 3 inch diameter reed bundles still cling to and hang down from the framework, as most of the reed bindings have broken, and most of the pitch has been weathered off by the mildly blowing dust and sand. No water is in the irrigation canals, not even a trickle of water, as they are now dry and parched, and are filled with cracked clay, and no greenery grows along them, because the water level in the river is now at a point 22 feet below the canal tie-ins. No one is working on the canals, as they have been abandoned and the banks have turned to sand, which has collapsed or blown into the dry canals, almost completely covering some of them. There are no crops, as the fields are also parched[433] and white with salt,[434] and sand blows across them and drifts in large scattered mounds. Nothing grows,[435] and dust and sand are mildly blowing everywhere. There

---

[431]M. Page 117, line 8 He commanded, and Anu and [Adad] guarded [the upper regions], and page 111, line 3 [Adad] above [made his rain scarce], and line 4 Below, (the river) was blocked up [and did not raise the (annual) flood from the Abyss], and page 79, line 1 Above, the rain was scarce, and line 2 Below, the (annual) flood did not [rise] from the abyss

[432]M. Page 117, line 10 The bolt, the bar of the sea, and line 11 Ea (God Enki) guarded together with [his] plants, and page 111, line 1 The bolt, [the bar of the sea], and line 2 [Ea] guarded [together with his plants].

[433]M. Page 111, line 5 The fields diminished [their yields], and line 6 Nisaba [turned aside her breast].

[434]M. Page 79, line 7 The black fields became white, and line 8 The broad plain was choked with salt, and page 111, line 6 [The black fields became white], and line 7 [The broad plain] produced salt.

[435]M. Page 79, line 4 The womb of earth did not bear, and line 5 Vegetation did not sprout [up], and page 111, line 8 [No vegetables] shot up, no cereals [grew].

are no signs of previous crops, as the fields have been long abandoned. The orchard trees are all bare, and now all that remains of the trees are only their dried out trunks. There are no signs of previous fruit, as they also have been long abandoned, and no one is seen in either the fields or the orchards tending them, as there is nothing to tend, for all the crops are gone. No sheep, goats, or dogs are seen in the streets or outside the homes, as most of them have been eaten, and those that are remaining are thin, and are guarded, and kept in the homes for protection. There are only very few of them left, and they are being very carefully rationed. The sheep and goats give no milk, as they are too weak, and all have been thoroughly plucked.

The pace is very slow, and the streets have very, very little adult activity, and are practically empty and neglected, with large mounds of wind blown sand everywhere. There is litter, but no weeds in the streets, and there are no children playing in them. The courtyards have large mounds of wind blown sand, as well as litter, but no weeds, and are unswept and look neglected. The cream colored plaster and sun-dried clay bricks on the outsides of the homes has crumbled to a high degree of deterioration, and the homes are also very dirty and in very poor condition, with large mounds of wind blown sand accumulating at the corners and in the doorways, and most of the clay has crumbled and been blown off of the clay coated nearly flat reed bundle roofs, and is sparsely visible, and the dried reed bundle roofs are full of large holes, and broken reed bundles hang down into the homes. The wood front doors are very dull from mildly blowing sand and lack of attention, and many of them show signs of broken and splintered wood. The small reed huts are in need of major repair, as they have been stripped of their 3 inch diameter reed bundles, with very little of their 6 inch diameter, triangular shaped, reed bundle framework remaining, for their reed bundles have already been used to patch roofs and as fuel in the fireplaces. The ziggurat and oval wall are in the same condition of neglect as the outsides of the homes, with large mounds of sand accumulating around them. The insides of the homes are unkempt and neglected, with unswept floors and large piles of wind blown sand beneath the large holes in the roofs. Large amounts of dust and sand are everywhere. The cream colored plaster is crumbling, and we notice very dull and broken furniture, and very soiled and unraveled window drapes that are almost rags, and there are no reed bundles in the fireplaces.

The people look totally emaciated and ravaged, very thin and frail, like walking skeletons, from the starvation and thirst caused by the renewed famine, which was brought on by the renewed drought. The intensity of the renewed famine and drought has overwhelmed all of mankind. Everywhere, the people are in various stages of severe suffering, with some lying in bed, too weak to walk, and others lying on the ground, close to the houses, or in the streets where they fell; dead, or in a near death state,[436] as the mildly blowing wind swirls sand around them. Their faces are altered by hunger,[437] and their cheek bones are protruding, forming severely hollow cheeks. Their faces look green,[438] from lack of nourishment,[439] and are encrusted with wind blown white dust and sand, like malt,[440] as there is no acceptable water to bathe in. Their appearance is unkempt, dusty and dirty looking, from lack of water, and they itch[441] from lice. The people that can, very slowly walk and are severely hunched[442] over, and their broad shoulders are now narrow,[443] and their long legs have become bent[444] and weak from the starvation.

Their clothing now consists of rags and shreds of cloth, worn, soiled and wrinkled, and unraveled at the seams, for their sources of clothing, animals and plants, have been almost eliminated. Some people have replaced their clothing with old cloth, that is wrapped around them, and we notice others removing the shreds of clothing from the dead. Their sandals have fallen apart, many with broken straps, and most of the people now are barefoot. The pendant earrings and bracelets are all in need of repair, and all have been removed, and the men's white turban-like hats and colorful head bands are soiled and unraveled, and all have also been removed. The women have no colorful head band ribbon in their hair, and wear no large crescent shaped earrings, but 60%

---

[436]M. Page 79, line 6 People were not seen [active].

[437]M. Page 79, line 11 The third year came, and line 12 [And] their features [were altered] by hunger, and page 111, line 14 [The people's features] were distorted [by hunger].

[438]M. Page 79, line 15 [Their] faces appeared green.

[439]M. Page 111, line 9 [Pestilence was laid on the peoples], (etc.).

[440]M. Page 79, line 13 [Their faces] were encrusted, like malt, and page 113, line 25 Their [faces] were overlaid [like dead malt].

[441]M. Page 79, line 10 For the second year they suffered the itch, and page 111, line 12 [When the second year arrived], [They suffered] the itch.

[442]M. Page 79, line 16 They walked hunched [in the street], and page 113, line 17 [They walked hunched] in the street.

[443]M. Page 81, upper line 17 Their broad shoulders [became narrow], and page 111, line 16 [Their broad shoulders] became narrow.

[444]M. Page 81, upper line 18 Their long legs [became short], and page 111, line 15 [When the fourth year arrived], Their [long] legs became short.

of the adults, and children 10 years of age in appearance or older, for there are no children under 10 years of age in appearance, continue to wear their symbol necklaces of the four main Gods, although they are also dirty, and many broken chains have now been replaced by unraveled pieces of string, but all their pendant cylinder seals have been long abandoned. They are very poorly groomed and dirty in appearance, with unkempt hair and natural beards.

We notice several people weakly pulling each other, fighting over the last remains of parched and dried grass and weeds, and eating it as food.[445a] The people are silent, apathetic, and lethargic, and they are now entirely too weak to caress, and there is now absolutely no child bearing,[445b] and there are now no mothers nursing babies, for there have been no births for the past 3,400 Earth years, nearly 1 Shar year, and almost no child bearing during the 21,600 Earth years, 6 Shar years, previous to that. Most of the clay utensils are broken, and no bird is being roasted, or fish smoke-dried, as they have all been eaten, and no other animal, ox or cow, is being roasted, for most of them, thin as they are, also have already been eaten, and the few remaining are also being very carefully rationed and guarded, and kept in the homes for protection. Mothers shut their doors, and do not open them to their older daughters, who watch their mothers go into their slightly more plentiful homes,[446] and mothers and younger daughters are closely watching each others small portions of meat and dried weeds on scales,[447] as they very carefully ration everything. There is no sesame coated bread, due to the lack of grain. The people drink muddy greenish color river water, for there is no beer or wine, and there are nearly empty vases and drinking cups with muddy greenish color river water everywhere.

60% of the adults continue to pray to the various personal Gods and Goddesses, sitting cross legged and bowing their heads in front of the niche altar in the wall of their homes, but they have no offerings to give, as there is very little food. They are very somber, with no merrymaking or music, as the musical instruments have been set aside long ago, and are now in a neglected,

---

[445a]M. Page 79, line 9 For one year they ate couch-grass.

[445b]M. Page 111, line 9 [Pestilence was laid on the peoples], [So that the womb was constricted and gave] birth [to no child].

[446]M. Page 113, line 18 [When the fifth year arrived], Daughter watched the mother's [going in], and line 19 [But the mother would not] open her door [to the daughter].

[447]M. Page 113, line 20 [The daughter] watched [the scales of the mother], and line 21 The mother watched [the scales of the daughter].

dusty and broken condition. There are no laboring workers visible, and the people are very sparsely seen outside their homes in the city streets. Children 10 years of age in appearance or older, for there are no children under 10 years of age in appearance, are in the same condition as the adults, and they do not play their games, but lie about their homes without vigor. There are no children in the streets, as they are all kept in their homes for their own protection; some of which are to be eaten as food, as many of the people are now turning to cannibalism, and eat their own smaller children, as they have already eaten those of their neighbors, for they have already eaten nearly all the animals, and almost exhausted their supply of other foods. The younger daughters that can, try to help their mothers, each especially watching the other very carefully, so they do not eat the portions which are not theirs. The clay school tablets have been long abandoned. The hunger and thirst of all mankind is increasing each day. The conditions for mankind are very desperate, and all mankind is now living on the brink of death.[448]

*

At an intersection of streets, in daylight, just outside the gate in the oval wall, amidst the constant and forceful wind, which mildly blows the dust and sand, and obscures our vision, we observe Ziusudra, who now appears to be 99 years old.

Several Days Later

He is very slowly walking with an elder, and both are weak, frail and hunched over, their faces are covered with white dust and sand, and both are clothed in rags, but continue to wear their symbol necklaces, with Ziusudra's being a gold God Enki symbol necklace. Ziusudra also wears his gold winged sphere symbol of Nibiru, and his gold bracelet. Reaching his later years, he shows signs of aging, among which are wrinkles on his face, and his black tight-curly hair and beard are turning gray, but all the aging signs are somewhat

---

[448]M. Page 79, line 14 [And they were living] on the verge of death, and page 113, line 26 [The peoples] were living [on the edge] of death.

concealed by the white dust and sand. Near them are several decomposing dead people, adults, lying in the street, with accumulating sand beginning to cover them, some stripped of their clothing, and others partially dismembered, being used as food. As they very slowly walk, they look through the blowing dust and sand, and see an adult fall in the street, in a near death state, next to a neglected corner house and near its side door composed of splintering wood. The elder shouts to Ziusudra, "Look Ziusudra!"

They both stop and look, and the splintered wood side door opens, and two severely hunched over people clothed in rags emerge, and slowly struggle to drag the vulnerable and unresisting dying adult into their house. The splintered wood door is then very quickly shut and bolted.

Ziusudra confesses to the elder, "I have continually appealed to God Enki, and made whatever offerings I could,[449] but he has not responded. He warned me of this severe resumption of the famine, and of the Gods involved, all at the command of God Enlil, but I did not believe it would be this intense," and then Ziusudra looks into the eyes of the elder, and sadly admits, "God Enki has not come to our aid." The elder asks, "Has he forgotten us?" Ziusudra reassures, "No, I do not believe God Enki has forgotten mankind. Our former plans for relief shall not succeed or help us this time, for there are too many Gods involved in this drought and famine, and we can not concentrate our obedience to one God, without angering the other Gods, and we have very few offerings to give. Even God Enlil, himself, guards that the drought and famine not be relaxed. We must wait for the advice of God Enki. As we have waited a long time, we must continue to wait." The elder inquires, "But we have suffered for twenty-five thousand of our Earth years, nearly seven Shar years! How long must we wait?" Ziusudra resolves, "I shall continue to wait, and pray to God Enki."

Then both men hear screams, and look at another neglected corner house, and witness a thin and weak 10 year old in appearance boy clothed in rags, wearily run out of the house, through the courtyard and into the street. The boy is screaming, "No . . . . no . . . . no . . . . ." Expressions of puzzlement are on the face of Ziusudra and the elder, as they watch the boy. Then, we witness

---

[449] M. Page 77, line 2 To the temple of his god, and line 3 Daily he set his foot, and page 117, line 12 Now Atra-ḫasīs (Ziusudra), [whose god was Ea] (Enki), and page 143, line 145 At that time, the king Ziusudra, the anointed [one], and line 146 He made offerings to Enki.

the boy's weak, frail and starving father, also clothed in rags, hunched over and staggeringly run out of the house, through the courtyard and into the street, pursuing the boy, snatch the boy and slowly declare, "I shall serve you up for food, my son, for I have already served up our daughter for dinner, and consumed my neighbors' children."[450]

Ziusudra and the elder look at each other in disbelief and disgust, as the defenseless child is slowly dragged through the courtyard and into the house, screaming, "No . . . . no . . . . no . . . . ." Then, the dull and broken wood door is quickly shut and bolted, and we continue to hear the boy's muted screams coming from inside the house, "Nooooo, nooooo, nooooo." We zoom to the door and listen, and the muted screaming continues for a moment, and then suddenly, the muted screaming stops. We slowly scan to the back of the house, and notice a pile of bones, some from animals, including dogs, and some of which are the skeletal remains of several 9 year old in appearance children, as the wind blown sand accumulates within them.

*

Through the obscuring constant and mildly blowing dust and sand filled air, in daylight, we see the outside of the house of Ziusudra. It looks like the others, with large mounds of wind blown sand everywhere, and in a very neglected condition.

Several Hours Later

Inside the house, we again see the large central room where the councils were held. Both the room and furnishings are neglected, and are in very poor condition, with crumbling cream colored plaster, unswept floors, large piles of wind blown sand accumulating beneath the large holes in the roof, and with broken reed bundles that hang down. Dust and sand are everywhere. The large

---

[450]M. Page 113, line 22 [When the sixth year arrived], [They served up] the daughter for dinner, 23 They served up [the son for food], and line 24 [Stomachs were filled with their own children], One [house] (neighbor) consumed another (since daughters were more expendable than sons, the conditions were so desperate that even their sons were expendable: extremely desperate conditions).

wood council chair for Ziusudra is dull and broken, as are some of the wood benches that were arranged in a circle, which has now somewhat lost its circular shape. We look through a doorway, and observe Ziusudra, sitting cross legged, bowing his head, praying in front of the niche altar in the wall, where he has placed a few small crumbs of dried sesame coated bread as an offering. He is weeping, and his tears streak down through the white dust and sand encrusted upon his face, and flow into his beard, as he prays for God Enki to intervene. Ziusudra slowly emotionally pleads, "I beg of you, God Enki,[451] intercede, and effect relief for mankind. We suffer greatly. Return, and give us guidance. You have always had my obedience. I have always followed your command. Do not fail me now. Every day, I weep, bringing oblations in the morning.[452] I prostrate myself in reverence and humility each day, as I pray at your temple.[453] Appear to me, and save mankind!"

*

Obscured by constant and mildly blowing dust and sand, we see a star-filled night sky with a crescent Moon, the symbol of God Enki, and then we look down to the moonlit gently sloping hills of the Abzu. As in Shuruppak, all vegetation has disappeared, and the large mounds of wind blown sand are everywhere, drifting among the bare remains of the tall bushes and dried out tree trunks.

ABZU
That Night

We look down further, and as the wind swirls the dust and sand, we

---

[451]M. Page 77, line 7 He swore by words of the god, and line 9 He swore by words of Enki.

[452]M. Page 77, line 2 To the temple of his god, 3 Daily he set his foot, 4 Every day he wept, 5 Bringing oblations in the morning, 11 Daily oblations to the temple of his god, and line 12 There he was seated, he wept, and page 117, line 13 Every day he wept to his god, and line 14 He used to bring offerings each morning, and page 143, line 146 He made offerings to Enki.

[453]M. Page 143, line 147 With humility (and) well chosen words, in reverence [held], and line 148 Every day he stood constantly present at [his] temple.

see the outside of the biological laboratory craft of God Enki, one of the three craft close together in a triangle, and linked together by the clear glass-like passageways. They are in the clearing, with the sand swept remains of the forest in the background. The craft appear as they had before, except there are accumulations of sand around each of their 4 support pods and the base of each ramp. We slowly zoom toward the craft, go up the ramp, and through the air flow door, a curtain of moving opaque air, which prevents any outside elements from entering the craft. We observe God Enki, alone, speaking to himself in front of a computer which has flashing display lights.

We zoom in further, over his shoulder, and see the computer, with an 8 inch square screen, a telepathic communicator, which is on but shows an out of focus picture. God Enki is reiterating to himself, and simultaneously touching the heat sensitive numbers on the computer console. The flashing lights dance on the display, as he programs the computer, "I am prevented, by the decree of God Enlil, to directly interfere with the famine."

He pauses, and then we again see a mischievous smile, and a sparkle in the eyes of God Enki, as he resolves, "But I may be able to, indirectly, using the telepathic communicator, through a dream subterfuge." We look at the screen on the telepathic communicator, and the picture begins to focus, and we see the face of sleeping Ziusudra intermittently appear on the screen. God Enki now leans over, and speaks into the computer, and slowly instructs, "Ziusudra, go to the river tomorrow night. I shall . . . . ," and the words of God Enki fade away, and we do not hear his full instructions.

*

Although obscured by constant and mildly blowing dust and sand, we see the crescent Moon, the symbol of God Enki, in the star-filled night sky, and then we look down to the moonlit courtyard and house of Ziusudra.

SHURUPPAK
The Same Night

Inside the house, we observe Ziusudra sleeping in his bed, clothed in rags, and wearing his gold God Enki symbol necklace, gold winged sphere symbol of Nibiru, and gold bracelet, tossing as he dreams. There is a blurry, cloud-like border surrounding the face of God Enki in the dream of Ziusudra.[454] The voice and face of God Enki are intermittent and jumbled, and Ziusudra cannot fully understand them, as he sees and hears, ". . . . go to the . . . . tomorrow night . . . . make offerings at the river . . . . face the river . . . . face the bank . . . . . " Ziusudra very slowly tosses and slowly awakens, but continues to lie down, and has a puzzled look upon his face, as he recalls the dream.[455] After a few moments of thought, he resolves, "I shall go to the river tomorrow night, and take offerings to God Enki." Then, Ziusudra closes his eyes and goes back to sleep.

✻

Still obscured by the constant and mildly blowing dust and sand, we again see the crescent Moon, the symbol of God Enki, in the star-filled night sky. We look down and see the very narrow and low, 3 feet deep water level of the stagnant, muddy greenish color Euphrates River, 32 feet below its normal level, here at Shuruppak. Then, as we look up and scan the river bank, amidst the blowing dust and sand, we notice Ziusudra, clothed in rags, but wearing his gold God Enki symbol necklace, and the other adornments that he had previously worn, bathed in the soft but obscure moonlight. He is sitting cross legged, bowing his head, praying on the west bank of the nearly dried out river.[456] Before him, on the sand, he has placed a small clay dish with a few

---

[454]M. Page 143, line 149 It was not a dream, coming out and spea[king aloud] (it was not a dream), and line 150 Conjured by heaven and underworld [alike], and page 113, line 27 [The discerning one], the man Atra-ḫasīs (Ziusudra), 28 Kept an open ear [to his lord], Ea (Enki), 29 [He (Ziusudra) spoke] with his god (God Enki, through the dream), and line 30 [And] Ea (Enki) spoke with him (Ziusudra, through the dream. This kind of communication, through a medium similar to a dream, may be an example of a pure form of mental telepathy, and early mankind may have been bestowed with the slight ability to use this power. Additionally, if the Gods programmed or gave some degree of this power to early mankind, they themselves would certainly possess it fully. Therefore, if Ziusudra did not require the aid of an agent, like a telepathic transmitting device, then God Enki would absolutely not have employed the use of an agent either, like the telepathic communicator computer. However, some agent, similar to that which I have described, may have been used by God Enki to increase the sensitivity, and consequently the receptibility of Ziusudra to such a power).

[455]M. Page 77, line 8, and line 10 Giving [attention] to dreams.

[456]M. Page 113, line 31 [He sought] the gate (the river) of his god.

small crumbs of dried sesame coated bread as an offering.

The Next Night

Ziusudra weeps as he faces the river and prays, and again his tears streak down through the white dust and sand encrusted upon his face, and flow into his beard, as he pleads, "God Enki, answer my words! Help us! We suffer greatly! Our time of need is upon us!"[457]

From a short distance, we observe Ziusudra as he continues to pray at the river bank, and as the night lengthens, he begins to fall in and out of sleep and dreams.[458] Then, later that same night, near the river bank, although obscured through the dust and sand filled air, we notice a craft rapidly approaching, with its brilliant glowing firefly-like light emitting from the underside, light up the landscape, and hover in the dust and sand filled air, near where Ziusudra is praying, while falling in and out of sleep and dreams. Ziusudra dreamingly sees the craft through the dust and sand filled air, and he very slowly bows down, and then prostrates himself, and then he very slowly stands, hunched over, and looks up at the craft, and watches as the craft descends nearby and prepares to land, as the glowing light begins to diminish, and the 4 pods like cattle hoofs extend 6 feet down, and the craft begins to land. Then, as he

---

[457]M. Page 77, line 13 His bed toward the river he put, and page 113, line 32, and page 77, line 14 On the bank he was seated, he wept, and page 117, line 15 When the river [also was quiet], and page 113, line 33, and page 77, line 15 [The night] was still young, and page 117, line 16 The night was still young, and page 77, line 16 In [the night] he finished calling Enki, and page 117, line 17 In the night he finished calling Enki, and page 77, line 17 His [god was] seen through the river, and line 18 Addressed [his god] of the river, and page 117, line 18 He spoke to Enki through the river, and page 77, line 19 Let the river take [offerings] and bear away, and page 117, line 19 May [you] take these offerings, and page 77, line 20 Let it [take them] away, and page 117, line 20 May it be established under Enki, and page 77, line 21 To bear [away] my [offerings] to Enki, and line 22 May he see [our suffering], and page 79, line 23 May he see our suffering, and page 117, line 21 Let Ea (God Enki) see our suffering (the immediately preceding lines seem to indicate Ziusudra placed floating meager offerings in the river, and appealed to the river to dispatch the offerings to the abode of God Enki in the Abzu, in order to communicate the suffering of mankind to him).

[458]M. Page 79, line 24 In the night I saw dreams, and page 117, line 22 In the night I saw dreams, and page 79, line 25 After he gave offerings, and page 117, line 23 After he gave offerings, and page 79, line 26 Facing the river at night, and page 117, line 24 Facing the [river at night], and page 79, line 27 On the bank seated facing the river, and page 117, line 25 Facing the river seated on the bank, and page 79, line 28 To the Apsû he dreamed, and page 117, line 26 To the Apsû he dreamed.

further awakens, he realizes it is not a dream, and his eyes fill with tears of joy at the sight of the craft, and again he begins to weep, and again his tears streak down through the white dust and sand encrusted upon his face, and flow into his beard, but he does not walk toward the descending craft, as he previously had done.

Then the craft lands upon the windy and sand swept Earth, and the glowing light further diminishes, and then completely expires, and after a moment, a side of the craft extends down, forming a 4 feet wide ramp and doorway. Then, God Enki[459a] walks through the opaque air flow door and down the ramp of the craft. He is alone, and is wearing the same red robe and adornments that he had previously worn, and he is without a helmet or body regulator, as the constant wind mildly blows the dust and sand around him. He smiles as he runs to Ziusudra, who again very slowly bows down, and prostrates himself, and then very slowly stands,[459b] and with little success, struggles to straighten his hunched over body. Ziusudra, who continues to weep, is overwhelmed, delighted and filled with joy. God Enki stops a few feet away from Ziusudra, and his face saddens, as he stares for a moment, and closely observes the poor condition of Ziusudra. Then, he moves closer to Ziusudra, and they exchange greetings, and then warmly embrace, as the mildly blowing dust and sand swirls around them.

After their long embrace, as he stops weeping, and in a relieved tone, Ziusudra exclaims, "I knew you would come, God Enki." Then God Enki apologetically advises, "Ziusudra, this famine is not with my approval. I have been under the close personal watch of God Enlil, who has also appointed two Igigi Gods as guards to closely watch and assist me. That is why I have been unable to help." He pauses to again observe the condition of Ziusudra, and then he asks, "Did you see me in your dream?" Ziusudra answers, "Yes, but I did not fully understand the dream." Ziusudra again begins to weep, and his tears again streak down through the white dust and sand encrusted upon his face, and flow into his beard, and we observe a saddened and repulsed God Enki, as Ziusudra emotionally reports, "Mankind suffers greatly. For one of your Shar years, three thousand six hundred of our Earth years, we ate dried weeds. The second Shar year, we suffered the itch. By the third Shar year, our

---

[459a]M. Page 79, line 29 Enki heard [his words], and page 117, line 27 [Ea] (God Enki) heard his (Ziusudra's) words.

[459b]M. Page 123, upper line 3 (see footnote 335).

features were altered by hunger, our faces encrusted, like malt. We were living on the verge of death, our faces appeared green. The fourth Shar year, our long legs became short, our broad shoulders became narrow, we walked hunched in the street. When the fifth Shar year arrived, mothers watched daughters, daughters watched mothers, that neither had excessive food. When the sixth Shar year arrived, some people served up their daughters for dinner, others served up their son for food. Stomachs were filled with their own children. One house consumed another. Now, during the seventh Shar year, the people are living on the edge of death."[460]

Then suddenly, after a brief somber pause, God Enki, who had looked very sad and horrified while hearing the description of the effects of the famine, again begins to mischievously smile, and again has a sparkle in his eyes, as he resolves, "I have a plan.[461] I shall slightly increase the level of the smaller seas, to half their lowered level, causing the rivers to rise. There shall then be released, an abundance of fish. I shall place the blame on my two guards, the Igigi Gods, whose negligence in my absence, caused the level of the smaller seas and of the rivers to rise, to half their lowered level. This trick shall work, I am certain. Now I must go,[462] before I am noticed by God Enlil."

We see time passing through a series of alternating cameo views of only several very thinly scattered dark clouds passing through the obscure sky, and obscure sunrise to sunset sequences, and sprinkled throughout the passing time, we see this plan also working.[463] Within the constant and forceful, but mildly blowing dust and sand, we observe the Euphrates River surging, with rising muddy water, flowing in from the slightly muddy smaller sea, the level of which is being increased to half its lowered level, with clear water from the larger unaffected sea, and very many fish jumping in the muddy river water. The river is almost half-full of muddy water, to the 14 feet level, and thick with fish, but the upper banks are still parched. For the surviving people, the

---

[460]M. Page 113, line 1, 2, 3, 4, 5, 6, 7, 8, 9, 10, and line 11, and page 115, line 12 They served up the son for food, Stomachs were filled with their own children, 13, 14, and line 15 (Ziusudra reports the sufferings of mankind to God Enki).

[461]M. Page 115, line 16, 17, 18, 19, 20, 21, 22, 23, 24, 25, and line 26.

[462]M. Page 115, line 27.

[463]M. Page 79, line 30 And [instructed] the water-monsters [as follows], 31, 32, 33 Go, the order obey, and line 34, Ask not, and page 117, line 28 And brought the [water monsters as follows], 29, 30, 31 Come, the order obey, and line 32 Ask not, and page 119, line 33, 34, 35, 36, 37, 38, 39, 40, 41, 42, 43, and line 44.

pace is slow, and there are many famished and still somewhat hunched over slowly moving people, clothed in rags, at the river collecting the fish, using various means. Some people are standing in the muddy water catching the fish with their bare hands, while others net them, with old worn nets, and others use old unraveled reed baskets to scoop them out of the muddy water. There are no birds flying in the air, as the dust and sand filled wind continues to blow them away. Reeds are beginning to sprout along the edge of the river. The skeletal remains of the reed bundle framework of the large and small, rectangular shaped, pitch coated reed boats are as they previously appeared, still in need of repair, and still scattered along the upper parched river banks. There is no water in the dry irrigation canals, and they are still parched and filled with cracked clay, and there is no greenery growing along them, because the water level in the river is still 11 feet below the canal tie-ins. No one is working on the canals, as they are still abandoned, and filled with wind blown sand, which continues to almost completely cover some of them. There are no crops, as the fields are still parched and white with salt, and sand still mildly blows across them and drifts in large scattered mounds. Beyond the river the drought still exists, and nothing grows, as the dust and sand are still mildly blowing. Only at the river is there relief from the drought and famine, for there is still no rain, mist, or dew. There are no signs of previous crops, as the fields are still abandoned, and the orchard trees are all still bare, and all that remains of the trees are still only their dried out trunks, except some of the trees next to the river are beginning to show signs of life, and no one is seen in either the fields or the orchards, tending them, as there is still nothing to tend. No sheep, goats, or dogs are seen in the streets or outside the homes, as almost all of them have now been eaten, and the few remaining thin animals are still kept in the homes for protection, with the sheep and goats still giving no milk, and all still thoroughly plucked.

The streets are beginning to show signs of activity, especially of slowly moving people going to and returning from the river with large amounts of fish, but there are still large mounds of wind blown sand everywhere. There is still litter in the streets, but no weeds, and there are slowly moving children beginning to again play their games in them; but all the surviving children are 10 years of age in appearance or older, for there are no children under 10 years of age in appearance. The courtyards still have large mounds of wind blown sand, as well as litter, but no weeds, and they still look neglected. The cream colored plaster and sun-dried clay bricks on the outsides of some of the homes is somewhat being repaired, now that the physical energy exists to utilize the muddy river water, and some of the homes are beginning to be coated with a cream colored wash, but there are still large mounds of wind blown sand at the corners and in the doorways, the size of which are now beginning to be somewhat reduced, as the constant wind continues to mildly blow the dust and sand; but the nearly flat reed bundle roofs are still in need of repair, due to the lack of adequate reeds, with fresh clay being applied to the old reed bundles, and to the very sparsely visible clay coating that remains, with some people taking the remaining old reed bundle framework from the small reed huts, and patching the holes in the roofs. The wood front doors are still very dull from lack of attention, and many of them still show signs of broken and splintered wood. The small reed huts have only a few small pieces of broken reed bundle framework remaining to suggest that they ever existed. The ziggurat and oval wall are in the same condition of repair as the outsides of the homes, with large mounds of wind blown sand, which have been somewhat reduced in size.

The insides of the homes are beginning to show signs of improvement, with swept floors, and dust and sand being reduced. The crumbling cream colored plaster is being repaired, but the wood furniture is still very dull and broken, and some of them have been partially repaired with old wood from other broken furniture. The window drapes are still very soiled and unraveled, and in rags, and there are a few small reed bundles in the fireplaces, taken from the newly sprouting reeds growing along the edge of the river. The people, who were very thin and frail from starvation and thirst caused by the renewed drought and famine, are slowly showing signs of recovery, but most of them stay in their homes for protection against the constant mildly blowing dust and sand. Everywhere, people are in various stages of recovery, and their appearance is also beginning to slowly improve. They are beginning to leave their beds, and slowly walk about the inside of their homes, and their faces are slowly returning to a more normal look, with protruding cheek bones and hollow cheeks disappearing, and their green looking faces are slowly returning to a normal skin color. Their faces are no longer encrusted with white dust and sand, like malt, as they now have ample muddy river water to bathe in. The posture and walk of the people slowly show signs of improvement, and their severely hunched over walk is improving, and their shoulders begin to broaden, and their legs begin to straighten from their bent position. Amidst the mildly blowing dust and sand, there are no dead people visible in the streets, as they have either previously been dismembered and eaten, or still lie temporarily buried surreptitiously under the large mounds of sand, which sometimes shift and only partially reveal their concealed ingredients, for the dead have yet to be officially buried.

The people's clothing still consists of rags and shreds of cloth; worn, soiled and wrinkled, and unraveled at the seams, and some people still have old cloth wrapped around them as a substitute for clothes, as the animals and plants formerly used for making cloth, are not yet available. Their sandals are still falling apart, many with broken straps, and most of the people are still barefoot. Their pendant earrings and bracelets, although still in need of repair, are being worn, as are some of the men's soiled and unraveled white turban-like hats and colorful head bands. The women are beginning to wear their colorful, but soiled and unraveled, head band ribbon in their hair, and wear old, but broken large crescent shaped earrings. Most of the adults and children, now 80%, wear symbol necklaces of the four main Gods, still with unraveled pieces of string as chain substitutes, but their pendant cylinder seals are still abandoned. They are beginning to look groomed and clean in appearance, with kept hair and natural beards. The people are beginning to converse, and caress each other, but again less than normal, and with some restraint, and no mothers are nursing babies, as none are yet born, but again some women are becoming pregnant. The broken clay utensils are being replaced, and no bird or animal is being roasted, but fish are in abundance, and are being freely eaten, and a few people attempt to smoke-dry them over very small fires in the fireplaces, while most people eat them raw, or nearly raw. The fish are everywhere, very numerous, and are the only food source. Fish are piled high on the dishes at meal time. Mothers again open their doors and welcome their older daughters to their homes, and mothers and younger daughters each prepare large amounts of fish for meals, without regard to the size of portions for each person. There is no sesame coated bread, due to continued lack of grain. The people drink muddy river water, for there is still no beer or wine, and there are nearly half-full vases and drinking cups with muddy river water everywhere,

Most adults, now 80%, pray to the various personal Gods and Goddesses, sitting cross legged and bowing their heads in front of the niche altar in the wall of their homes, after they place many fish as offerings. The people are beginning their merrymaking ways again, and the long neglected, dusty and broken musical instruments are being cleaned and repaired, and again played, but only to a limited degree. There are now some slowly laboring workers, reducing the large mounds of sand from around the city streets and their homes, with temporary success, amidst the mildly blowing dust and sand. The children, all of which are 10 years of age in appearance or older, for there are no children under 10 years of age in appearance, are in the same stage of recovery as the adults. They are beginning to play their games, both in their homes and in the streets, and have created new games which utilize the large mounds of sand, as they are no longer kept in their homes for protection, and to be eaten as food. The younger daughters help their mothers prepare the fish for meals, and they each voluntarily share their amounts. The children are beginning to again use the clay school tablets for reading and writing pictographs. The conditions begin to improve, and the appearance of mankind quickly improves, as the drought and famine ease and subside, but only near the river. Away from the river, the drought and famine affecting the land and mankind is not broken, and does not end, for there is still no rain, mist, or dew, and the wind continues to mildly blow dust and sand everywhere. Mankind is only partially recovered, and the only relief lies at the river, providing fish as food, and muddy river water to drink. Slowly, life for mankind begins to improve to an improvised but considerably less than normal state, and mankind is again saved, but again, conditions have generally declined.

*

As the constant and mildly blowing dust and sand filled air obscures the daylight, still with only several very thinly scattered dark clouds, again we see the inside of the oval walled city of Nippur, the city of God Enlil, as it appeared before, with its colorful 7 tiered ziggurat in the center, with the temple Ekur and the antennas atop it, as the seventh tier.

NIPPUR
.03 Shar Years Later
or
100 Earth Years Later
8,900 B.C.

We scan, and amidst the constant and mildly blowing dust and sand, we notice partially recovered mankind, all with black tight-curly hair and brown eyes, working at a slow pace in the courtyard, reducing the mounds of wind blown sand as they accumulate, pampering the ziggurat, and tending the few remaining thin sheep in the small sheepfold nearby. They are all very dusty and dirty, and noticeably tired, but not sweating. All wear God Enlil symbol necklaces. There are 2 Igigi Gods, 1 male and 1 female, crisp and clean, both without helmets or body regulators, and 2 Anunnaki God officers, 1 male and 1 female, crisp and clean, both helmeted and one wearing a body regulator, supervising the work. There are eight craft grouped in an area, near the ziggurat, one of which is the assembly craft. We notice 4 Gods and 2 Goddesses, all helmeted and four wearing a body regulator, quickly walking toward the ramp.

Inside the assembly craft, again our eyes scan, and it appears as it did before, a smaller version of the assembly on Nibiru, complete with a smaller dome viewing monitor in the center, which shows mankind in the present, at a slow pace, amidst the constant mildly blowing dust and sand, being partially recovered near the river, and with no recovery beyond the river, from the reimposed drought and famine: catching and eating fish, and drinking muddy river water, and somewhat returning to normal behavior of being defiant, assertive and disobedient; and young and older Gods and Goddesses again freely caressing young humans[464] with straight blond hair and blue eyes, and black and brown variations of wavy and tight-curly hair, as the constant and mildly blowing dust and sand swirls around them. All the main Gods, including God Anu, are present. Three of the 7 great Anunnaki Gods are seated behind the slightly raised section, and are quietly conversing, as are the 20 Gods and 10 Goddesses. God Anu, God Enlil, God Enki, and Goddess Ninhursag sit at

[464]T. Middle page 334 through page 336.

the slightly raised section for the four main Gods, four of the 7 great Anunnaki Gods. All the Gods are watching the viewing monitor, and again some are pointing at it in disgust. Their communication helmets and body regulators are on the ledge that forms the back of the white pew-like molded circular seats. The 4 Gods and 2 Goddesses rush in through the opaque air flow door, and quickly remove their helmets and body regulators, and sit down. God Nusku, the Anunnaki God officer and personal servant of God Enlil, is standing next to the air flow door, and he is wearing a communication helmet.

God Anu rises, and the conversations quickly fade to silence, and all the Gods bow their heads to him, and then he motions to his right, to God Enlil, and then God Anu sits down. Then God Enlil rises, and all the Gods bow their heads to him, all except God Anu. He first looks at God Anu, and then he scans the assembly, looking for assurance, and then God Enlil sternly addresses the assembly, "In this assembly of the Gods, I, God Enlil, supreme commander and ruler of the Earth, have brought you to my presence.[465] I have two important matters to reveal," and now God Enlil reflects, "first, all of the Gods, and we great Anunnaki Gods, decided together on a rule.[466] To increase the famine, and decrease the numbers of mankind. God Adad, under the close watch of God Anu," and we observe God Adad, and then God Anu, who have concerned looks on their faces, "guarded the upper regions,[467] and withheld 'all' forms of rain from mankind, and caused it to be released at the north and south axis of the Earth." He pauses, and all the assembly members look at the viewing monitor, and for the balance of his address, we alternate between the assembly, the viewing monitor, and God Enlil.

We see the present on the viewing monitor, which shows the polar regions, still completely covered with low dark clouds, being blanketed by continued very heavy snow, which is still rapidly accumulating to an extremely great depth; upon the flat ice sheet areas at the north axis, and upon the frozen

---

[465](Preceding this assembly, there seems to have been a messenger dispatched by God Enlil, across the wide sea to God Enki, and then sent back to God Enlil by God Enki, who repeated their claims; and also preceding this assembly, the two Igigi God guards assigned to watch God Enki were brought before God Enlil, probably to determine if God Enki was using the Igigi Gods to subvert the decree of God Enlil. However, for continuity I have combined all these actions into the assembly itself).

[466]M. Page 81, line 14, and line 15, and page 83, line 28 [All we] great Anunnaki, and line 29 Decided together on a rule, and page 85, line 23, and line 24.

[467]M. Page 81, line 16, and page 83, line 30 Anu and Adad guarded the upper [regions], and page 85, line 25, and page 119, line 2, 9, and line 16, and page 121, line 32.

mountainous areas at the south axis. Then God Enlil continues, "I, personally, guarded the middle Earth,[468] the surface air currents, the winds, and prevented birds from reaching mankind," and we continue to see the present on the viewing monitor, which shows a bird being blown through the dust and sand filled air, and away from the people, by the mild but forceful and constant wind, against its direction of flight, and against its will. Again God Enlil continues, "God Enki," and we observe God Enki, who is calm and collected, "was to guard the seas, together with his plants,[469] mankind, and prevent mankind from obtaining fish from the seas, through gravity control, by increasing the gravity over the smaller seas, near the people, thus reducing the water level of the smaller seas and of the rivers, and driving the fish away from the people, and preventing any fish from being taken from the smaller seas and the rivers." We see the past on the viewing monitor, which shows amidst the constant and mildly blowing dust and sand, the water level of the slightly muddy smaller seas near the people dropping, and then the water level of the muddy rivers dropping and surging toward the lowered slightly muddy smaller seas, with fish jumping in the surging waters, as they are carried away from the people, and toward the center of the larger unaffected clear seas.

God Enlil then turns around to the ledge behind him, where there is a flat rectangular object, again concealed under a white plastic-like cloth. He unveils and reveals the object, a one foot long full-sized fish, upon a clear glass-like tray.

We observe the concerned faces of the Gods in the assembly, all with puzzled looks, as they look at each other, and then at the fish.

God Enlil looks at the fish, and then turns back toward the assembly, "This punishment has also failed." Then he pauses, and scans the assembly, and then he looks at God Anu, looking for assurance. He now looks at God Enki, and he turns and takes the tray from the ledge, and holds it up toward God Enki, so he can clearly see the fish, and now God Enlil, in an angry tone accuses, "God Enki, you loosened the yoke of oppression, and established freedom from the drought and famine. You let loose an abundance of fish for

---

[468]M. Page 81, lower line 17, and page 83, line 31 I guarded the lower earth, and page 85, line 26, and page 119, line 3, 10, and line 17 [Sin and Nergal] did guard the middle earth, and page 121, line 33.

[469]M. Page 119, line 4 [That the bolt], the bar of the sea, 5 [You should] guard together with your plants, 11, and line 12.

the people!"[470]

Again we observe some of the Gods in the assembly, as they look at each other, some angry and some smiling in approval, and then we observe God Enki, calm and collected, with an apathetic look upon his face.

Now God Enlil contemptuously asks, "God Enki, what fish were left in the wide seas, that the people did not catch?" Again he pauses, and looks at the viewing monitor, and we see the past events leading to the present on the viewing monitor, which shows, amidst the constant and mildly blowing dust and sand, the rivers surging, with rising muddy water, flowing in from the slightly muddy smaller seas, the level of which is being increased to half their lowered level, with clear water from the larger unaffected seas, and the almost half-full muddy rivers thick with jumping fish, and the many hunched over very slowly moving people catching the fish, using various means, and eating the fish and partially recovering from the famine, while the drought still exists away from the rivers; and presently, of young and older Gods and Goddesses again freely caressing young humans with straight blond hair and blue eyes, and black and brown variations of wavy and tight-curly hair, as the dust and sand mildly swirls around them.

Then we observe some of the Gods in the assembly, as they look at each other in anger, while others smile in approval.

God Enlil returns the fish tray to the ledge, and sits down, and God Enki rises as God Enlil sits down, and now smiling he apologetically responds, "I did guard the seas, but in my absence, my two guards, the Igigi Gods, the guards who God Enlil assigned, I left in charge, and through negligence, they lost control, and the smaller seas and the rivers rose to half their lowered level, and a myriad of fish escaped. I laid a penalty on them, and punished them!"[471] Then God Enki looks at God Enlil, smiles, and sits down.

---

[470]M. Page 81, lower line 18 Where Enki [went], 19 He loosed the yoke [and established freedom], and lower line 20 He let loose [abundance for the peoples], and page 119, line 6, and line 13, and page 81, lower line 21 He established [relief in/from the heat (drought) of the sun], and page 83, line 32, and upper line 1, 2, and upper line 3, and page 85, line 27, 28, 29, and line 30.

[471]M. Page 119, line 18 [The bolt], the bar of the sea, 19 [I did] guard together with my plants, 20 When [control] escaped from me, 21 Releasing a myriad of fish, one myriad escaped, 22 Guards and I (then) got together and it (the problem) disappeared, 23 And they (the guards) broke half of [the bolt], and line 24 [After] I had killed (severely punished) the guards of the sea, and page 121, line 25 I laid [penalty] on them and punished them, 26, 27, 34, 35, 36, 37, 38, 39, 40, 41, 42, and line 43.

God Enlil quickly rises, and is now very angry, and looks at God Nusku, his personal Anunnaki God officer servant, who stands next to the air flow door, and shouts as he gives him an order,[472] "God Nusku, the two Igigi Gods waiting outside, the two guards I appointed to watch and assist God Enki; let the Anunnaki God officers bring them to me! Let the Anunnaki God officers send the guards into my presence!"[473] As God Enlil remains standing, God Nusku rushes out of the assembly through the opaque air flow door, and many of the assembly members quietly debate among themselves, some for and some against; God Enki or the Igigi Gods, being responsible for negligence in guarding the seas.

One charges, "The guards were negligent," and another responds, "God Enki was negligent." Another accuses, "The guards let loose the fish," and another responds, "God Enki let loose the fish."

Then, after this controversy and emotional display continues for a brief moment, God Nusku leads the 2 Igigi Gods, both males, without helmets or body regulators, followed by 4 Anunnaki God officers, 3 males and 1 female, all helmeted and two wearing a body regulator, through the opaque air flow door, and into the assembly and up to God Enlil. They all bow their heads to him, and God Nusku steps aside, and the 2 Igigi Gods are now standing directly before God Enlil, with the 4 Anunnaki God officers standing behind them. God Enlil heatedly and loudly inquires,[474] "You are the two guards that I assigned to watch and assist God Enki?" In unison, both Igigi Gods calmly and meekly answer, "Yes, God Enlil." Then God Enlil looks at them with piercing eyes, and more heatedly and loudly again inquires, "Did you, through negligence, lose control of the seas, causing the smaller seas and the rivers to rise?" Again in unison, both Igigi Gods calmly and meekly answer, "No, God Enlil," and then God Enlil shouts, "You may go!" The 4 Anunnaki God officers, 2 Igigi Gods, and God Nusku all bow, and then turn, and begin to walk toward the air flow door. Then, God Enlil turns and angrily stares at the seated God Enki, and we observe a calm and collected God Enki, with a broad smile upon

---

[472]M. Page 83, line 22, and line 23.

[473]M. Page 83, line 24 Let them bring [to me] the Igigi (the guards), and line 25 Let them [send] them into [my] presence (the Igigi Gods were suspect, because it was known that they empathized with God Enki, for he relieved them of their toil, and if they took a side, it would be with God Enki, for without mankind to labor for them, they again would bear the yoke of oppression).

[474]M. Page 83, line 26 They brought [to him] the Igigi (the guards), and line 27.

his face, as many of the Gods in the assembly look at each other and quietly resume the debate, some charge, "The guards," others respond, "God Enki," and this also continues for a brief moment.

Then God Enki begins to laugh, very quietly and to himself at first, and then out loud, as he exclaims, "I am fed up with sitting in this assembly."[475]

Some of the Gods in the assembly broadly smile, as they look at God Enki laughing, and a few even laugh with him, but somewhat subdued.

God Enlil angrily shouts above the laughter, "God Enki broke the rule! There is slander in his hand!"[476] As he scans the assembly, again looking for assurance, he looks at God Anu, and then at God Enki.

God Enki, still seated and still laughing, consolingly responds, "Nooooo, no."[477]

Again God Enlil shouts, now with a scowling look upon his face, as he points to God Enki, and accuses, "He broke the rule of the assembly!" As he glares at God Enki, the assembly is in an uproar, as some Gods softly laugh, while others continue to debate who is to blame for breaking the rule and releasing the fish to mankind.

Some charge, "God Enki," while others respond, "The guards."

This uproar lasts for a moment, and then slowly quiets, and then God Enlil looks at the viewing monitor, and we see the past on the viewing monitor, which now shows the revolt of the Igigi Gods, led by God Wêila, who have surrounded the oval walled compound and temple of God Enlil at midnight, as they wave their arms, and wave and push their tools, some flaming, up and down in the night air. Then God Enlil loudly reflects, with the assembly again quiet, as he points to the viewing monitor, "You raised a cry for mankind!"[478] We observe the three great Anunnaki Gods, the commanders of the military, sitting next to each other behind God Enlil, all with a stern look upon their face. Then God Enlil continues to reflect, as he now points to God Enki, and in a mild tone which builds, "You selected genes from God Wêila,[479] the Igigi

---

[475]M. Page 83, line 15, 16, 17 [Enki] got fed up with sitting, and line 18 [In] the assembly of the gods laughter overcame him.

[476]M. Page 85, line 19 Enki broke the rule, there is slander in his hand.

[477]M. Page 85, line 21 In the assembly of the gods they argued, and line 22 Together they argued, Enki and Enlil.

[478]M. Page 85, lower line 32 [You] raised a cry [for mankind].

[479]M. Page 85, line 33 You slaughtered (obtained a sample from the donor) [a god] together with [his personality] (good qualities: see footnote 184 and 185).

God who led the revolt, and you sat and created mankind!"[480] He pauses, and looks at the viewing monitor, and we again see the past on the viewing monitor, which shows the inside of the biological laboratory craft of God Enki, with glass-like flasks and beakers containing bubbling chemicals of different colors, and God Enki seated at the circular computer-like microscope, with its flashing display lights and one foot diameter viewing screen, fertilizing the egg, and then seated at the gene selector computer with flashing display lights, showing the DNA ladder with 7 genes opposite 7 other genes; a total of 14 genes, upon the one foot square screen. We observe God Enki, who has stopped laughing and now displays a proud smile upon his face.

Then God Enlil continues to reflect, as he now first points to Goddess Ninhursag, and then he glances at and points to the viewing monitor, which again shows the revolt of the Igigi Gods, as he looks at her, "She imposed your toil on man!"[481] Again he pauses, and looks at the viewing monitor, and we continue to see the past on the viewing monitor, which now shows the first 14 birth-Goddesses, being implanted in the creation chamber of the medical craft. We observe Goddess Ninhursag, sitting next to God Enki, and she also has a proud smile upon her face. Then God Enlil again points to God Enki, and he loudly declares, "People you created! Freedom you created! Problems you created! People bring problems!"[482]

Again he pauses, and scans the assembly, and then he looks at God Anu, again looking for assurance, and then sternly, and in a softer tone, he tauntingly remarks as he again looks at God Enki, as the viewing monitor returns to the present, and shows the many people at the muddy rivers, at a slow pace catching the abundant jumping fish, by various means, amidst the mildly blowing dust and sand, "Through genetic engineering you programmed the functions and guidelines for mankind.[483] Let us turn to you.[484] Let us bind God Enki first, by an oath."[485]

God Enki now has a puzzled look upon his face, as God Enlil sarcastically continues, "Water, you provided for the people. What an appropriate gift, for

---

[480]M. Page 85, line 34 [You] sat and created.
[481]M. Page 85, lower line 31 [She imposed] your toil [on man].
[482]M. Page 85, line 35 People bring problems.
[483]M. Page 85, line 36 You determined on a rule for mankind.
[484]M. Page 85, line 37 Let it (the assembly) turn to you.
[485]M. Page 85, line 38 Let us bind prince Enki [first] by an oath.

we shall also give them water. We shall let water destroy mankind, with a flood!"

God Enlil remains standing, and he pauses, and scans the assembly members, as the Gods gasp and grimace, and some look at each other in disbelief, and respond, "Nooooo, nooooo," while others smile in approval.

The assembly quickly becomes silent, then God Enki, who now has a stern and serious look upon his face, rises, and petitions the assembly,[486] "Why shall you bind me with an oath, first?[487] Am I to lay my hands on my own people?[488] On my creation? The flood that you are commanding of me,[489] whose plan is it? I do not know.[490] Am I to give birth to a flood?[491] I shall not!" He now turns, and points to God Enlil, and continues, "That is the task of God Enlil![492] Let him choose! I shall not cause a flood to destroy mankind!" God Enki then sits down, with God Enlil still standing.

There is silence in the assembly, as the Gods look at God Enki with expressions of empathy, and then God Enlil again sternly addresses the assembly, "This is the second, and the most important matter confronting us at this assembly." We now simultaneously hear the words of God Enlil, and see them, shown as hologram-like pictures of the past and future on the viewing monitor, and we alternate between the assembly, the viewing monitor, and God Enlil, for the balance of his descriptive speech. Then God Enlil continues, "As you all know, our heavenly planet, Nibiru, which crosses through the solar system of planets, and whose symbol is the cross, shortly, shall again be orbiting the star we call the Sun, and passing in the vicinity of this small planet, Earth, being nearest to the north axis area. Normally, during each orbit, the gravitational and electromagnetic effects of Nibiru, being the largest planet in this solar system, causes water displacement, a pulling up, or bulging of the waters, to occur upon the Earth and be drawn across it, in the form of mild flooding in the central regions, and a more pronounced bulging of the waters,

---

[486]M. Page 85, line 40, and line 41.

[487]M. Page 85, line 42 Why will you bind me (God Enki) with an oath [first]?

[488]M. Page 85, line 43 Am I to lay my hands on [my own peoples]?

[489]M. Page 85, line 44 The flood that you are commanding [me].

[490]M. Page 85, line 45 Who (whose plan) is it? I [do not know].

[491]M. Page 85, line 46 Am I (God Enki) to give birth to (bring forth/deliver/endorse) [a flood]?

[492]M. Page 85, line 47 That is the task of [Enlil], and page 87, line 48 Let him [and Adad] (the storm God) choose.

severe flooding at the north and south axis, with the effects being greater at the north axis, because it is nearer to Nibiru. The air is also affected in much the same way, and the surface air is drawn across the Earth toward Nibiru, in the form of mild winds. The gravitational and electromagnetic force fields also cause agitation to take place throughout the Earth, as a result of the molten core being pulled toward the surface, toward Nibiru, which results in moderate tremors with fissures, and volcanic eruptions which emit lava, and thick clouds of smoke, ash, and steam, as well as increased temperatures upon the surface of the Earth and of the air. These same force fields of Nibiru simultaneously cause a pure water mist, which is being excessively evaporated by the increased frictional heat of the molten core acting upon the surface waters,[493] to be attracted and drawn up from the Earth and appear as a rainbow, and then condense, causing a profuse moderate rain to fall upon the Earth."

"But this orbit, Nibiru shall have an abnormal effect upon the Earth, and shall cause destruction to life upon the Earth, in the form of an unprecedented, devastating flood!" He pauses, and scans the assembly, looking at the Gods, and we observe the intense faces of the assembly members, anxiously waiting for his words. Then God Enlil elucidates, "I have just received a final computer report, confirming previous reports, without a doubt. Because of the acute closeness of Earth to Nibiru during this orbit, and the location of the other nearby planets, as has occurred on previous orbits with varying degrees of disturbance and affliction, the Earth is unusually vulnerable to the gravitational and electromagnetic force fields of Nibiru, which shall drastically influence the Earth, and intensify and multiply the normal effects of Nibiru upon the Earth."

"Giant waves of water[494] shall be pulled up and over the surface of the

---

[493]X. Upper page 105 Two celestial bodies were driven near to each other. The interior of the terrestrial globe pushed toward the exterior. The earth, disturbed in its rotation, developed heat. The land surface became hot. Various sources of many peoples describe the melting of the earth's surface and the boiling of the sea, and middle page 380 Furthermore, we found that excessive evaporation of water from the surface of the oceans and seas, a phenomenon that was postulated to explain excessive precipitation and formation of ice covers, was caused by extraterrestrial agents (see footnote 38).

[494]M. Page 87 , line 51 Let Errakal (an epithet for God Nergal, a son of God Enki) [tear up] the mooring poles (and pull out the masts of boats. Even the son of God Enki was to assist in the destruction), and line 52 Let [Ninurta] go and make [the dykes] overflow (God Ninurta, a son of God Enlil, was to assist in the destruction).

Earth, primarily toward the north, toward Nibiru. These gigantic waves of water shall be accompanied by unstable waves of air, forcefully driven winds,[495] and heavy rain storms, which shall cause sudden climatic changes, as they are also drawn primarily toward the north, toward Nibiru. And as the molten core of the Earth is drawn toward Nibiru, severe tremors shall simultaneously shake and fissure the Earth, as all the volcanoes erupt with huge amounts of molten lava, dense clouds of smoke and ash, and extreme amounts of steam, causing greatly increased temperatures. But it shall be the gigantic waves of water that cause the destruction, as they sweep across the lands toward the north,[496] causing a great deluge."

Again he pauses, and scans the assembly, looking at the Gods, and again, we observe the intense faces of the assembly members, including the face of God Enki and Goddess Ninhursag, as they all stare at the viewing monitor, and gasp, "Nooooo" and "Ohhhhh."

Then God Enlil warns, "This shall be far more serious than the water displacement which normally occurs with each orbit of Nibiru, every three thousand six hundred Earth years. This orbit shall have extreme consequences for the Earth."

God Enki jumps up and anxiously requests of God Enlil, "Is there any way of preventing this?" He remains standing, waiting for a reply, and then Goddess Ninhursag jumps up and eagerly inquires of God Enlil, "Can we do anything?" She also remains standing, waiting for a reply, and God Enlil consolingly answers both questions, "We can do nothing to prevent it. It is beyond our control." Then God Enki and Goddess Ninhursag sink into their seats, as they look at each other through tearful eyes.

God Enlil instructs, "This is a perilous circumstance for the Gods. You should all immediately assemble your craft, and go to the city of Sippar, our mission control center, where God Utu," and God Enlil looks at God Utu in the assembly, and we observe God Utu, and then God Enlil continues, "who is

---

[495]M. Page 87, line 49 Let Šullat and [Ḫaniš] go [in front] (two Gods associated with violent storms, especially violent winds).

[496]X. Middle page 89 especially in the north, and upper page 90 gigantic waves, and middle page 95 (quoting Job 26:7 through 14) He (Nibiru) stretcheth out (pulls) the (waters toward the) north over the empty place; He bindeth up the waters in his thick clouds; He hath compassed the waters with bounds; The pillars of heaven tremble and are astonished at his reproof; his hand hath formed (created the Earth, from out of) the crooked serpent (erratic Tiamat); Lo, these are parts (examples) of his ways (power/strength, similar to these).

in command there, shall prepare the craft for departure and orbiting, during the period of flooding. At Sippar, you shall wait for my command to ascend."

God Enlil now smiles, as he confidently proclaims, after he again scans the assembly members, looking for assurance, "This flood shall achieve for me, what I have been unable to do with the cooperation of the Gods, and the physical forces which I command. To destroy mankind!"[497]

Again he pauses, and again we observe the faces of the Gods, some scorn in disbelief, and others smile in approval, others gasp and grimace, and others beam with joy.

Then God Enlil commands,[498] "This information, the impending flood, we shall keep secret, and withhold from mankind." God Enlil now looks at God Anu, as he pauses, again looking for assurance, and then he continues, "I, God Enlil, having been given the divine decree of supreme power and function of God Anu on Earth, by God Anu, president of the assembly, to rule over the Earth, decree that we completely lift the existing drought and famine, immediately. In addition, we shall all take an oath of silence concerning the flood. Come, all of us, and take the oath to keep the flood a secret from the people, from mankind."[499] God Enlil then looks at God Anu, for approval, and then he sits down. Then God Anu stands, and raises his right hand, and pledges, "By this oath, I am bound,"[500] and he remains standing, and he looks at God Enlil. Then God Enlil stands, and looks at his two sons, one of which is God Ninurta, and they also stand. All three raise their right hand, and together pledge, "By this oath, I am bound,"[501] and they remain standing.

Then slowly, all the Gods in the assembly stand, all except God Enki and Goddess Ninhursag, some very reluctantly, and raise their right hand, and

---

[497]V. Genesis 6:7 And the LORD (God Enlil) said, I (God Enlil) will destroy man (mankind) whom I have created (by decree) from the face of the earth (similar to this).

[498]M. Page 121, line 44 Enlil opened his mouth to speak, and line 45 And addressed the assembly of all the gods.

[499]M. Page 121, line 46 'Come, all of us, and take an oath to bring a flood' (as with the previous disciplinary measures, which were brought about under the direction and control of the Gods, from the context of this statement and the statement in footnote 497, the flood itself may have been brought about and been within the sphere of control of the Gods. However, if the flood was under the control of the Gods, the chances are greatly increased that the catalyst for the flood, Nibiru, may not be a natural celestial body, but may be an artificial satellite, energy force, or the manifestation of a concept far beyond our present comprehension; as I further explain under 'May Not Be A Planet' on page 374 in my epilogue).

[500]M. Page 121, line 47 Anu swore first (following their order of authority).

[501]M. Page 121, line 48 Enlil swore, his sons swore with him.

together also pledge, "By this oath, I am bound," and they remain standing. Now, all the assembly members look at God Enki and Goddess Ninhursag, who are the only ones seated, and a long moment of silence passes.

Then Goddess Ninhursag slowly stands, her eyes full of tears,[502a] and she slowly raises her right hand, and slowly declares, "With regret, and against my will," she pauses, and looks at the standing assembly members, and then she slowly and softly pledges, "By this oath, I am bound," and she remains standing, and looks down at God Enki, who sits to the right of her. All the assembly members look at God Enki, who is now the only one seated. Some are looks of empathy, while others are looks of anger, and another long moment of silence passes.

Then God Enki slowly and reluctantly stands, with tears developing in his eyes, and he scans the assembly members, and then looking at God Enlil, he softly accuses, "God Enlil, by commanding total destruction, you place an evil deed on the people."[502b] He pauses, and looks at the viewing monitor, which has now returned to the present, and shows the many people, amidst the mildly blowing dust and sand, at a slow pace catching the abundant jumping fish in the muddy rivers, by various means, and eating the fish, and their state of partial recovery from the reimposed drought and famine. He then looks at the standing assembly members, who are also looking at the viewing monitor, and in a loud tone, God Enki slowly declares, "By this assembly decision, I am also bound to silence." His eyes now fill with tears, as again he pauses, and then he slowly and softly professes, "My lips agree, but my heart does not." He slowly raises his right hand, and almost in a whisper, and with a tear running down his cheek and into his beard, he pledges, "By this oath, I am bound," and he remains standing.

After another long moment of silence passes, God Enlil, now with a satisfied look and smile upon his face, commands, "Let us clasp hands together." All the Gods clasp hands, and then God Enlil demands, "Your oath to me."

Together, all the Gods declare, "The decree of God Enlil is binding."[503a]

[502a]M. Page 143, line 140 Then Nin[tu (an epithet for Goddess Ninhursag) feared her] crea[tures (her creation = mankind) destiny], and line 141 Holy Inanna (an epithet for Goddess Ninhursag) we[pt] because of the (peril of the) people.

[502b]M. Page 87, line 34 The gods commanded total destruction, and line 35 Enlil did an evil deed on the peoples.

[503a]M. Page 143, line 143 An (Anu), Enlil, Enki (and) Ninhursag, and line 144 The gods of the universe, had [taken an oath by] the names of An (Anu) and Enlil, and page 133, upper line 1, and upper line 3; and F. page 80, line 14, 15, 16, 17, 18, and line 19.

*

We see a very clear star-filled night sky with a crescent Moon, the symbol of God Enki, and then we look down to the moonlit gently sloping hills of the Abzu, which have now fully returned to normal, with their lush greenery of tall and thick bushes and trees everywhere.

ABZU
.03 Shar Years Later
or
100 Earth Years Later
8,800 B.C.

We look down further, and we see the outside of the biological laboratory craft of God Enki, one of the three craft close together in a triangle, and linked together by the clear glass-like passageways. They are in the clearing, with the normal forest in the background. They appear as they had before the drought, as the Abzu has now fully recovered from the drought and famine, and has returned to normal. We slowly zoom toward the craft, go up the ramp, and through the opaque air flow door, and we observe God Enki, alone, again speaking to himself in front of a computer which has flashing display lights.

We zoom in further, over his shoulder, and see the computer, with an 8 inch square screen, the telepathic communicator, which is on but again shows an out of focus picture. God Enki is reiterating to himself,[503b] and simultaneously touching the heat sensitive numbers on the computer console, as he previously had done. The flashing lights dance on the display, as he programs the computer,"I am bound, by oath, not to directly inform mankind of the impending flood."

He pauses, and then we again see a mischievous smile, and a sparkle in the eyes of God Enki, as he resolves, "But again, as I did before, I may be able to inform Ziusudra, indirectly, using the telepathic communicator, through a

---

[503b]M. Page 143, line 142 Enki bethought himself (of the situation, even though he had taken an oath).

dream subterfuge. I shall program the computer to a slightly different wave length this time." We look at the screen on the telepathic communicator, and the picture begins to focus, and again we see the face of sleeping Ziusudra intermittently appear on the screen. God Enki now leans over, and speaks into the computer, and slowly instructs, "Ziusudra, gather reeds and build a boat. Take your wife, your family, and your friends. Take animals. Save yourself. Annihilation is upon you. Water shall destroy all mankind." The eyes of God Enki now fill with tears, and one tear rolls down his cheek and into his beard, as he despairs at his own words.

*

We see the crescent Moon, the symbol of God Enki, in the clear, star-filled night sky, and then we look down to the moonlit courtyard and house of Ziusudra. It appears normal and is now fully recovered, restored, and well kept, as it was before the drought and famine.

SHURUPPAK
The Same Night

Inside the house, sleeping in his bed, clothed in his purple robe, and wearing his gold God Enki symbol necklace, gold winged sphere symbol of Nibiru, and gold bracelet, and tossing as he dreams, we observe 99 year old in appearance Ziusudra. His signs of aging, a wrinkled face, and his black tight-curly hair and beard, which are turning gray, are now plainly visible. There is a blurry, cloud-like border again surrounding the face of God Enki in the dream of Ziusudra.[504] The voice and face of God Enki are again intermittent and jumbled, and again Ziusudra cannot fully understand them, as he sees and hears, " . . . . build . . . . wife . . . . family . . . . friends . . . . animals . . . . save yourself . . . . annihilation . . . . water . . . . ." Ziusudra tosses

---

[504]M. Page 143, line 149 It was not a dream, coming out and spea[king aloud] (it was not a dream), and line 150 Conjured by heaven and underworld [alike], and page 113, line 27, 28, 29 [He (Ziusudra) spoke] with his god (God Enki, through the dream), and line 30 [And] Ea (Enki) spoke with him (Ziusudra, through the dream), and page 133, line 4 King (Ziusudra) hears [Ea] (Enki), and line 5 The [King] (Ziusudra) had Ea (Enki) in his heart (revered God Enki: see footnote 454).

and abruptly awakens, and partially sits up, and has a puzzled and startled look upon his face, as he recalls the dream.[505] Then, after a few moments of thought, he questions himself, "What does it mean?" He pauses for a moment, and then he declares, "We do not suffer now, the drought and famine have been lifted." Then, Ziusudra lays back and closes his eyes and goes back to sleep.

*

We see a bright blue daylight sky with normal clouds, and then we look down, and from a distance, we observe the outside of the oval walled city of Shuruppak, and it appears as it had appeared before the drought and famine, with its colorful 300 foot high central 7 tiered ziggurat, composed of solid stages, platforms, or tiers, with ramp-like connecting stairs, progressively reduced in height and width, and each coated with a different color plaster, white, black, purple, blue, vermilion, and silver, and gold colored temple, the communication and control center of the Gods, for their exclusive use only, as the seventh tier, at the top, with antennas atop it, and the rectangular homes sprawling around the wall for one mile in all directions from the ziggurat and temple compound, and the wide river to the east, close to the city, with clear water in it, and the irrigation canals veining through the area. Ziusudra, the loyal servant of God Enki, is high Priest and King. The city and the people, most of which have survived the drought and famine, are back to normal and fully recovered. Everything, including the pace, is again back to normal, and there are no remaining signs of the drought and famine; but there are no Gods in the city, among the people, neither Anunnaki Gods, nor Igigi Gods, for they are all at Sippar, their mission control center, preparing to leave the Earth.

The Next Day

We zoom in and see many cameo views. The wind is normal, being variable, it ranges from almost calm to very light, and there is normal rainfall.

---

[505] M. Page 77, line 8, and line 10 Giving [attention] to dreams.

The Euphrates River is wide, and slowly flowing, again with clear water, and at a full, normal water level of 35 feet. There are tall, lush, thick reeds along and near the river. People are fishing in the river with flawless nets, and there are fish jumping in the water. People are also catching birds with bow and arrow. Again there are large and small, rectangular shaped, freshly pitch coated reed boats in the river, which are entirely constructed of reed bundles, and are a common and familiar sight. The canals are filled with water, at the normal 10 feet water level, with clear water, and are irrigating the crops. Again the people are digging with wood hoes and wood shovels in the canals, keeping them open and maintaining them. They scoop and shovel mud from under the water, and pull it up over the top of the bank, and some people are using reed baskets, with a rope support which they grasp at their chest, to haul the mud away on their back, to be fully utilized as clay. The fields and orchards again have black soil and lush crops of grapes, figs, dates, olives, apples, pears, pomegranates, almonds, walnuts, pistachios, corn, wheat, barley, and other grains, as well as lentils, cucumbers, lettuce, onions, beans, and other vegetables. The fields are weedless, with the adults, and children 10 years of age in appearance or older, tending the crops. The sheep and goats are again fat, and the children are tending, herding, and milking them. The adults are plucking the sheep and collecting the wool, as others spin and weave the wool into cloth, as they do with flax, for making linen.

The streets are again clean and well kept, and are filled with adults, with the Gods being conspicuously absent, and children 10 years of age in appearance or older are playing games or playing with their fat pet dogs. The streets are clean and even, and are not paved with clay bricks, but are of smooth compacted sand. The rectangular courtyards, surrounded by the 7 feet high sun-dried clay brick walls, coated with cream colored plaster, and a cream colored wash, are again clean, and are paved with baked clay bricks. Again the outsides of the houses are clean and well kept, in excellent condition, and look new, with fresh cream colored plaster and cream colored wash over the sun-dried clay brick walls, and with freshly clay coated nearly flat reed bundle roofs. The wood front doors are again highly polished. The small rectangular reed huts, with reed doors, used as storage buildings, and entirely constructed of reed bundles, which are attached to the back of the houses, are again in excellent condition, clean and well kept. Inside, the houses are clean and orderly, with pale red sun-dried clay brick floors, and cream color plastered walls. The furniture is made with highly polished wood, and consists of chairs, and rectangular tables and benches, all in excellent condition. The window draperies are again like new, clean, and well sewn. The U-shaped open top baked clay brick fireplaces, one in each kitchen area, are full of reed bundles, which are the common fuel.

Most adults, 80%, continue to pray, sitting cross legged, as they bow their heads, after placing offerings of large round loaves of sesame coated bread at the wall niche altar in their homes, or at any of the eight smaller temples near the base of the ziggurat. They pray to various personal Gods, and implore, "God Anu," "God Enlil," "God Enki," "Goddess Ninhursag," and "GODS OF THE UNIVERSE™,"[506] and also they petition, "accept this offering."

The people's clothes are again exceptionally clean, crisp and well sewn, and beautiful, and consist of: white tunic-like knee length skirts for working in; and white plain toga-like full length robes for everyday; and colorful full length robes with edging, for special events, which are only draped over one shoulder. The Priests, members of the council, and special individuals clothes consist of colorful full length robes with edging for everyday and special events, like the special events clothes of the people, except their robes cover both shoulders. All adults again wear leather wedge-like sandals with straps. Some people wear a flawless wire earring with stone pendants, and some wear an intricately detailed wood or silver bracelet on their wrist, and some men wear an impeccable, white turban-like cloth hat or colorful cloth head band. Some women wear a faultless, wide colorful cloth head band ribbon in their hair, and large crescent shaped earrings on special occasions.

Most adults, and children 10 years of age in appearance or older, 80%, continue to wear a 4 inch diameter clay, wood, gold, or silver symbol pendant of the four main Gods, hanging around their neck on a silver chain necklace, like the Gods wear; which are the highly worshipping people, especially the people that give offerings and pray. They again have long legs and broad shoulders, and walk without any signs of their previously hunched over posture, and all are again well groomed and clean, with combed hair and natural shaped beards. All the people in this region still have white pigment skin, with full faces and fair features, and look like Caucasian mankind of today. Most of the people have black hair, some tight-curly, others wavy, and others straight; all

---

[506]M. Page 143, line 143, and line 144 (see footnote 274).

of this type have brown eyes. Other people have brown or blond variations of hair; with blue, green, or brown eyes. All hair is of medium length on both men and women. Again people converse with each other, and display affection toward, and freely caress each other, but no mothers are nursing babies, for at the present time, none have been born; however, some women appear to be 4 months pregnant. There are no children younger than 10 years of age in appearance. (This is due to the different aging rate between Ziusudra and mankind today. The body processes of early mankind were genetically retarded, to give them longevity, and Ziusudra's life span was approximately 36,000 years, as ours, for ease of computation, is approximately 100 years. 360 of Ziusudra's years equal 1 of our years today, and a full term pregnancy, or 9 months today, is equal to 270 of Ziusudra's years. But the animals in Ziusudra's time, age like the animals of today, for they were not genetically retarded. That is why there are many animals, because they have had 200 years to replenish after the lifting of the drought and famine, but the humans have only had .56 of our years to replenish, or approximately 7 of our months.)

They are again using flawless pale red and cream decorated clay utensils, tableware, and vases, and use clay cups for drinking, and reed baskets of various sizes, all without handles, and they use flint knives. They again roast birds, ox and lamb on wood skewers, and smoke-dry fish in the U-shaped open top baked clay brick fireplaces, one located inside, in each kitchen area, or in the other located outside, in back of their houses. They again eat large round loaves of sesame coated bread, which are baked in dome top ovens built into the rear portion of the fireplaces, and again drink the clear river water, beer, and red or white wine, and again there are full vases and drinking cups with all three everywhere. Mothers open their doors and welcome their older daughters to their homes, and younger daughters, all 10 years of age in appearance or older, are helping mothers prepare the meals, and with the household duties.

There are no small children, under 10 years of age in appearance, to play with the toys in the courtyards, or behind the houses, only older children, who play games. All the children are again learning, by using the 3/4 by 3-1/2 by 5-3/4 inches long clay tablets for reading and writing with pictographs. Again the people are joyous, partying, singing, dancing, and merrymaking, and some adults and children play musical instruments, like the tambourine, harp, drum, and flute. Many of the people again sign their name and status, by using personalized pictograph cylinder seals, which some again wear as a pendant around their neck. Again the people are everywhere in the city. All the signs of the drought and famine have disappeared, and everything has again returned to normal. Mankind is still comparatively intelligent and knowledgeable, and is again independent, disobedient and defiant, and assertive to other people. These are the normal and fully recovered conditions that we find in Shuruppak, the next day after the dream, in which God Enki appeared to Ziusudra.

*

We observe Ziusudra, within the bright sunlight, with a wrinkled face and graying black tight-curly hair and beard, clothed in his special full length purple robe with gold edging, which covers both shoulders, and his tall, purple fez-like hat, and wearing his gold God Enki symbol necklace, and his gold winged sphere symbol of Nibiru on his robe, above the left breast, and his gold wire earring with gold pendants on one ear, and his gold bracelet, and carrying his gold scepter and gold ring, as symbols of his authority as high Priest and King of Shuruppak.

The Same Day

He is walking near the river, in back of a house and near its attached small reed hut. He is admiring the crops, and observing the adults and children who are working in the fields and orchards. There is an open area near him, and at short intervals, he looks up, waiting expectantly for God Enki, for Ziusudra is uncertain of what he saw in his dream the previous night. As he walks and waits, he again questions himself, "What was the meaning of the dream last night? Was it a message from God Enki, or was it just a dream?"

Then he looks at the adults and children working in the fields and orchards, and then, out of the corner of his eyes, Ziusudra sees a craft rapidly approaching, glistening in the bright sunlight, with its brilliant glowing firefly-like light emitting from the underside. From a distance, we watch the craft hover over the open area, and then descend and prepare to land, as the glowing light begins to diminish, and the 4 pods like cattle hoofs extend 6 feet down, and the craft begins to land. We observe Ziusudra, who is smiling and pleased to see the descending craft, as he quickly walks toward it.

Then the craft lands, and the glowing light further diminishes, and then completely expires, and after a moment, a side of the craft extends down, forming a 4 feet wide ramp and doorway. Then, God Enki walks through the opaque air flow door and down the ramp of the craft. Again he is alone, and is wearing his full length, from elbow to ankle, red robe with gold trim at the bottom, neck, and elbow, 5 inch wide gold waist band, gold wedge-like sandals with straps, gold winged sphere symbol of Nibiru on his robe, above the left breast, gold watch-like bracelet instrument, and his 4 inch diameter identifying gold symbol pendant, hanging around his neck on a gold chain necklace, and he is without a helmet or body regulator.

At the same time, Ziusudra arrives to greet him, and he bows down, and prostrates himself, and then he stands,[507] and they exchange greetings, and then warmly embrace.

After their long embrace, Ziusudra exclaims and then questions,[508] "Welcome God Enki. You appeared to me in a dream last night.[509] I did not understand it, for we do not suffer now. Teach me the meaning of the dream, the reason, that I may seek its outcome."[510] God Enki looks toward the reed hut, and motions to Ziusudra, and they both begin to walk toward it, and as they walk toward the reed hut, God Enki replies, "I sent you a message in the dream, but again, you did not fully receive it. Ziusudra, as my loyal servant, you have lived thirty-six thousand Earth years. You are most obedient, and your lineage, your ancestors and family, have not intermarried to degenerate the Gods.[511] Many of the other people have defiled themselves.[512] Because of your programmed genetic evolution, through the reproductive knowledge I sympathetically granted mankind, your intelligence shall progressively increase, but your succeeding generations shall revert to live only very short lives, like the creatures from which you were created. I bestowed upon you

---

[507]M. Page 123, upper line 3 (see footnote 335).

[508]M. Page 89, line 1, 2, 11, and line 12, and page 123, upper line 4.

[509]M. Page 123, line 11 In a dream I have seen your face.

[510]M. Page 89, line 13 Teach me the meaning [of the dream], and line 14 [The] reason that I may seek its outcome, and page 123, line 12 Please tell me your [plan].

[511]V. Genesis 6:9 Nō´-ăh (Ziusudra) was a just man and perfect in his generations, and Nō´-ăh (Ziusudra) walked with God (was pious/obedient to God Enki, similar to this: see footnote 311).

[512]V. Genesis 6:11 The earth also was corrupt before God (God Enki), and the earth was filled with violence, and 6:12 for all flesh (mankind) had corrupted his (God Enlil's) way (prohibition on intermating between Gods and mankind) upon the earth (similar to these).

kingship, a ruler of your people, here in the city of Shuruppak. You are my trusted servant. With humility and reverence you make offerings each day.[513] I am under oath, not to divulge a secret of the Gods to mankind."[514]

God Enki and Ziusudra have now walked to within 10 feet of the reed storage hut, and God Enki points to the hut, and we see a close up view of the storage hut, which has a closed door. The hut and door are both made of 3 inch diameter reed bundles, and are attached to a 6 inch diameter, triangular shaped, reed bundle framework, which is not visible from outside the hut. Then God Enki asks and commands,[515] "You say, 'What am I to seek?'[516] Go into this reed hut,[517] and stand at the left side wall, and listen[518] closely," and both walk to the reed hut door, and Ziusudra opens the door, which swings to the right, and it remains open. Then he enters the reed hut, and turns to the left as he walks through the doorway, and after God Enki watches Ziusudra enter the reed hut and turn, God Enki walks close to the outside of the reed hut, and stands in front of the left side wall,[519] to the immediate left of the doorway. We observe Ziusudra in the reed hut, with his ear only 6 inches from the left side wall, and we now plainly see the 6 inch diameter, triangular shaped, reed bundle framework.

God Enki looks around, and sees that no one is close enough to hear, and then God Enki speaks to and instructs the reed wall,[520] "Wall, listen to me.[521] Reed wall, reed hut, observe all my words. Pay attention to my

---

[513]V. Genesis 6:8 But Nō´-ăh (Ziusudra) found grace (was favored and loved) in the eyes of the LORD (God Enki, similar to this).

[514]V. Genesis 6:17 And, behold, I (God Enki), even I (God Enki), do bring (took the oath of secrecy to withhold the information from mankind) a flood of waters upon the earth (similar to this.)

[515]M. Page 89, line 15, and line 16.

[516]M. Page 89, line 17 You say, "What am I to seek?"

[517]M. Page 143, line 151 In the ki-ur (reed hut?), (away from) the gods, (there was) a wall [of reeds].

[518]M. Page 143, line 152 Ziusudra hea[rd], standing by its side, and page 89, line 18 Observe the message that I will speak to you, and page 127, line 2 Your dream last night I will explain; and F. page 80, line 10 Namely, a secret of the gods will I tell thee.

[519]M. Page 143, line 153 He stood at the left of the side-wall [speaking].

[520]M. Page 123, line 13 [Ea] (God Enki) opened his mouth to speak, and line 14 [And addressed] the reed-hut, and page 133, line 12 He (God Enki) repeated their words (the words of the Gods in the assembly) to the wall; and F. page 80, line 20 Their (the assembly members) speech he repeated to a reed hut.

[521]M. Page 89, line 20 Wall, listen to me!, and page 123, line 15 Wall! Wall!

instructions,"[522] and God Enki pauses, and then he reveals, "Ziusudra, man of Shuruppak, son of Ubar-Tutu,[523] listen. Waters of a flood shall seize all the peoples together.[524] From the south it shall come. On all dwellings, over all cities, the storm shall sweep.[525] The destruction of the seed of mankind is near."[526] We observe a startled and wide eyed Ziusudra, as he carefully listens to each word, and then God Enki continues, "The final sentence, the word of the assembly, by oath,[527] was to keep the secret from mankind. These words, the oath, was taken by God Anu, God Enlil, God Enki, and Goddess Ninhursag, GODS OF THE UNIVERSE,™ under a decree issued by God Anu and God Enlil.[528] The overthrowing of the kingship of mankind now approaches."[529]

Again we observe Ziusudra inside the reed hut, now very excited, who now has his ear pressed to the wall, and then God Enki instructs, "Prepare yourself, before the flood sets out. Assemble all that are close to you.[530] Abandon your house, together build a big boat.[531] Build a hexagon shaped boat, a boat composed of six equilateral triangles, six decks let it contain."[532]

Reed-hut! Reed-hut!, and page 143, line 154 Side-wall, I want to talk to you, [hold on] to my word; and F. page 80, line 21 Reed hut, reed hut! Wall, wall!

522M. Page 89, line 21 Reed wall, observe all my words!, and page 123, line 16 Wall listen, pay attention to me!, and page 133, line 14 Wall, hear all my words!, and page 143, line 155 [Pay atten]tion to my instructions; and F. page 81, line 22 Reed hut, hearken! Wall, consider!; and W. page 243, Volume 1 And it was revealed to Noah (Ziusudra, similar to this).

523F. Page 81, line 23 Man of Shurippak, son of Ubara-Tutu! (see footnote 302).

524M. Page 127, line 3 [Waters of a flood] will seize all the peoples together.

525M. Page 143, line 156 On all dwellings, over the capitals the storm will [sweep]; and V. Genesis 6:17 and every thing that is in the earth shall die (similar to this).

526M. Page 143, line 157 The destruction of the descent of mankind [is near].

527M. Page 143, line 158 The final sentence, the word of the assembly [by oath], and line 159 The word spoken by An and En[lil and Ninḫursag].

528M. Page 143, line 143, and line 144 (see footnote 503a).

529M. Page 143, line 160 The overthrowing of the kingship [of] mankind, Now approaches.

530M. Page 127, line 4 Prepare yourself before the flood sets out, and line 5 Assemble all that there are close to you.

531M. Page 89, line 22 Destroy (abandon) your house (house = Kingdom), build a boat, and page 127, line 6 Together build a big boat; and F. page 81, line 24 Tear down (abandon your) house (Kingdom), build a ship!; and V. Genesis 6:14 Make thee an ark of gō´-phĕr wood (gō´-phĕr could mean a gufa (round)-like boat, similar to this. A similar Hellenic word of interest is κοῦφιος, which means hollow, an enclosed empty space, which is phonetically like go-pher; as I further explain on page 298 in my comments and under 'Architecture' on page 362 in my epilogue).

532M. Page 129, line 1 [Six] decks let it contain (thinking along the thoughts of style and habit of scribal continuity of repeating words and blending one line into the next, the

From inside the reed hut, Ziusudra anxiously inquires, "I have never built a boat like that before. Draw a design on the ground, that I may see the design, and build the boat,"[533] and Ziusudra looks at the ground, in the open doorway.

Then God Enki bends down, pulls and breaks a 10 inch long reed from the hut wall, and using it as a stylus, draws a top, side, and perspective view of a hexagon shaped boat, a 6 sided figure, containing 6 equilateral triangles, divided into 6 floors, and a slightly tapered downward from the center roof, and a slightly tapered upward from the center bottom, with the perimeter height equal to one-quarter of its diameter.

As God Enki draws with the reed on the ground[534] in the doorway, he continues to instruct, "Spurn property, disregard goods, and save life![535] The boat which you build,[536] let its structure be made entirely of reeds,[537] for flexibility. Its width and its length shall be equal, measured accurately.[538] Let it like a circle be,[539] below water its two-thirds.[540] Let it be completely roofed over,[541] above and below,[542] with a strong covering,[543] and seal it so the Sun shall not see inside it.[544] Thoroughly caulk the boat with pitch,[545] above and

---

possibility exists that this line, line 1, could have been originally rendered [Six] sides let it contain, or in other words, a hexagon shape, for the following line, line 2, is where 'like a circle' appears, which was the profound clue, or stimulus and then impetus that caused me to discover the shape of the boat).

533M. Page 129, line 11 Atra-ḫasīs opened his mouth to speak, 12 And addressed Ea (God Enki), [his] lord, 13 'I have never built a boat like that before (because this was a special/unusual/uncommon/unique boat), 14 Draw the design on the ground, and line 15 That I may see [the design] and [build] the boat.'

534M. Page 129, line 16 Ea (God Enki) drew [the design] on the ground.

535M. Page 89, line 23 Spurn property and save life; and F. page 81, line 25 Abandon (your) possessions, seek (to save) life!, and line 26 Disregard (your) goods, and save (your) life!

536M. Page 89, line 25 The boat which you build; and F. page 81, line 28 The ship which thou shalt build.

537M. Page 127, line 7 Let its structure be [made] entirely of reeds.

538M. Page 89, line 26 Its width to be equal [to] its length; and F. page 81, line 29 Its measurements shall be (accurately) measured (followed), and line 30 Its width and its length shall be equal; and P. upper page 46 If the texts mention the "length" and "width" of such a figure (triangles) it is only from the context that we can determine the exact meaning of these two terms, and upper page 47 (regular hexagon, and equilateral triangle).

539M. Page 129, line 2 [Let] it like a circle be (a hexagon is like a circle. Since the system of measurement at that time was sexagesimal, based on the unit of 60, which I believe is actually based on the unit of 6, which is an expression of water; as I further explain on page 298 in my comments and under 'Water' on page 344 in my epilogue, and the precise number of decks (6) is given above, then I believe the shape of the boat was a hexagon (6); partially because of the sexagesimal (60/6) system of measurement, and my belief that the shape of the boat itself was expressed in terms of water (6), but mainly

below the water, both inside and out.[546] Let the pitch be tough, and so give the boat strength.[547] The tackle should be very strong.[548] Truly, let it be a má-gur-gur boat, with the name, The Life Saver,"[549] and he finishes drawing the hexagon shaped boat on the ground, and we see a close up view of the three drawings.

Then, as Ziusudra stares at the drawings on the ground, he affirms, "God Enki, my Lord, what you commanded, I shall build."[550]

---

because of the measurements given, and primarily because of this line, which is the only surviving clay tablet clear reference to the shape of the boat, for certainly a hexagon is like a circle; as I further explain on page 280 in my comments).

[540]F. Page 83, line 79 Below water its two-thirds (if this two-thirds figure is based on the externally visible displacement/buoyancy of one-third of the height of the boat, then it indicates how the completed empty boat was to appear or appeared in the water, and it may not have any relevance to the number of decks inside the boat, but I am very suspicious of it. However, if it is a clue and is relevant to the number of decks in the boat, it could imply and thus identify the number of decks in the boat as a number divisible by three, for two of three decks would equal four of six decks, four-sixths reduced to two-thirds; were to be or were below the water line).

[541]M. Page 89, line 29 Roof it over like the Apsû (the boat is to have a well braced faultless roof, like the roof of the mines in the Abzu. However, this line could also have another meaning, which would be a subtle clue to the angle or degree of taper of the roof: the roof of the boat is to have a slightly tapered roof, like the high nearly flat plateau of Ethiopia, or like the gently sloping land of southeastern Africa — the Abzu/Apsû: see footnote 77); and F. page 81, line 31 Cover it [li]ke the subterranean waters (are covered: totally covered by the ground).

[542]M. Page 89, line 31 Let it be roofed over above and below (the water: the boat is to have the same type of roof on both the top and the bottom, and because of the slight tapering, the 6 equilateral triangles comprising the top and bottom will become 6 isosceles triangles, giving the top and bottom a slight pyramid shape, and thus more strength).

[543]M. Page 127, line 9 Completely roof it over with a strong covering.

[544]M. Page 89, line 30 So that the sun shall not see inside it (no sunlight is to enter the sealed boat, as no sunlight enters the mines of the Abzu).

[545]M. Page 129, line 4 Thoroughly caulk the [boat].

[546]V. Genesis 6:14 and shalt pitch it within and without with pitch (similar to this).

[547]M. Page 89, line 33 Let the pitch be tough, and so give (the boat) strength, and page 129, line 3 Let [the pitch] be strong above and below (the water).

[548]M. Page 89, line 32 The tackle should be very strong.

[549]M. Page 127, line 8 Truly let it be a maqurqurrum-boat (=má-gur-gur boat) with the name, The Life Saver, and page 171, note 205; and G. page 52, note 12 magurgur = "a boat driven by the wind"/"a boat which can be closed by a door"/"a house boat"; and F. page 232 magurgur = "a very great ship," "a giant boat," and page 233 The Old Testament word for the ark is tēbâ. Outside of the biblical flood story, this term occurs only in Exodus 2:3,5, where it is applied to the reed vessel in which the infant Moses was saved. Egyptian ḏb't = "chest," "box," "coffin." In the Septugint and the New Testament — κιβωτός = "box," "chest," "coffer"; and T. page 351 Hebrew - teba = "sunken", and a submersible boat - a submarine, and page 352 MA. GUR. GUR. - "a boat that can turn and tumble."

[550]M. Page 129, line 17 [Ea] (God Enki) my lord, what you commanded, I (Ziusudra)

Then God Enki resumes, "Into the boat which you shall build,[551] send your barley, your goods, your property,[552] your wife, your family, your relatives, and the skilled people.[553] Supplies[554] heap up. Also creatures of the steppe, all the wild creatures of the plains that eat grass, and birds of the heavens,[555] I shall send to you, and they shall wait at your door.[556] The fat and clean animals and birds, those without disease,[557] you shall catch and put into the boat," and then we observe Ziusudra, inside the reed hut, again with his ear pressed to the wall, and his eyes fixed on the drawings upon the ground in the doorway. Then God Enki instructs, "Observe the appointed time, of which I shall inform you.[558] On that day, as the first shimmer of morning beams forth, and the appearance of the weather changes, and you hear thunder roaring in the clouds, enter the boat, bolt the door and seal it with pitch."[559]

From inside the reed hut, again Ziusudra anxiously inquires,[560] "But what shall I answer the city, the people, and the elders,[561] when they inquire as to the boat's purpose?"

God Enki responds,[562] "So that you do not alarm the people, for you shall need their labor, this is what you shall tell the elders[563] . . ."

---

shall build; and F. page 81, line 32 When I understood this, I said to Ea (God Enki), my lord, 33 [Behold], my lord, what thou hast thus commanded, and line 34 [I] will honor (and) carry out; and V. Genesis 6:22 Thus did Nō´-ăh (Ziusudra) according to all that God (God Enki) commanded him, so did he (Ziusudra, similar to this. It is not only the boat/ark of Ziusudra/Noah, it is also and foremostly the boat/ark of God Enki. Also, the Bible in Genesis 7:6 and 7:11 reports that when the flood began, Noah was six hundred years old. Could the number six have been a reference to the shape of the boat (see footnote 539), and was it transposed to be the age of Noah, a convenient age; or could the 600 year age of Noah be a transposed reference to the 600 foot perimeter of the boat: is it just coincidence, or do you believe, as I believe, that there is a linkage, and a cryptic message, whether intentional or not by the Biblical scribes, within the age of Noah).

[551]M. Page 127, line 10 [Into the boat which] you will make.

[552]M. Page 129, line 7 [Send up into] it your barley, your goods, your property.

[553]M. Page 129, line 8 [Your wife], your kith, your kin, and the skilled workers; and V. Genesis 6:18 and thou shalt come into the ark, thou, and thy sons, and thy wife, and thy sons' wives with thee (similar to this).

[554]M. Page 127, line 12 Supplies heap up; and V. Genesis 6:21 (supplies/food, similar to this).

[555]M. Page 127, line 11 [Send (to you I will) all the] wild creatures of the steppe, (and) birds of the heavens, and page 129, line 9; and F. page 81, line 27; and V. Genesis 6:19 (all wild creatures/all flesh), and 6:20 (birds/fowls, similar to these).

[556]M. Page 129, line 10 [I] will send to you and they will wait at your door.

[557]V. Genesis 7:2 (creatures/beasts), and 7:3 (birds/fowls, similar to these).

[558]M. Page 129, line 5 [Observe] the appointed time of which I will inform you.

[559]M. Page 129, line 6 Enter [the boat] and close the boat's door.

[560]M. Page 133, line 9 I knew (of) the counsel (secret) of the great gods, and line 10 I knew of their oath, though they did not (directly) reveal it to me.

As we quickly zoom away, God Enki is speaking to the reed wall of the hut, and we do not hear the answer, but we stop at a short distance, and we observe God Enki speaking to the reed hut, but we cannot hear his words of instruction.

*

It is bright daylight, and we see the outside of the large house of Ziusudra, and the gate leading to the courtyard, which is within the oval wall at Shuruppak, near the colorful ziggurat. The courtyard and house are beautiful, and look as though they were new, with a fresh cream colored wash, and a highly polished wood front door.

The Next Day

The city elders, 15 males and 5 females, all wearing their symbol necklaces, are gathering for a council meeting. Some are being brought by ox drawn four wheel cart, and others arrive on foot. They are beautifully clothed in their colorful full length robes with edging, which cover both shoulders, and all appear very healthy. They are going through the gate[564] and courtyard, and into the house.

Inside the house, we see the large nicely furnished central room, imitating the assembly of the Gods, which is again beautiful, and also looks as though it was new, with highly polished wood benches arranged in a circle for the elders, and a large highly polished wood chair for Ziusudra. Ziusudra, clothed in his special full length purple robe with gold edging, which covers both shoulders, and his tall, purple fez-like hat, and wearing his gold God Enki symbol necklace, and his gold winged sphere symbol of Nibiru on his robe, above the left breast, and his gold wire earring with gold pendants on one ear, and his gold bracelet, and holding the symbols of his authority as high Priest and King, his gold

[561]F. Page 81, line 35 [But what] shall I answer the city, the people, and the elders?

[562]F. Page 81, line 36 Ea (God Enki) opened his mouth and said, and line 37 Speaking to me, his servant.

[563]F. Page 81, line 38 Thus shalt thou say to them; and M. page 91, line 38 Atra-ḫasīs (Ziusudra) received the command.

[564]M. Page 91, line 39 He assembled the elders to his gate (house).

scepter and gold ring, is welcoming the group, and they quickly sit on the benches and freely speculate about the purpose of the meeting.

Ziusudra then sits in his chair, and the elders quiet and look at him, and Ziusudra declares,[565] "I have gathered you in council, because my God, God Enki, does not agree with your God, God Enlil.[566] God Enki and God Enlil are angry with one another,[567] because of me. They have expelled me from my house,[568] Shuruppak. I have learned that God Enlil hates me,"[569] and we observe the surprised and puzzled faces of the elders. Then Ziusudra reveals, "Since I reverence God Enki, he told me of this matter.[570] Therefore, I cannot dwell in your city.[571] I cannot set my feet, nor turn my face, within the domain of God Enlil.[572] With the Gods angry[573] with each other, this is what God Enki told me to do,"[574] and again we observe the expectant faces of the elders, waiting for his words.

Then Ziusudra elucidates, "With your assistance, I am to build a boat, and go down to the Abzu, and dwell with God Enki, my Lord.[575] When God Enlil sees me departing, on you he shall rain down plenty,[576] bringing an abundance of birds, and a profusion of fishes,[577] and bringing a large harvest and wealth to you.[578] To begin this new prosperity, God Enlil shall instruct God Adad, the God in charge of rain distribution, to cause a gentle, light, wheat-rain, to softly rain down upon you in the evening."[579]

---

[565]M. Page 91, line 40 Atra-ḫasīs (Ziusudra) opened his mouth, and line 41 And addressed the elders.

[566]M. Page 91, line 42 My god [does not agree] with your god.

[567]M. Page 91, line 43 Enki and [Enlil] are angry with one another.

[568]M. Page 91, line 44 They have expelled me from [my house] (Kingdom).

[569]F. Page 81, line 39 [I have le]arned that Enlil hates me.

[570]M. Page 91, line 45 Since I reverence [Enki], and line 46 [He told me] of this matter.

[571]M. Page 91, line 47 I can[not] live in [your city]; and F. page 81, line 40 That I may no (longer) dwell in yo[ur ci]ty.

[572]M. Page 91, line 48 I cannot [set my feet on] the earth (within the domain) of Enlil; and F. page 81, line 41 Nor turn (show) my face to (within) the land (domain) of Enlil.

[573]M. Page 91, line 49 With the gods angry.

[574]M. Page 91, line 50 [This] is what he told me to do.

[575]F. Page 81, line 42 [I will therefore g]o down to the apsû (Abzu) and dwell with Ea (God Enki), my [lor]d.

[576]M. Page 89, line 34 I (He) will rain down upon you here (I believe I (God Enki) is a scribal error, and the word should be He, or the name should be either God Enlil or God Adad); and F. page 81, line 43 [On] you he (God Enlil or God Adad) will rain down plenty.

An elder eagerly exclaims, "We shall gladly assist you with this task!" Another elder shouts, "Let the task begin immediately[580] Ziusudra!" Then another elder affirms, "For with your departure, it shall rain down upon all our crops, and we shall have a large harvest, and increase our wealth!" We observe all the elders nod their heads in approval, as they discuss with each other, with wide smiles upon their faces, the benefits of their greater wealth, and then we observe the face of Ziusudra, who is forcing a half smile, as he scans the group of elders.

Again we see the elders, conversing and smiling, as they continue to give their approval with a nod of their heads.

*

It is middle afternoon, on a bright and clear day at the Euphrates River, at the far northeast edge of the city of Shuruppak, the boat building area of the city.

Approximately January 21st

We scan and see a very busy scene, with people, many people, approximately 700 people,[581] working at an almost quick pace, wearing white tunic-like knee length skirts, symbol necklaces, and colorful cloth head bands, and without ear wear or bracelets, all actively engaged in preparing reeds on the west bank of the clear and calm flowing Euphrates River, with orchard trees to the northwest, and with the common and familiar large and small, rectangular shaped, freshly pitch coated reed boats nearby in the river. There are both older people,[582] and children all older than 10 years of age in

---

[577]M. Page 89, line 35 (Bringing) An abundance of birds, a profusion of fishes; and F. page 81, line 44 [Abundance of b]irds, profusion of fishes.

[578]F. Page 81, line 45 Bringing a large harvest-wealth.

[579]F. Page 81, line 46 [In the evening the leader] of the storm (God Adad), and page 82, line 47 Will cause a wheat-rain (a helpful rain) to rain down upon you.

[580]F. Page 82, line 48 As soon as [the first shimmer of mor]ning beamed forth (the next morning), and line 49 The land was gathered [about me] (people and construction materials were assembled).

[581]M. Page 91, line 14 The poor man [brought what was needed] (his labor).

[582]M. Page 91, line 10 The elders gathered reeds.

appearance, males and females, including some women still appearing to be 4 months pregnant, who are cutting the reeds that grow in abundance along the river banks and in marshy areas nearby, and placing them in stacks. These reeds are then being carried by older people to nearby reed workers, who pound the reeds flat on a stone, using another stone as a mallet.[583] These flattened reeds are then being lashed together,[584] bound, into small 3 inch diameter bundles, using reeds as tackle, and the unequal ends are being chopped off by the carpenters, using a flint axe,[585] to a final length of 5 feet, which equals three, 20 inch cubits. These small 3 inch diameter by 5 feet long reed bundles are then carried away by other stronger people.

Then, we scan further and see a large, 6 sided, hexagon shaped boat, taking shape[586] in the Euphrates River. There are to be a total of 6 decks,[587a] of which the first and second decks have been completed, and the third deck is being built. The boat is composed of 6 equilateral triangles per deck, with each side of each triangle being 100 feet long,[587b] thus giving the framework of the boat a radius of 100 feet, and a diameter of 200 feet,[588] and a perimeter of 600 feet, and a framework, deck to deck height of 8 feet 4 inches on center, and a side hull eventual framework height of 50 feet,[589] and a center axis, or hub eventual framework height of 53 feet 4 inches. The standard unit of measurement being the 20 inch cubit, the framework must equal even cubits, thus the framework of the boat has a radius of 60 cubits, and a diameter of 120

---

[583]M. Page 91, line 12 The reed-worker [carried his stone].

[584]M. Page 91, line 16 They brought reeds together.

[585]M. Page 91, line 11 The carpenter [carried his axe].

[586]F. Page 82, line 56 On the fifth day [I] laid its framework (after four days of bitumen and reed preparation), and line 59 I 'laid the shape' of the outside (and) fashioned it (built the boat: 'laid the shape' could be a variant of footnote 539 'like a circle').

[587a]F. Page 82, line 60 Six decks I built into it (probably a variant of footnote 532, and thus could be interpreted as Hexagon decks I built into it), and line 61 (Thus) dividing (6, the framework members of the 6 decks) into seven (splitting/increasing 6 members/decks into 7 members/levels).

[587b]F. Page 82, line 57 One ikû was its floor space (each deck contained 25,980 square feet; as I further explain on page 284 in my comments), one hundred and twenty cubits each (triangle) was the height of its walls (120 cubits = 200 feet; as I further explain on page 285 in my comments).

[588]F. Page 82, line 58 One hundred and twenty cubits measured each side (to side) of its deck (120 cubits = 200 feet); and V. Genesis 6:15 The length (perimeter) of the ark shall be three hundred cubits, the breadth (radius) of it fifty cubits (similar to this); and M. page 135 (Berossus, according to Polyhistor) He did not disobey, but got a boat built, five stades long and two stades wide.

[589]V. Genesis 6:15 and the height of it thirty cubits (similar to this).

cubits, and a perimeter of 360 cubits, and a framework, deck to deck height of 5 cubits on center, and a side hull eventual height of 30 cubits, and a center axis, or hub eventual height of 32 cubits.

The boat is approximately one-third complete, with workers swarming everywhere, using small, rectangular shaped, shuttle boats to work from. The boat is under construction upwards, floating in the river, and is held in place by a 6 inch diameter braided reed hawser, tied to a beam, which is tied to a tree on the west bank, to the north, upstream from the boat, with approximately two-thirds, or 12 feet of the boat below the water line, and approximately one-third, or 6 feet above the water line.[590] Three of the small 3 inch diameter reed bundles are being lashed together with other reeds, using them as tackle, while alternately staggered, to form an equilateral triangle framework of 6 inch diameter reed bundles or beams, like spokes in a wheel. Each of the 6 equilateral triangles has been divided into a ground or floor plan of 9 primary trapezoid shaped sections, with these sections looking somewhat like open compartments,[591] by truss bracing beams composed of triangles, each extending vertically from floor to ceiling, and connecting each equilateral triangle and thus supporting each deck, with each spaced 6 cubits, or 10 feet apart on center, except the nearest to the center axis, or hub, which is 6 feet 7 inches on center from the hub. They extend from the hull toward the center axis, or hub, somewhat like a spider's web, with vertical and horizontal beams dividing each triangle truss. These vertical triangle truss bracing beams with dividing beams, also compose the perimeter or hull of the 6 equilateral triangles, and provide the primary framework for the hull. Diagonal truss bracing beams also extend around the perimeter of the boat, 5 feet on center, and brace the hull against the first vertical triangle truss bracing beam, which is 10 feet from the hull.

---

[590]F. Page 83, line 79 Below water its two-thirds; and V. Genesis 6:16 with lower, second, and third stories shalt thou make it (similar to this. I believe the Biblical account given here is a variant of and thus derived from the preceding Babylonian description given in footnote 540. Another variant appears in) Josephus IV, Jewish Antiquities, Books I-IV, page 37 He constructed an ark of four stories (etc.). This ark had stout sides and roof so as not to be overwhelmed from any quarter and to defy the violence of the waters. (A larger number of decks means a stronger vessel, and an increased or doubling of capacity of the vessel, thus enhancing its primary utilitarian purpose).

[591]F. Page 82, line 62 Its ground plan I divided into nine (sections; as I further explain on page 290 in my comments); and V. Genesis 6:14 rooms shalt thou make in the ark (similar to this. This very well could be a variant of line 62).

Numerous beams, also 6 inches in diameter, but one cubit, or 20 inches on center, further brace the boat. These beams extend horizontally up the 6 sides of the boat, to form the horizontal hull bracing beams, and provide the secondary framework for the hull, and connect the hull to the vertical triangle truss bracing beams; and they also extend outward from each equilateral triangle, 20 inches on center, and provide the primary framework for the floors and ceilings. To this framework of beams, the 3 inch diameter by 5 feet long reed bundles are being perpendicularly lashed, woven with reeds, and staggered, as are their ends, and alternately reversed to fit together, to form the 6 inch thick hull and floors. Reed bundle ramps, 5 feet wide by 6 inches thick, also composed of the 3 inch diameter reed bundles, and braced with 20 inches on center, 6 inch diameter beams, connect each floor next to the center axis, or hub of the boat, and their access openings also permit ventilation within the boat. The center axis, or hub of the boat, is composed of 19 solid, 6 inch diameter reed bundles, alternately staggered, which form 6 equilateral triangles, and together have a diameter of 30 inches.[592]

Then we observe Ziusudra, clothed in his special full length purple robe with gold edging, which covers both shoulders, and wearing his gold God Enki symbol necklace, and his gold winged sphere symbol of Nibiru on his robe, above the left breast, and his gold bracelet. We do not see Ziusudra wearing his tall, purple fez-like hat, or wearing his gold wire earring with gold pendants on one ear, or carrying his gold scepter or gold ring, from this time forward.

He is directing the workers,[593] who are lashing the reed bundles together, "Bind those tighter," as he rubs his hands with ointment.[594] Then Ziusudra admits to a worker near him, "My hands are not accustomed to this labor," as he turns and instructs another worker, "have the strong bring whatever else is needed,"[595] and the worker bows, turns and runs, following the order of his high Priest and King.

We scan further and see smoke rising from a U-shaped furnace, containing a large 4 feet 2 inches, 2-1/2 cubits in diameter baked clay pot, for

---

[592](See page 301 in my comments for detailed construction probabilities notes).

[593]M. Page 91, line 17 The boat was built, and line 18 Atra-ḫasīs (Ziusudra) directed them.

[594]F. Page 83, line 75 I[nto] ointment I put my hands.

[595]F. Page 82, line 55 The strong brought [whatever else] was needful.

melting bitumens, which are used to make pitch.[596] A workman shouts, as he mixes the pitch with an 8 feet 4 inches, 5 cubits long wood pole, "Bring more asphalt and bitumens to fix the pitch," and other workers quickly respond, and bring chunks and small clay pots of raw bitumens, and he dumps some of them into the large pot and stirs it. The furnace is an elongated half-circle of baked clay bricks, 6 feet 8 inches, 4 cubits high, with the pot, which has a pouring spout, resting on a long wood pole hinge, that passes through two opposite holes near the lip of the pot. There is a blazing fire beneath the pot. The pitch is poured into smaller clay pots, which are being carried to the boat by the children,[597] all older than 10 years of age in appearance, in a net sling suspended from a wood pole, which two children place on their shoulders.[598a] The pitch is then being applied as a caulk,[598b] but only to the inside and outside of the reed boat hull, as the work progresses upward, above the water line. The pitch is being applied from inside the boat, first to the outside of the hull, and then to the inside of the hull. It is applied by first pouring it over the reeds, at the top of the work area, and then by mopping and rubbing it into the reeds until they are saturated, with a mop-like applicator, composed of a medium length wood pole and cloth rags, and it gives a shiny black appearance to the boat.

Again we observe Ziusudra, as he instructs the workers applying the pitch, "It must be sealed above and below the water line, inside and out," and we see a close up view of the workers, as they lash and caulk the boat. One worker exclaims to another, "Greater prosperity and wealth are coming, with the departure of Ziusudra," as he lashes the small reed bundles to the beams. Another worker shouts to the reed carriers on the river bank, "Bring more reeds!" Then he declares to a worker next to him, "It is to be made entirely of reeds. We are to build a total of six decks, and each of the six sided decks is to be made up of six triangles. Each deck is to be two hundred feet across, and its

---

[596]F. Page 83, line 65 Six shar of pitch I poured into the furnace, and line 66 (And) three shar of asphalt [I poured] into it.

[597]M. Page 91, line 13 [The child carried] the pitch; and F. page 82, line 54 The child [brou]ght pitch, and page 83, line 67 Three shar of oil (probably a lighter/thinner grade of pitch) the basket-carriers brought, and line 68 Besides a shar of oil (pitch) which the saturation consumed.

[598a]N. Page 86, ILL. 48 (left center register).

[598b]F. Page 82, line 63 I drove water-stoppers into it (a type of caulk, besides the pitch, probably a single reed or a few reeds twisted or braided together, or possibly a rope or cloth caulk was used to fill gaps in the reeds: see footnote 545).

ground plan is to be divided into nine sections, and a connecting ramp between each deck at the center, and then roof it over," as he lashes the 6 inch reed bundles together, using reeds as tackle, to form pie shaped sections of beams. Another worker responds to the previous worker, "Its shape is to be like a circle, with a nearly flat roof and bottom, which are to be slightly tapered from the center," as he applies pitch to the inner side of the hull.

Again, we scan further and see an overall view of the busy construction area, which has a festive air about it, with many young sheep,[599] called lamb, and many young ox, called bullock,[600] being turned and roasted over a long pit of open fires, and some workers resting and refreshing themselves, by drinking plenty of beer, and red or white wine,[601] from pale red and cream decorated clay cups, and also directly from similarly decorated clay vases. Other workers are freely eating roast lamb and bullock, vegetables and fruits, and freely drinking, as they rest at rectangular wood tables and benches, near the roasting fires. Also near the roasting area, at one of the tables, as a worker prepares to eat a large portion of meat, which he grasps with both hands, he remarks to another worker seated across from him, "Everyday, Ziusudra prepares ox and sheep for us," and the other worker responds, as he drinks red wine directly from a clay vase, "Each day is like a New Year's Day feast.[602] We are drinking beer, and red or white wine, as if it were river water,"[603] and again he freely drinks, and then we see a close up view of the red wine being poured from the vase, directly into his mouth.

We see time passing through a series of alternating cameo views of normal clouds passing through the sky, and sunrise to sunset sequences, and sprinkled throughout the passing time, we see the progressing construction of the boat.

*

It is middle afternoon, on another bright and clear day at the clear and calm flowing Euphrates River, at the far northeast edge of the city of Shuruppak.

---

[599]F. Page 83, line 71 Sheep I killed every day.

[600]F. Page 83, line 70 Bullocks I slaughtered for [the people].

[601]F. Page 83, line 72 Must (grape juice, not yet fermented into wine), red wine, oil (olive oil for pouring over food), and white wine.

[602]F. Page 83, line 74 They made a feast as on New Year's Day.

Approximately April 23rd 3 Months Later

We scan and see the completed hexagon shaped boat, with the common and familiar large and small, rectangular shaped, pitch coated reed boats nearby in the river. It is thoroughly coated with shiny black fresh pitch,[604] and slightly tapered downward from the center of the roof, as we scan completely around the comparatively enormous hexagon shaped boat. The enhanced finished exterior dimensions are: height at the side hull is 51 feet 6 inches, and at the center axis it is 54 feet 10 inches. The enhanced finished exterior diameter is 202 feet, due to the 6 inch secondary beams and the 6 inch hull skin: a total of 12 inches on each of the 6 sides are added to the framework dimensions; and due to the one-half beyond center of the 6 inch bracing beams, or 3 inches, and the 6 inch hull, 9 inches on the top and bottom are added to the framework dimensions. Normally we add onto a framework, but because we do not know exactly how thick the hull was, for practical reasons, I shall use the framework dimensions, given previously, for the completed boat, which may actually be the completed dimensions, for the secondary beams and hull itself may have been included within the dimensions given for the framework. The boat rides about 16-1/2 feet above the water line,[605] with a nearly 6 feet high by 4-1/2 feet wide doorway, which has an inward tapering perimeter, the only opening in the boat, located at the top of a side, near the center; the bottom of the doorway being about 9 feet above the water line. It is still tied to the tree on the west bank of the river, to the north, upstream from the boat, and securely held in place by the 6 inch diameter braided reed hawser, which now passes through the hull of the boat, near the top of the side, next to the doorway, and into the boat through a 6-1/2 inch diameter outward tapering hole. There is a long ramp, 5 feet wide by 6 inches thick, also composed of 3 inch diameter reed bundles, and braced with 20 inches on center, 6 inch diameter reed beams, but not coated with pitch, leading from the bank to the doorway of the boat,

---

[603]F. Page 83, line 73 [I gave] the workmen [to drink] as if it were river water.

[604]F. Page 83, line 78 Sealed above and below (probably a variant of footnote 542: see footnote 541).

[605]F. Page 83, line 79 Below water its two-thirds (see footnote 540).

and being slightly reduced in width to fit into the doorway, it also serves to secure the boat to the bank, and prevent drifting.

Ziusudra is standing with his pilot at the base of the ramp, on the river bank, instructing the workers, who are beginning to load the boat. The workers carry clay pots and jars in a net sling suspended from a wood pole, wood chairs, rectangular wood tables and benches, clay utensils, large reed baskets of grain and food supplies, mostly dried, including large quantities of smoke-dried fish, large pale red and cream decorated clay vases of fresh drinking water and wine, most of which are sealed, and 3 inch diameter by 20 feet, 12 cubits long wood punting-poles[606] for poling and steering the boat. The workers are carrying household possessions of Ziusudra up the ramp, through the doorway, and into the boat.[607a] Ziusudra directs the workers carrying clay pots of pitch[607b] up the ramp, "Store that pitch near the door."

Then, there is a very light Earth tremor, and Ziusudra notices it, as he looks up to the bright blue sky. He then fearfully looks at the workers, to see if they also noticed. The workers pause for a second, but do not pay serious attention to it, as it quickly stops. We observe Ziusudra, who now has a relieved look upon his face, and again instructs the workers, who are carrying grain in large reed baskets, "Store the grain in its place," and then Ziusudra turns to his pilot, and declares to him, "after several months of building, finally the boat is completed. Difficult was the task.[608] I shall entrust you, as the pilot, to guide the boat safely,"[609] and the pilot responds, "You shall be safe with me, my King." Then Ziusudra professes to the pilot, "Whatever I have, silver,[610] gold,[611] all my possessions, are to be loaded," as the workers carry beautifully carved wood chests up the ramp. Ziusudra then turns and instructs several workers, who wait nearby with clay oil lamps and clay jars of oil,[612] "Load the lamps and their oil, also bring more barley and wheat, and water and wine,

---

[606]F. Page 83, line 64 I provided punting-poles and stored up a supply (since there was a pilot, it must be presumed the boat was to be navigable at a later date).

[607a]F. Page 83, line 80 [Whatever I had I] loaded aboard her; and M. page 91, line 29 Bringing whatever he had, he loaded onto the boat, and page 123, lower line 2 [All he] (had he) put inside.

[607b]F. Page 83, line 69 Two shar of oil (pitch) [which] the boatman stowed away (for stopping/plugging leaks).

[608]F. Page 83, line 76 [Finally] the ship was completed, and line 77 Difficult was [the task] (if the boat was built in just seven days, it would not have been a difficult task).

[609]F. Page 84, line 94 For the navigation of the ship to the boatman Puzur-Amurri,

and lash them in their place," and the workers bow, and quickly respond. One worker curiously remarks to another worker, as they bring more reed baskets of grain, and prepare to walk up the ramp, "Such a strange boat, it has no openings, except for the door," as they look at each other, with puzzled expressions upon their faces.

*

From space, we observe Nibiru approaching the Mars area, and passing between Mars and Jupiter, on a plane of orbit that is similar to the 17 degree angle of Pluto.

Then, also from space, we observe the planets Jupiter and Mars, and then we zoom in for a close up view from the air, close to the surface of each planet, and we notice slight tremors.

12 Hours Later

At pre-dawn, we see a silhouette of Shuruppak, then we scan and see the hexagon shaped boat in the river. There is no Moon, as we look up to the northern sky, where we observe Nibiru, which is clearly visible against the star-filled sky. Nibiru, with its two rings of clouds, within which orbit its four moons, appears as a radiating and glowing cross in the sky, bathed in a sunset red glow. We look down, and there is a light Earth tremor, which goes unnoticed, for it is very early in the morning, pre-dawn, and all the people are still soundly sleeping. Then we notice some of the animals, which are now beginning to sense the approach of Nibiru, as they nervously look around and up to the sky, but they are not yet frightened enough to run, and again we look up at Nibiru.

*

---

and line 95 I intrusted the mighty structure with its goods.

[610]F. Page 84, line 81 Whatever I had of silver I loaded aboard her; and M. page 93, line 30 Whatever he had of silver, he loaded.

[611]F. Page 84, line 82 Whatever I [had] of gold I loaded aboard her; and M. page 93, line 31 Whatever he had of gold, he loaded.

[612](Lamp oil, probably olive oil as in footnote 601, or possibly pitch related oil as in footnote 597, was used to fuel the lamps).

It is middle morning, on another bright and clear day, as we look up into the sky.

April 30th
7 Days Later

Then we look down slightly, and follow a group of birds as they enter our field of vision, and excitedly fly through the air, with Shuruppak in the near distance, to the east. Then, we scan toward the far west, the direction from which the birds are flying, and we see the swooping craft of God Enki in the air, glistening in the bright sunlight, with its brilliant glowing firefly-like light emitting from the underside.

Inside the craft, we observe God Enki, alone, looking down out a porthole at many running animals and the flying group of birds. In the background, the small viewing monitor shows a close up view of the scene below. Then from the ground, beneath the craft, we watch the craft swooping low to the ground, buzzing and herding the animals and birds, and then from the air, away from the craft, we watch the craft of God Enki buzzing and herding the frightened animals and birds, out of the fields and steppe to the far west, and toward an open area of Shuruppak near the boat, at the west edge of the Euphrates River. He is using a high frequency, computer sound source for the herding, which sounds like high-pitched whistling, and is almost inaudible to human ears, to frighten the animals and birds toward the river and the boat. Then, the craft of God Enki zooms away toward the boat, and we follow it, and watch it hover, descend and prepare to land near the river, in the open area near the boat, as the glowing light begins to diminish, and the 4 pods like cattle hoofs extend 6 feet down, and the craft begins to land.

We observe a smiling Ziusudra, appearing as he had previously, running toward the descending craft.

Then the craft lands, and the glowing light further diminishes, and then completely expires, and after a moment, a side of the craft extends down, forming a 4 feet wide ramp and doorway. Then, God Enki walks through the opaque air flow door and down the ramp of the craft. Again he is wearing the same red robe and adornments that he had previously worn, and he is without a helmet or body regulator.

At the same time, Ziusudra arrives to greet him, and he bows down, and prostrates himself, and then he stands,[613a] and they exchange greetings, and then warmly embrace; and as God Enki and Ziusudra embrace, in the background we notice some of the animals and birds arriving and gathering in the open area and in the orchards, near the river and the boat. Their frightened run and flight quickly changes to a calm approach, as they near the area.

After their long embrace, God Enki reveals to Ziusudra, "The flood shall come on the seventh day.[613b] For the definite hour, on the sixth night, watch for the signal by God Utu[614] from his city of Sippar, to the northwest, where all the Gods and craft are gathered; and when the rain begins, enter the boat and seal the door."[615] Then Ziusudra anxiously inquires, "What shall I do now God Enki, my personal God," and God Enki commands, "The time has arrived to go into the boat.[616] Be prepared, and notice the days. Take your water clock and fill it for seven days, on this, the tenth day of the second month. Load as many of the clean animals and birds, in pairs,[617] as many as shall fit," and they both turn to look at the animals and birds in the background, as God Enki cautions, "do not load the diseased animals." He then pauses, and looks at Ziusudra, and a very long moment passes, and then God Enki concedes, "I must now go to Sippar, and wait for the fateful hour." Then Ziusudra affirms, "I shall do as you command,[618] God Enki," and again God Enki and Ziusudra warmly and now tightly embrace, as their eyes fill with tears.

Then, after their warm and emotional embrace, which both realize may be their last, God Enki turns and quickly walks toward the ramp of his craft, as he struggles to restrain his emotional tears. We watch him go up the ramp

---

[613a]M. Page 123, upper line 3 (see footnote 335).

[613b]M. Page 91, line 37 He (God Enki) announced to him (Atra-ḫasīs/Ziusudra) the coming of the flood for the seventh night (night could be a scribal error or misnomer; as I further explain under 'The Beginning Of The Flood' on page 316 in my epilogue); and V. Genesis 7:4 For yet seven days, 7:10 And it came to pass after seven days, and 7:11 in the second month, the seventeenth day of the month (similar to these).

[614]F. Page 84, line 86 Shamash set for me a definite time (Shamash/Utu = brightness/light = sunrise/Sun God).

[615]F. Page 84, line 87 When the leader of the sto[rm] causes a destructive rain to rain down in the evening (pre-dawn), and line 88 Enter the ship and close thy door.

[616]V. Genesis 7:1 And the LORD (God Enki) said unto Nō´-ăh (Ziusudra), Come thou and all thy house into the ark (similar to this).

[617]V. Genesis 7:8 (similar to this).

[618]V. Genesis 7:5 And Nō´-ăh (Ziusudra) did according unto all that the LORD (God Enki) commanded him (similar to this).

and through the opaque air flow door, and after a moment, the ramp retracts to form a side of the craft, the underside begins glowing, the craft rises, the 4 pods like cattle hoofs retract, the glowing increases and becomes a brilliant glowing firefly-like light as it ascends, and the craft glistens in the bright sunlight, and we follow it until it quickly disappears into the distant bright blue sky.

Then, we slowly look down, and see the animals and birds arriving and gathering in the open area and in the orchards, to the northwest of and near the boat, and we observe Ziusudra watching them assemble, as his emotional tears streak down his face, and flow into his beard. (At this time, March 21st is New Year's Day, for it is the day of Vernal/Spring Equinox. Allowing for 30 days per month, the 1st day of the 2nd month is April 21st; therefore, the 10th day of the 2nd month is April 30th.)

*

Inside the boat, we scan and see the clay oil lamps, securely lashed to the vertical beams, providing soft illumination within the interior.

Several Hours Later

With workers walking by, who continue to load the boat with many various articles, we take a closer look throughout the boat, and we notice that the boat is entirely made of reeds, and none of the interior truss beams, floors, ceilings, or ramps are pitch coated, only the inner side of the hull, including the top deck ceiling and the bottom deck floor are pitch coated. The finished floor to ceiling beam height for all decks is 7 feet 4 inches throughout, at the side hull and at the center axis, with the top and bottom deck being 7 feet 4 inches at the side hull, and 9 feet at the center axis, due to the 1 foot 8 inch, 20 inch, or 1 cubit outward taper of the top and bottom from the center axis. Each deck is divided into 6 equilateral triangle shaped sections by the framework of the hexagon shaped boat, and each section is further divided into 9 trapezoid shaped sections by the vertical triangle truss bracing beams, with vertical and horizontal beams dividing each triangle truss, which extend from floor to ceiling, supporting each deck, and a staggered 5 feet wide by 6 inches thick connecting ramp between each deck, next to the center axis, or hub. The interior floor space of each of the 6 sections now looks like a forest of reed beams, due to the framework, with all of the 9 vertical divided triangle truss bracing sections built into the pie shaped sections being trapezoid shaped, now looking like open compartments. Being 6 stories, or decks in height, there are 36 equilateral triangles, 6 per deck, making up the hexagon boat, all radiating out from the 30 inch diameter center hub, ramp, and ventilation area [619]

Again we notice the workers walking by, carrying a variety of articles to their proper place, and with few exceptions, smaller personal possessions to the sixth, the uppermost, or main deck, and larger personal possessions and supplies to the fifth deck, and lashing them with long 1/2 inch wide leather strips, to the 6 inch diameter vertical, horizontal, and diagonal reed beams which make up the framework of the boat. We notice lashed large clay vases of drinking water and wine, most of which are sealed, and lashed large reed baskets of grain on the fifth deck, with a few being lashed within the main deck; also within the main deck, there are woven reed mats, placed among the smaller personal possessions, upon the reed bundle floors in all of the compartment-like areas designated for people to sleep in.

Then we observe Ziusudra, standing inside the boat, next to the nearly square doorway, and a nearly square reed bundle door, fully pitch coated, which has inner protruding reed fastening loops, as does the inner doorway, and also has an inward tapering perimeter, to which has been applied a 1 inch thick layer of thick pitch, and in the background through the doorway, we notice the ramp, now somewhat closer to the water. He is instructing some of his relatives and friends, of which some women still appear to be only 4 months pregnant, as they lash everything in place. All of his relatives have tight-curly black hair, except for older relatives whose hair and beards are turning gray, like Ziusudra's, and all have brown eyes, also like Ziusudra's; but some of his friends do not have these traits, as they have variations in hair and eye color. Ziusudra directs them, "My relatives, my friends, lash everything securely."

The wife of Ziusudra, and their grown children; their daughter and their three sons, now enter the boat through the doorway. Ziusudra embraces them and instructs, "My wife, my daughter, my three sons, see that our possessions

---

[619](See page 301 in my comments for detailed construction probabilities notes).

are all loaded and secure," and his children move into the background, and begin checking their possessions, while his wife checks the lashings nearby, on some of the vertical and diagonal truss beams.

Ziusudra then turns and walks to the water clock, which is securely lashed with 1/2 inch wide leather strips to a vertical reed beam near the doorway. He takes a clay vase of water, and fills the water clock to the 10th day,[620] and he sets the marker peg on the 17th day, and he places the spool marking the month, on the 2nd month, and we see a close up view of the water clock, as Ziusudra informs and instructs his wife, who is now watching him set the clock, "We shall be leaving in seven days. We shall now move into the boat, and prepare our places. Give notice to our daughter and our three sons, our relatives,[621] our friends, the pilot, the boatmen, and all the craftsmen,[622] to complete the loading of their supplies, possessions, and tools, and to move into the boat." Then his wife affirms, "I shall do as you command, my husband," and she turns and walks toward the others, to inform them to move into the boat.

*

We observe Ziusudra, near the 5 feet wide reed bundle ramp, now noticeably closer to the water, on the west bank of the clear and calm flowing Euphrates River, on another bright and clear day.

Several Days Later

---

[620]M. Page 91, line 36 He opened the water-clock and filled it.

[621]F. Page 84, line 84 After I had caused all my family and relations to go up into the ship; and V. Genesis 7:7 (similar to this. Incidentally, in another account of the flood, a Hellenic legend, which I believe refers to the same event and hero, the name of the hero is Deucalion, an epithet, an identification of succession and rank, which I interpret as meaning; deuca = 10 and lion = king, or the 10th King/Ruler, which Ziusudra/Deucalion certainly was. In this story, the name of the hero's wife is given as Pyrrha, whereas in the other Hellenic/Greek accounts and the Sumerian/Babylonian accounts, the name of the wife is not given. It could be that the name Pyrrha is also an epithet, and not the actual name of the individual).

[622]F. Page 84, line 85 I caused (etc.), all the (local) craftsmen to go (into it).

He is inspecting and selecting the animals to be loaded. The workers are herding[623] the animals[624] from the open area and the orchards, past Ziusudra and other workers who are inspecting and selecting them. There are sheep, goats, and oxen, plus many other varieties of animals common to that area, being paraded before Ziusudra, and we notice in the background, many varieties of birds[625] also common to that area, perched on the orchard trees, waiting for the workers to take them to Ziusudra. The animals that are selected are being led up the ramp[626] by other workers. Ziusudra is picking only the healthy and strong animals, and rejecting the sick and weak animals. He instructs the workers who assist him in selecting, as other workers walk up the ramp with large bundles of grass and large flat clay bowls of drinking water, "Load only clean,[627] healthy animals. The strong and fat[628] animals, not the sick and weak animals," and then he points to several healthy oxen in an approaching group, "herd these to the third and fourth lowest decks, and bring drinking water and a supply of grass for the animals to eat." The workers bow and obey as he commands, by separating the approaching group of oxen, and leading the selected healthy oxen up the ramp, as the other workers load more large drinking bowls and more large bundles of grass.

Now a variety of birds are being brought to Ziusudra, some carried on workers' arms, others in their hands, and Ziusudra instructs, "Load the doves, swallows, and ravens, and the other healthy birds into the first and second lowest decks, and bring drinking water and large baskets of seeds," and the workers bow and obey, as they bring the birds before Ziusudra for his

---

623M. Page 93, line 34 (clean and fat animals) He caught [and put on the boat].

624F. Page 84, line 85 I caused the game of the field, the beasts of the field, (and) all the (local) craftsmen to go (into it); and M. page 93, line 36 The cattle [of the field], and line 37 The wild [creatures of the steppe]; and F. page 84, line 83 Whatever I had of the seed of all living creatures (etc.: whatever animals he had available to him of all the animals on Earth; not all the animals on Earth); and V. Genesis 6:20 (cattle/cattle, similar to this).

625M. Page 93, line 35 The winged [birds of] the heavens.

626F. Page 84, line 83 Whatever I had of the seed of all living creatures [I loaded] aboard her; and M. page 93, line 38 [The creatures] he put on board, and line 39 Soon after the moon disappeared (which I interpret to mean soon after the end of the month. This corresponds to several days after April 30, probably within the first 4 days of May, for it would be wise to load the animals and birds on the latest possible date); and V. Genesis 7:9 (similar to this).

627M. Page 93, line 32 Clean (healthy animals) [he sought]; and V. Genesis 7:2 (clean/clean beast), 7:3 (clean/fowls also = clean), and 7:8 (clean/clean beasts and fowls, similar to these).

628M. Page 93, line 33 Fat (strong animals) [he sought].

inspection, as they did previously with the animals. Ziusudra picks and selects only the healthy birds, and rejects the sick and weak birds, as the other workers load large drinking bowls and large flat reed baskets of seeds.

Meanwhile, inside the boat, the animals and birds are being led down the reed bundle ramps, and into their compartment-like areas. The birds are put into the lowest decks, the first and second, because their light weight prevents them from damaging the hull, which is the first deck, and as soon as they are let go within the first deck, they perch upon the exposed horizontal floor beams, horizontal hull beams, and horizontal and diagonal truss reed beams that support the boat, with their large flat reed baskets of seeds, and their large flat clay bowls of drinking water, both neatly placed about them. The first deck hull floor, being the bottom of the boat, has been entirely coated with pitch, in addition to the hull sides, and it slightly tapers downward toward the center, all with exposed bracing beams.

The animals are put into their decks, the third and fourth, with their large grass bundles, and their large flat clay bowls of drinking water, both neatly placed about them. Only the hull sides of these decks are coated with pitch, and the floors are flat and level with reed bundles.

Then, again outside the boat, we observe the relatives and friends of Ziusudra, as they are gathering and going up the ramp, a total of 350 people, most of whom carry additional personal items, including small storage chests. As Ziusudra greets them, he calls to a group of playful teenagers, some with their pet cats and dogs, and directs, "My beautiful nieces and nephews. Go into the boat," and they bow and obey. We notice potters with their wheels; teachers with their clay tablets; metal workers with partially finished silver and gold pieces, and their embossing tools; leather workers with animal hides and their tools; and others skilled in special crafts, all carrying their supplies and tools, as they make their way to the ramp, where Ziusudra nods his head and exchanges words with them, welcoming them to the boat.

*

We see the water clock in the boat, securely lashed to the vertical reed beam near the doorway. It indicates the 16th day, and the spool is on the 2nd month, with only one more day until it reaches the marker peg, which is set on the 17th day.

May 6th

We look at the doorway, which faces toward the west, and we notice the bright indirect early morning sunlight, as it enters the boat, and illuminates the interior area beyond the doorway.

It is a bright and clear day, middle morning, on the same day, and we observe Ziusudra standing on the west bank of the clear and calm flowing Euphrates River, next to the 5 feet wide ramp leading to the shiny black pitch coated hexagon shaped reed boat. Being fully loaded, the boat now rides only 10 feet above the water line, at the hull, and 40 feet below the water line, with the bottom of the doorway being about 2 feet above the water line. The common and familiar large and small, rectangular shaped, pitch coated reed boats are nearby in the river. Ziusudra is supervising the few remaining personal items that are now being loaded, as well as large reed baskets with fresh fruit, additional sealed large clay vases of fresh drinking water, and additional bundles of grass for the animals.

The pilot is approaching Ziusudra at the ramp, followed by two workers who carry a few of his possessions; included among them are navigational instruments. The pilot holds a sextant-like device in his hand, and Ziusudra welcomes and questions him, "A warm greeting to my pilot this morning. I see you have your instruments with you. Do you have all the instruments for a safe trip? Be sure to bring all of them," and the pilot consolingly responds, "You need not worry, my King. The trip to the Abzu, to dwell with God Enki, shall be a pleasant one," and he walks up the ramp, followed by the two workers.

Then Ziusudra begins gathering the workers, as he commands, "All of you, all the workers who have labored building and loading the boat, gather near me," and all the workers that are in the area quickly gather before Ziusudra, and he announces, "Tell all the other workers, whoever assisted in this task. Tonight we shall have a farewell banquet,[629] to celebrate our departure tomorrow. Make the banquet preparations, and bring your families and your friends. For with our departure, for you a new prosperity shall begin," and the workers spontaneously beam and cheer their approval. Ziusudra forces a half-smile, for he notices a light Earth tremor as he concludes the announcement,

---

[629]M. Page 93, line 40 Together he invited his people, and line 41 Treating them to a banquet.

and he looks up to the bright blue sky. The light Earth tremor is noticed by the workers, but again not taken seriously, as they give it only a second of their time, for they are preoccupied with the farewell banquet announcement, and the thought of greater prosperity, and they continue to cheer their approval.

*

It is dusk, the Sun is setting, and we look across the clear sky, to the opposite horizon, but we do not view the Moon, as the Moon is not yet above the eastern horizon.

6:00 p.m.

We look down and notice clay oil lamps, with steady flames in the calm air, hanging from tree branches by 1/2 inch wide leather strips. As we look down further, we view the farewell banquet scene on the west bank of the river, with the orchard trees to the northwest; between the open area near the river and the orchards, among scattered trees and a few large bushes. There is soft music floating in the air, being played on tambourines, harps, drums, and flutes. We scan and observe musicians playing their instruments, and many lambs and bullocks being turned on wood skewers while roasting over a long pit of open fires, with reed bundles as fuel. There are 100 rectangular wood tables and 200 rectangular wood benches, one gathered close to each side of the tables, and 100 woven reed mats spread upon the ground, around the tables. Upon the tables and mats are set pale red and cream decorated clay cups, vases, and dishes. The banquet meal has not yet begun, as the animals are not quite thoroughly roasted.

The people are still gathering, wearing their special events clothes: colorful full length robes with edging, only draped over one shoulder; with Priests, members of the council, and special individuals robes covering both shoulders. Some people wear their flawless wire earring with its stone pendants; and some people wear their intricately detailed wood or silver bracelet; and some men wear their impeccable white turban-like cloth hat; and some men wear their colorful cloth head band. Some women wear their faultless, wide colorful cloth head band ribbon in their hair; and some women wear their large crescent shaped earrings. Most of the adults, and children 10 years of age in appearance or older, 80%, wear their clay, wood, gold, or silver symbol necklace of the four main Gods, and some wear their personalized pictograph cylinder seal, as a pendant around their neck. All adults wear their leather wedge-like sandals with straps. The people are standing throughout the banquet area, as they all drink beer, and red or white wine from the clay cups and directly from the clay vases. It is a very joyous event, and the people are all eager to celebrate, as they laugh and drink with each other, and after a few short moments, several groups of 6 to 8 younger people hold hands, forming small lines, and begin to circle dance, taking three bouncing steps to the right and then one bouncing step to the left, as the merrymaking begins, with the leaders of each line of dancers either waving a handkerchief from a raised hand, or snapping their fingers as they dance.

Then we see the hexagon shaped boat, with the relatives and friends of Ziusudra walking down the ramp from the boat, also with their colorful special events robes, symbol necklaces, pendant cylinder seals, head and ear wear, and bracelets; joining the festivities, as friends embrace each other, and bid them a safe trip to the Abzu, and a joyous and unsuspecting farewell.

Now we observe a worker, turning a lamb over the fire, as he declares to another worker, who is watching him while drinking red wine directly from a vase, "This shall be a fitting farewell to Ziusudra, our Priest-King," as he turns the lamb, and the other worker attests, "And a fitting welcome to our new prosperity and wealth," as he drinks more wine from the vase, now holding the vase high, and pouring the red wine into his wide open mouth. Then both workers notice several light Earth tremors, but they, as most of the merrymakers, do not pay attention, and give only a passing thought to them, as they, like all the other people, are now totally preoccupied with the festivities.

From the river bank, again we see the boat, tied to the west bank of the calm flowing Euphrates River. We look across the water, scanning from the boat to the river bank, and we notice a barely discernable mist beginning to develop and blanket the water, barely rising above it, and the level of the water, which is beginning to steadily increase. At this time, the water level is only several inches higher than normal at Shuruppak.

Then from space, we observe the planet Nibiru, approaching closer to its perihelion, its closest distance to the Sun, and simultaneously closer to its perigee, its closest distance to the Earth, in its normal orbit, and in all its glory, with its two rings of clouds, within which orbit its four moons, appearing as a radiating and glowing cross, bathed in a sunset red glow, and with star-filled space surrounding the planet.

Then we see the Earth from space, and we zoom in for a close up view, and from the air, amidst increasing ground and air temperatures, we view the mountainous areas far to the east and north of Shuruppak, which are beginning to be shaken by the increasing frequency and intensity of the Earth tremors, as they increase from very few light Earth tremors, to moderate Earth tremors, which now cause fissures and volcanic eruptions to occur, with huge amounts of molten lava, dense clouds of smoke and ash, and extreme amounts of steam spewing forth, as the Earth begins to groan.[630]

From the ground, in the wilderness, as temperatures of the ground and air increase, and as the ground begins to fissure, we notice the underground waters, including steam, beginning to come up to the surface, as the very few light Earth tremors increase, and become moderate, and a mist begins to develop, with a rainbow forming in the sunlit distance, and it appears rain is imminent. The animals are becoming more and more frightened by the increased frequency and intensity of the Earth tremors, and they begin to run in all directions, as the vibrations of all the distant tremors are fully detected, transmitted through the ground to their feet.

From the ground, in the desert, as temperatures of the ground and air increase, and as the ground begins to fissure, we notice the underground waters, including steam, beginning to come up to the surface, bubbling up through the sand, and forming small pools of water, as the very few light Earth tremors increase, and become moderate, and a mist begins to develop, with a rainbow forming in the sunlit distance, and it appears rain is imminent.

From sea level, in the far northern hemisphere, where the bulging effect of the increase in water is much more pronounced, amidst a developing fog, we notice the level of the water at the seashore, which is beginning to steadily increase, as it begins to cover small boulders in the water.[631]

From the air, close to the Earth, we see the Earth and successively both

---

[630]X. Upper page 78 The earth, forced out of its regular motion, reacted to the close approach of the body of the comet (Nibiru): a major shock convulsed the lithosphere, and the area of the earthquake was the entire globe, and page 110 through upper page 114 (quoting Pliny, circa 23 A.D. - 79 A.D.) "earthquakes are preceded or accompanied by a terrible sound" (theophany/cacophony, similar to these, but at an earlier time); and Z. page 137 through page 154, and especially lower page 145 sudden changes in the earth's spin could trigger earthquakes.

[631]X. Page 85 through Page 90 The ocean tides are produced by the action of the sun and to a larger extent by that of the moon. A body larger than the moon or one nearer to the earth would act with greater effect.

polar ice caps, with their surrounding waters beginning to bulge, as Nibiru begins to pass closer to the nearest area of the Earth, the north pole area, creating a primary pulling up of the waters toward the north, a bulge of waters, and a secondary pulling of the Earth away from the waters of the south and a concurrent floating away from the Earth of the waters of the south, also a bulge of waters, like the tidal effect upon the Earth caused by the Moon,[632] except far more severe, with the north pole area pulling effect much more intense than the south pole area pulling effect.

From over the water at the equator, more than 2,000 miles south of Shuruppak, where the Earth and its waters are normally 13 miles higher than at the poles, due to the centrifugal force of the Earth, we observe the gravitational and electromagnetic effects of Nibiru, which have just overcome and neutralized the forces of the Earth; as they now begin to transcend the forces of the Earth, and quickly pile the equatorial bulge waters, and the waters of the southern hemisphere, nearly to the Tropic of Capricorn, into 20[633] huge sloping, rolling purple waves,[634] each 300 feet high,[635] which are being forced to recede downward from the 13 mile high bulge of the Earth at the equator, and primarily receding and simultaneously being pulled toward the north, toward Nibiru,[636] with a secondary receding of the waters toward the south. The series of 20 huge sloping, rolling waves, each being 300 feet high, are followed by 21 rippling, sweeping waves, each being 100 feet high: all increasing in speed, spaced at short intervals, and quickly moving toward the northern hemisphere. They contain approximately an 8,100 feet cumulative

---

[632]Z. Page 29 through page 43, and especially illustration on page 34.

[633]M. Page 95, line 12 Its might came upon the peoples [like a battle array], and page 125, line 19; and F. page 85, line 110 ('like a battle array' implies a rank, formation, rows, or series of waves, as in a military formation); and X. middle page 101 the channels of water were seen, and middle page 332 (waves, a) recession of the waters from the equatorial bulge (moved poleward, similar to this).

[634]M. Page 97, line 5 What? Have they given birth to (brought forth) the [rolling] sea? (They could refer to the assembly, but could be interpreted as; What? Were they given birth (brought forth/created) by the rolling sea? In this context, they would be a reference to the waves); and X. lower page 177 "purple waves" (similar to this).

[635]Z. Page 155 through page 169 (similar to this); and V. Psalm 33:7 He (Lord/Nibiru) gathereth the waters of the sea together as an heap: he layeth up the depth in storehouses (similar to this); and W. upper page 244, Volume I waves like mountains (similar to this); and X. lower page 177 "the sea [that] rises up like a wall" (similar to this).

[636]X. Lower page 85 but the celestial body nearby would disturb this poleward recession, drawing the water toward itself.

depth of water.[637] The first rolling wave has a severe destructive and crashing crest, with the other 19 rolling waves having more subtle and whooshing crests, which are not as severe, and the following 21, 100 feet high rippling waves, are just sweeping waves, all without crests. We see the huge purple waves, just as they begin to roll and gain speed. We view this scene from the air, and from the sea level, and from the crest of the first wave level, and then we see the first 20 huge sloping, rolling lead waves, each 300 feet high, sweeping across the open sea, racing toward the north, with the 21 smaller 100 feet high rippling, sweeping waves following. The 20 huge sloping, rolling lead waves make a hissing, whistling-like sound, with the lead wave crest also making a crashing sound, and the other 19 crests making a whooshing sound.

The air[638] is also being affected, and forms into waves, like the water, and wildly undulating extremely warm masses of air, laden with extreme amounts of water vapor, which additionally contributes to the adverse conditions, have quickly formed into very dark clouds with violent storms,[639] containing very dense darkness, and fierce lightning and thunder, and violently howling wind driven torrents of heavy rain, with a somewhat less intense raging storm preceding them, as they all precede the waves across the sea,[640] and are causing drastic temperature changes; both now traveling 200 miles per hour[641] across the open sea, leaving the equatorial area, and the southern hemisphere, the region of the weaker pulling effect of Nibiru, and entering the northern hemisphere, the region of the strongest pulling effect of Nibiru. Now, in addition to the waves being pulled north by the stronger gravitational and electromagnetic effects of Nibiru upon the waters of the northern hemisphere,

---

[637]X. Page 85, note 1 (quoting Lalande, circa 1795 A.D.) who computed that a comet (or a celestial body/Nibiru) with a head as large as the earth, at a distance of 13,290 lieues, or about four diameters of the earth, would raise ocean tides 2,000 toises, or about four kilometers high (2.486 miles or 13,126 feet, similar to this).

[638]Z. Page 115 through page 123 Just as tides cause currents in the oceans, they induce currents in the atmosphere, in the form of winds.

[639]X. Page 82 through middle page 83 (the effects of a close passing/orbit of a celestial body/Nibiru) all contributed to produce hurricanes of enormous velocity and force and world-wide dimensions.

[640]M. Page 97, line 25 Came the deluge (rain), the storm (winds), [the flood] (waves); and F. page 85, line 128 The wind blew (winds), the downpour (rain), the tempest (strong winds), (and) the flo[od] (waves) overwhelmed the land (the three components of destruction: rain, winds, and waves); and V. Genesis 7:11 the same day were all the fountains of the great deep broken up, and the windows of heaven were opened (similar to this).

[641]Z. Middle page 158 waves that race through the open sea at speeds of 500 or 600 miles per hour (similar to this).

the waves are being propelled by their own momentum, thus extremely compounding the bulging effect of the waters in the north regions, the regions nearest to Nibiru during the period of passing. (Thus, the most recent major ice age is about to come to an abrupt end.)

*

As we look across the sky at the farewell banquet in Shuruppak, now in full progress, we hear the music, louder now, which combines with the din of the people, which has also increased. It is dark now, and the sky is clear and filled with stars. The Moon has not yet appeared above the horizon.

6:30 p.m.

We look down and scan from the west, and view the festivities near the river's edge, with the enormous shiny black pitch coated hexagon shaped reed boat, and the common and familiar large and small, rectangular shaped, pitch coated reed boats, all silhouetted in the background, against the eastern sky, which is slightly illuminated by the yet to appear Moon. Then we notice the clay oil lamps, hanging from the tree branches by their leather strips, with their steady flames in the calm air, which appear to have increased their illumination, due to the absence of both the Sun and Moon.

We observe the workers, approximately 700 people, the people who labored and built the boat, now with their families and friends, approximately 1,650 people, the people of Shuruppak, merrymaking at the banquet, and we also observe the 350 relatives and friends of Ziusudra, who have come off the boat to enjoy the feast. All of the people are enjoying themselves, both younger and older, a total of approximately 2,000, most of which are now seated 10 per table upon the benches, and on the mats, 10 per mat, and are now eating roast lamb and bullock, vegetables and fruits, and drinking more beer and wine;[642] but some people prefer to circle dance instead of eat, especially the younger people, with the leaders of some lines now bobbing up and down and hitting the ground with the palm of their hand, then raising their foot behind

---

[642]M. Page 93, line 42 Before he sent his family on the boat, and line 43 They ate and they drank.

themselves and hitting the side of the heel of their foot with their fingers, or raising their foot in front of themselves and backhanding their toes, and then spinning with raised arms and snapping fingers, all accompanied with exuberant shouts of "OPA" . . . . "OPA" . . . . "OPA", and "YES". . . ."YES" . . . . "YES", and "HEP" . . . . "HEP" . . . . "HEP".

From a distance, we observe Ziusudra, clothed in his special full length purple robe with gold edging, which covers both shoulders, and wearing his gold God Enki symbol necklace, and his gold winged sphere symbol of Nibiru on his robe, above the left breast, and his gold bracelet. Ziusudra is not eating or drinking as he walks past the people, offering farewell to them, and thanking them for their assistance in the construction of the boat.

Ziusudra stops at a table of people, and conscientiously submits, "Thank you for your labor in building the boat. Your Priest-King offers a grateful farewell to you, my subjects. I shall miss you deeply." One worker at the table invitingly replies, "Come join us, our King, that we may enjoy a longer farewell," as he lifts a cup of wine, and Ziusudra responds, "No, thank you for your kindness. I have no appetite for food or drink," and then he continues to walk. As Ziusudra walks away, another worker at the same table shouts, "Farewell Ziusudra, our Priest-King," and he lifts a cup of wine and drinks. Then, another worker at the same table remarks to his friend sitting across the table, as Ziusudra walks further away, "Look how saddened he is because he must leave our city, Shuruppak, and go dwell with God Enki, in the Abzu," and he picks up a piece of roast lamb which is on his dish, and eats it by holding onto the bone. The friend then comments, "Leaving us has caused Ziusudra, our Priest-King, to lose his appetite," and he also eats some of the roast lamb, in the same manner as his friend. Again we observe Ziusudra, as he makes his way through the banquet area. He nods his head and alternately waves his hands, as he bids farewell and thank you to the people of Shuruppak.

*

We view a 3/4 Moon in the lower eastern sky, as it begins to rise, surrounded by stars in the clear sky, as we hear the music and din from the banquet.

8:00 p.m.

We look down and scan, and circle dancing has overcome nearly everyone, as almost all of the people are participating. There are now many larger groups of people forming longer lines of circle dancers, and in addition to the bobbing up and down, hitting heels or toes, spinning, raised arms and snapping fingers, and exuberant shouts, males have taken to the center of each circle, facing each other, trying to outdo their competitors' dancing skills and techniques, all as the merrymaking totally engulfs the people. There is loud music, and the din of the people is almost equally as loud, now that most of the people have finished the meal, and freely drink, but a few people continue to nibble at the food. The landscape is softly lit by the 3/4 Moon, low in the eastern sky, which has risen above the horizon on its journey through the sky. The clay oil lamps hang from the trees by their leather strips, and their flames seem to be dancing to the music, as the winds begin to stir, and become very light, and now slightly blow at the flames.

Ziusudra is walking past another table, as he nods and waves farewell and thank you, and then walks past them. He feels a light-moderate Earth tremor, and he stops and looks around. Most of the people do not give it a second thought, as the beer and wine are now relaxing their attentive interest, and Ziusudra quietly and anxiously reiterates to himself, so that others do not hear, "The hour approaches," and then he gazes at the 3/4 Moon rising in the east, using it as a time reference, and then he looks away, toward the darker portion of the sky, to the northwest, toward Sippar. Again, he quietly worries to himself, "I do not yet see the signal of God Utu," as he turns and walks toward another group, resting at the tables from the circle dancing, as they drink more beer and wine.

As Ziusudra approaches the group, he notices his pilot, who is freely drinking and still nibbling at the roast lamb, and Ziusudra promptly walks to him and exclaims, "At last, my pilot, our farewell banquet. The food must have been very good," as the pilot takes a long, very long, drink of wine directly from a vase, and then the pilot remarks, "It was excellent. This is my second portion," and Ziusudra adds, "Enjoy yourself. Eat well. Eat until you can eat no more," and Ziusudra turns away and quietly cautions to himself, "for you may not eat for a few days." Then Ziusudra turns back toward the pilot, leans over his shoulder and commands, "You can take the remainder of your food into the boat. You can finish your meal there, but leave the wine to the others," and the pilot affirms, "I shall do as you command, my King," and he immediately sets the wine vase down on the table, adds more food to his dish, rises and bows to Ziusudra, and takes his dish with him, as he walks toward the boat ramp, and goes up the ramp and into the boat.

*

We view the rising 3/4 Moon, now 15 degrees above the horizon, surrounded by stars in the clear sky, and we hear the music and din from the banquet.

9:00 p.m.

We look down and observe Ziusudra, looking up at the Moon, using it as a time reference. Then, he looks to the northwestern sky, and sees the clearly visible stars. Ziusudra walks to a mat, with seven teenagers seated upon it who are entertaining themselves with hand games, and commands, "My nieces and nephews, the evening is late for you. It is time for you to sleep. Gather your cousins and go into the boat and sleep, for we shall have a long journey tomorrow." The teenagers eagerly affirm, "Yes, we shall Uncle Ziusudra, our King," and they immediately rise, bow, and run off, and gather their cousins, and we watch as twenty-five teenagers quickly assemble, and then run up the ramp and into the boat. Then, we scan from the boat, across the now nearly calm flowing water, to the river bank, and we notice the mist, now a very slight fog, which hugs the water, and the water steadily inching higher.

Ziusudra, who had been watching the teenagers run up the ramp, turns back toward the banquet, and suddenly stops, as he feels several moderate Earth tremors. He looks around, but the people are having a magnificently good time drinking and circle dancing, and they do not pay attention to the tremors, as they do not want to be distracted from their merrymaking. Then, as Ziusudra continues to walk among the people, we notice the roasting fires, as they burn lower.

Ziusudra now promptly walks to his wife, who is at a table with their grown children, their daughter and their three sons, and their children's friends, who are drinking as they rest from the circle dancing, and Ziusudra declares and then inquires of his wife, "Such a joyous occasion for my wife, my daughter, and my three sons. Did you eat and drink well, my family?" His wife responds, "We did Ziusudra. Come join us," and Ziusudra replies, "No, I must offer my farewell to all the people." Then, he leans down over his wife's shoulder, and privately, close to her ear, commands, "We must make ready to embark tomorrow, to our new home in the Abzu. The hour is late. Take our family and relatives into the boat, now, and make ready for tomorrow," and his wife affirms, "We shall do as you wish, my husband." Then Ziusudra continues privately to his wife, "Also instruct the craftsmen, the learned men, and our friends, everyone that departs with us, to go into the boat," and Ziusudra slowly walks away, as his wife nods her head approvingly, and immediately begins informing their grown children, and the others. We observe them, as they somewhat reluctantly accept the abrupt merrymaking interrupting and terminating command, but they go, some with food and drink, up the ramp and into the boat.[643]

*

We view the 3/4 Moon, now 30 degrees above the horizon, surrounded by stars in the clear sky, and we hear the music and din from the banquet.

10:00 p.m.

We look down and again observe Ziusudra, looking up at the Moon, which brightly illuminates his face and the surrounding area, as the circle dancing and merrymaking continue. Ziusudra has a concerned look upon his face, as he walks about bidding farewell to all the people. As he walks, he looks with glancing eyes to the northwest sky, at regular intervals, then he looks at the Moon, and then slowly turns his eyes to the northwestern sky, and again sees only stars. He walks through the banquet area impatiently, and then up the boat ramp, and into the boat, to look in on his family, relatives, and friends.

[643]V. Genesis 7:13 In the selfsame day entered Nō´-ăh (Ziusudra and his family, etc.), into the ark (similar to this).

Then he walks down the ramp, and after noticing the very slight fog, which continues to hug the water, and briefly staring at the water level at the river bank, which is steadily inching higher, he becomes more impatient, and then he paces back and forth: wherever he goes, he paces back and forth. He does not sit, for he is too upset, and he is filled with nervous anxiety.[644]

*

We view the 3/4 Moon, now 60 degrees above the horizon, surrounded by stars in the clear sky, and we hear softer music and a more subdued din from the banquet.

12:00 Midnight

We look down and scan, and view the banquet area, which is now illuminated by the bright Moon. The merrymaking continues, but at a slower pace, as some of the older people are tired, and they sit and watch the younger adults and teenagers circle dance. Hanging from the tree branches by their leather strips, the clay oil lamp flames are now dancing more than some of the people, as the winds increase from very light, and become light, and now additionally blow at the flames. As the people watch, they sit on the wood benches at the wood tables, and now slowly drink smaller amounts of beer and wine.

Again we observe Ziusudra as he paces, offering farewell and waving good-bye to all the people he meets. He continues to keep an eye on the northwest sky, as he apprehensively glances, and now stares for a minute at a time, even over the people's heads as he bids them farewell. As he looks up, he also notices the position of the Moon, as a time reference, and then his eyes slowly turn to the northwestern sky, but he still only sees the stars. Suddenly, Ziusudra feels another Earth tremor, as do the people, much stronger this time, a strong-moderate Earth tremor, and it lasts for a longer duration than the preceding tremors. As the musicians feel the tremor, their music becomes

---

[644]M. Page 93, line 45 But he was in and out: he could not sit, could not crouch.

irregular, and almost stops, as do the dancers, as they all take notice. But like before, as the tremor ends, the merrymaking resumes and the tremor is again disregarded, as the effects of the beer and wine consumption prevail. Then we observe the face of Ziusudra, relieved that the people had not become alarmed by the tremor, while he watches the circle dancing, as the merrymaking continues.

Then, from the air, sea, and first wave crest level, we see the 20 huge 300 feet high, rolling lead waves, followed by the 21 smaller 100 feet high, sweeping waves, and we hear the hissing and whistling sound of the 20 lead waves, and the crashing sound of the lead wave, and the whooshing sound of the other 19 lead waves, and see the very dense darkness and fierce lightning, and hear the thunder, and see and hear the violently howling wind driven torrents of heavy rain, in the very dark clouds with violent storms, with a raging storm preceding them,[645] as they sweep across the sea, racing toward the north at 200 miles per hour, carried by the stronger pulling effect of Nibiru upon the waters of the northern hemisphere, and propelled by their own momentum.

Then, from the air, land, and first wave crest level, we see the lead wave has now reached the Arabian peninsula, 900 miles south of Shuruppak, and 4.5 hours away, and quickly begins to cover the trembling and fissuring land,[646] as the waves have already done with the lower half of Africa. The lead wave appears as a sloping wall of purple water, and all the sounds are very loud, especially of the water crashing down upon the wave, beneath the crest of the lead wave.

*

We view the 3/4 Moon, now 75 degrees above the horizon, surrounded by stars in the clear sky, and we hear soft music and a very subdued din from the banquet.

---

[645]M. Page 123, lower line 4 The wind blew and brought the [storm], lower line 5 Adad (rain/storm God: see footnote 379) rode on (utilized) the four winds, [his] asses, and lower line 6 The south wind, the north wind, the east wind, the west wind, and page 125, line 7 The storm, the gale, the tempest blew for him, 8 The Evil Wind with the winds arose, 9 The south wind strongly arose at his side, and line 10 The west wind blew at his side (all blew together as one: see footnote 495).

[646]M. Page 125, line 11 They reached the land together.

1:00 a.m.

We look down and scan, and again view the banquet area, which is still illuminated by the bright Moon, where a few of the remaining youthful merrymakers and drinkers are trying to extend the evening. The music is now soft, and floats over the area, and the din has quieted to a very low level. The banquet is almost over, with the few remaining young people refusing to face the fact. All the older people have left, or are in the process of leaving, and everyone is tired, and filled with food, drink, and music. All that remains of the roasting fires are hot embers of reeds, as the flames have also departed. The clay oil lamp flames are the only banquet survivors, appearing as they had when the banquet started, hanging from the tree branches by their leather strips, with dancing flames, which are still blowing in the continuing light winds. Then, we again scan to the boat, and then across the nearly calm flowing water, to the river bank, and we notice the very slight fog, now being blown away as quickly as it forms, and the water steadily inching higher.

Then we observe Ziusudra, now with a very serious look upon his face, anxiously walking past the remaining merrymakers and drinkers, as he constantly paces. He is intensely observing their warm, youthful faces, as he continues to walk closer to them. They have abandoned the tables and benches, for the more comfortable woven reed mats. Some are lying on the mats, and are freely caressing. Ziusudra stops and watches them for a moment, and we see a close up view of the young people caressing, as they lay on the mats. The warmth of the young people, and their impending fate, affect and upset Ziusudra very deeply, and he walks a short distance from them, and stands next to a large bush, and then he hurriedly steps around to the other side of the large bush, and tries to vomit, and does, but only in a dry-heave manner. We view the bush, and hear the dry-heave sounds of Ziusudra, coming from the other side. Ziusudra quickly recovers, and steps from behind the bush, and softly confesses to himself, "My heart is broken. I am vomiting gall,"[647] as he resumes walking among the young people, troubled, and almost constantly glancing, intensely looking to the northwest sky. He also notices the Moon, as a time reference, and then his eyes slowly turn to the northwestern sky, where he still only sees the stars.

---

[647]M. Page 93, line 47 For his heart was broken and he was vomiting gall (dry-heave).

Meanwhile, inside the boat, we see a close up view of the water clock, securely lashed to the vertical reed beam near the doorway. Having almost reached the marker peg, the clock nearly indicates the 17th day, and the spool is on the 2nd month. As we scan, we notice the clay oil lamps, which are securely lashed to the vertical reed beams, softly illuminating the interior, and the securely lashed large clay vases of drinking water and wine, most of which are sealed, and the securely lashed large reed baskets of grain. Then we observe the family, relatives, and friends of Ziusudra, who are relaxing in their compartment-like areas of the sixth deck, the uppermost, or main deck, preparing their makeshift beds upon the woven reed mats, placed among their smaller personal possessions. Many of the teenage children are already sleeping, and the balance are preparing their beds. The adults quietly converse among themselves, as they reflect upon their enjoyable evening at the banquet, and the people who are staying in Shuruppak. Very faint music and din can be heard within the boat, from the musicians and few remaining young people at the banquet. Many of the clay oil lamps are now being blown out, and the remaining clay oil lamps are somewhat dimly illuminating their compartment like areas, as the people on the boat go to bed, wearing their colorful robes, and symbol necklaces, after removing their head and ear wear, pendant cylinder seals, and bracelets, which they carefully put away among their other small personal possessions.

Then, we see the animals in their compartment-like areas, the sheep, goats, oxen, and others, including their pet cats and dogs, sensing danger, and becoming restless and frightened, as they very slightly detect the water cushioned vibrations of distant severe Earth tremors. The animals are now awakening, some standing up from their lying down sleep position, while others awaken from their sleep position as they stand, with their large grass bundles of food, and their large flat clay bowls of drinking water, both neatly placed about them.

Then, we see the birds in their compartment-like areas, sitting upon the horizontal and diagonal reed beams, as if they were a perch. The birds begin to slowly flutter their wings, as they are also aroused from their sleep, with their large flat reed baskets of seeds, and their large flat clay bowls of drinking water, both neatly placed about them. It is a very quiet, nervous and anxious scene among the animals and birds, and the only sounds that can be heard are the very, very faint music and din from the banquet, for on the lower decks of the boat, the intensity level of all outside sounds is reduced, compared to the main deck. The securely lashed clay oil lamps which are lit, are somewhat dimly illuminating their areas.

*

We view the 3/4 Moon, now 90 degrees above the horizon, surrounded by stars in the clear sky, and we hear very soft music and a very, very subdued din from the banquet.

2:00 a.m.

We look down to Ziusudra, who is standing unnoticed on the ramp, near the doorway of the boat, noticing the water level at the river bank, which continues to steadily inch higher. He is bathed in bright moonlight, as he now scans over the banquet area, as the merrymaking is finally ending. Then the musicians stop playing, and the only sounds that can be heard are low levels of conversations, mostly good nights. There are a few drunken young circle dancers, who continue after the music stops, but they quickly realize the music has ended, and they dishearteneḋly stop dancing. Most of the young people have dispersed, and only a few remain. Young couples bid other young couples good night, and the musicians offer good night to each other, as they depart for their homes. All the young couples walk arm in arm, caressing, and some are giggling, as they walk toward their homes, located to the southwest of the banquet area. A few are too drunk to move, and they sleep on the woven reed mats. The hot reed embers left from the roasting fires have now burned out. The banquet area is now empty of merrymakers, and is now very quiet. Only the empty wood tables and benches, empty clay cups, vases, and dishes upon the tables, and a few sleeping young people upon the woven reed mats remain, and bear witness to the existence of the farewell banquet. Only the clay oil lamps show signs of life, as all otherwise is peaceful, as the light winds increase, and become mild, and now exceedingly blow at the flames, causing them to flicker, as the lamps, hanging from the tree branches by their leather strips, begin to sway.

We observe Ziusudra, anxiously looking at the 3/4 Moon, which has now reached the zenith of its journey through the sky, and then his eyes impatiently turn to the northwestern sky, waiting for the signal from Sippar, but again, he only sees the stars in the clear sky. His earlier glancing looks have now become long stares, and his face is filled with fear and anticipation. We zoom past Ziusudra, and into the shiny black pitch coated reed boat, where many of the clay oil lamps are out. It is quiet, and somewhat dimly illuminated, with all the people sleeping in their compartment-like areas. Then, after a brief moment, we zoom back outside, and again we observe Ziusudra on the ramp, near the doorway, anxiously watching the northwestern sky for the signal of God Utu. He notices the 3/4 Moon, which suddenly disappears behind scattered clouds quickly approaching from the south, and it instantly becomes very dark. Again he notices the Earth quiver, another strong-moderate Earth tremor, which has a very slight effect on the boat, because of the cushioning effect of the water, causing only a very slight sway, and very small ripples on the now slightly choppy river water around the boat.

Meanwhile, in the city, there is no reaction to the strong-moderate Earth tremor from the people in their homes. They are all sleeping, wearing their colorful robes, and symbol necklaces, after removing their head and ear wear, pendant cylinder seals, and bracelets; after becoming exhausted from the merrymaking at the farewell banquet earlier. The city streets are empty and quiet, as the mild winds blow through them, but the animals, the dogs, sheep, goats, oxen, and others, are frightened and tremble, and many begin to run aimlessly, as they now fully detect all of the distant severe Earth tremor vibrations, transmitted through the ground to their feet.

Then from space, we observe the planet Nibiru, approaching in all its glory, with its two rings of clouds, within which orbit its four moons, appearing as a radiating and glowing cross, bathed in a sunset red glow, and with star-filled space surrounding the planet.

Then we see the Earth from space, and we zoom in for a close up view, and from the air, water, land, and first wave crest level, we see the 20 huge 300 feet high, rolling lead waves, followed by the 21 smaller 100 feet high, sweeping waves, and we hear the hissing and whistling sound of the 20 lead waves, and the crashing sound of the lead wave, and the whooshing sound of the other 19 lead waves, and see the very dense darkness and fierce lightning, and hear the thunder, and see and hear the violently howling wind driven torrents of heavy rain, in the very dark clouds with violent storms, with a raging storm preceding them, as they sweep across the trembling and fissuring land,[648] racing toward the north.

Then, from the air, land, and first wave crest level, we see the lead wave has now reached the central lands of the Arabian peninsula, 500 miles south of Shuruppak, and 2.5 hours away, and quickly covers the land, as the waves have already done with the lower half of the Arabian peninsula. The lead wave appears as a sloping wall of purple water, and all the sounds are very loud, especially of the water crashing down upon the wave, beneath the crest of the lead wave.

*

We view a slightly illuminated cloud perimeter, among the scattered clouds that nearly fill the sky, concealing the 3/4 Moon, now 120 degrees above the eastern horizon, as it begins lowering itself in the western sky, with the clear sky and stars as a background between the scattered clouds. It is very quiet.

4:00 a.m.

We look down to Ziusudra, who is still standing on the ramp, near the doorway, amidst the mild wind, searching the northwestern sky for the signal from Sippar. It is very dark with the Moon behind the scattered clouds, which have now reached him, and are overhead. Then he slowly scans across the sky, toward the north, and between the scattered clouds, he notices Nibiru, now clearly visible, with its two rings of clouds, within which orbit its four moons, appearing as a radiating and glowing cross in the sky, bathed in a sunset red glow. After a long and thoughtful gaze, he slowly looks down from the sky to the banquet area, and again scans over it. Suddenly, Ziusudra sees a faint glow out of the corner of his eyes. He turns his head, and in amazement, he looks directly into the low northwestern sky, and off in the distance at approximately 110 miles, he notices a soft glow behind the orchard trees, close to the horizon. The orchard trees somewhat conceal the soft glow, and Ziusudra becomes very excited, and runs halfway down the ramp for a better, unobstructed view through the trees.

---

[648] F. Page 84, line 99, and line 100 Coming as heralds over hill and plain.

Again he stares, and fixes his eyes on the soft glow.

As Ziusudra looks toward the low northwestern sky, he immediately notices the soft glow is slowly ascending, and simultaneously increasing in brilliance, with the surrounding landscape at Shuruppak now dimly illuminated by the glow, miles away, which has now quickly become brilliant. Then Ziusudra exclaims, "The signal of God Utu, from Sippar. The Gods are ascending.[649] Their craft appear as torches,[650] lighting up the land with their brightness.[651] The definite time has arrived."[652]

We see a close up view of the group of ascending craft over Sippar, which is an extremely bright scene. There are fifty saucer shaped shuttle craft, each 75 feet in diameter, slowly ascending, with a brilliant light, like torch light, being emitted from the underside of each craft. The light is not from a flame, or burning process, it is a brilliant glowing light, like that of a firefly.[653]

Inside a craft, we observe young and older Gods and Goddesses, most of whom are members of the astronaut corps, and are also considered astronauts, Anunnaki Gods, Anunnaki God officers, and Igigi Gods, all of which are standing, looking out the portholes, or at the small viewing monitor, which shows a close up view of the brightly illuminated landscape below. All are

---

[649]F. Page 85, line 114 They fled (the Earth and) ascended to the heaven (space/sky) of Anu.

[650]F. Page 85, line 103 The Anunnaki (astronauts) raised (their) torches (craft); and V. Ezekiel 1:13 and like the appearance of lamps (similar to this).

[651]F. Page 85, line 104 Lighting up the land with their brightness; and V. Ezekiel 1:4 and a brightness was about it, and 1:13 and the fire was bright (similar to these).

[652]F. Page 84, line 89 That definite time arrived (see footnote 614).

[653]V. Ezekiel 1:4 a great cloud, and a fire infolding itself, and 1:13 their appearance was like burning coals of fire, (etc.), and out of the fire went forth lightning (similar to these. In this reference and those in the preceding footnotes, the use of the terms torches, lamps, fire, and burning coals would seem to imply a fire/flame/burning process, which should emit carbon particles or black smoke, however, no mention, either directly or indirectly, is made to smoke, and a torch/fire or lamp/flame without some degree of smoke would be very unusual. Therefore, a white vapor trail/steam/artificial cloud, and not black smoke, could also have been present. The observation and report could have been of a "cold" chemical light, such as a firefly, as I describe, with no vapor trail; or the observation by Ezekiel of a great cloud, fire, and accompanying lightning, could have been a laser, which is essentially similar to harnessed lightning; or it could have been, again because of the presence of a bright torch-like light, which may have been the same lightning phenomenon Ezekiel described, electrical energy produced by creating resonances by oscillation inside the Earth's ionospheric cavity, as described by Margaret Cheney in Tesla: Man Out Of Time, Chapter 14, which creates electricity surrounded by an artificial cloud, like natural lightning and a natural cloud which is formed around lightning during a thunderstorm).

now wearing their loose fitting flight attire of gold-like material, except the Igigi Gods, who are not considered astronauts, Anunnaki Gods. There is an Anunnaki God officer, a male, seated at a computer console, which has flashing display lights. He confirms to another Anunnaki God officer, a female, who is seated near him at another computer console, "We are instructed to stay in open communications with all the other craft," and the other Anunnaki God officer nods her head, as she touches several heat sensitive numbers on the console, and we hear other voices emitting from the computer, coming from the other craft.

*

We observe Ziusudra, halfway down the ramp, as he watches the glow of the ascending craft slowly fade into the sky, and the landscape again darkens, but does not become as dark as it was previously. He then turns, and looks to the east, over the shiny black pitch coated hexagon shaped reed boat, and across the slightly choppy Euphrates River, and he notices a shimmer[654] of light on the eastern horizon. Dawn is breaking over the horizon, daylight is arriving.

SUNRISE
May 7th

Ziusudra is now watching the sunrise, as the Sun rays pierce the dim sky, forming a rainbow in the distance that quickly begins to fade away and then disappears. He turns and scans the sky, and between the scattered clouds, he again gazes at Nibiru in the northern sky, and as he slowly turns to the south, he looks beyond the scattered clouds, and notices that beyond them, loom dark, thick clouds,[655] appearing as a solid line, which are still in the far distance, but are quickly approaching. We scan to the banquet area, which is still a quiet and peaceful scene, and we notice the flickering flames of the clay oil lamps, the only signs of life, swaying as they hang from the tree branches by their leather strips. Unexpectedly, a sudden gust of wind, as a harbinger of death,

---

[654]F. Page 84, line 96 As soon as the first shimmer of morning beamed forth.
[655]F. Page 84, line 97 A black cloud came up from out the horizon.

snuffs out all the flames, and now there are no signs of life at the darkened banquet area.

Then we scan back to Ziusudra, and we see a close up view of his face, and suddenly a drop of rain falls upon his cheek. He notices it, as he reaches up and touches it with his finger, and wipes it away, as if he could forestall his imminent fate. We see another drop, and another, as a soft sporadic rain begins to fall, and the drops of rain streak down his face, and flow into his beard. Ziusudra looks puzzled, and filled with fear, as he looks at the dark, thick, solid line of clouds, that are rising and quickly approaching in the distance from the south, beyond the scattered clouds overhead, for he now faintly hears the hissing, whistling, and crashing sound of the lead wave. He anxiously turns his ear toward, and then looks at the dark clouds.

The weather suddenly changes,[656] as the mild south winds increase, and a warmer breeze begins to blow. He now sees fierce lightning and faintly hears thunder[657] in the quickly approaching dark, thick, solid line of clouds, which are now in the near distance, and are becoming darker. The rain increases and becomes moderate, and is now constant, adding to the hissing, whistling, and crashing sound of the lead wave, and the thunder, which have also increased. The south winds begin to blow with more force, and become even warmer, as Ziusudra loudly and anxiously declares to himself, "The appearance of the weather has changed.[658] The destructive forces are upon us,"[659] as he now quickly looks to the east, again over the boat, where the Sun rays are now brighter in the eastern sky, and it is slightly lighter. Then, he again quickly looks to the south, where the thick, solid line of clouds have now become even darker in the sky. Suddenly there is a strong Earth tremor, which almost shakes Ziusudra off the widely swaying reed bundle ramp. He struggles to keep his balance, and slowly staggers up the ramp, and just as the tremor ends, he staggers into the slightly swaying boat, which is cushioned by the now choppy water.

Inside the somewhat dimly illuminated boat, we hear the sounds of the

---

[656]F. Page 84, line 91 I viewed the appearance of the weather.

[657]M. Page 93, line 49 Adad (the storm God, in charge of rain, lightning, and thunder) roared in the clouds; and F. page 84, line 98 Adad thunders within it (within the black cloud).

[658]M. Page 93, line 48 The appearance of the weather changed.

[659]F. Page 84, line 90 In the evening (at dawn) the leader of the sto[rm] (God Adad) caused a destructive rain to rain down.

rain upon the roof of the boat, as if it were a drum, and as Ziusudra rushes by, we again see a close up view of the water clock, securely lashed to the vertical reed beam near the doorway, which indicates: the 17th day, exactly at the marker peg, with the spool on the 2nd month. Ziusudra quickly wakes up the pilot, boatmen, and the other people, some of whom have already been aroused by the slight swaying of the boat. Ziusudra reveals, as he excitedly shouts a command to them, "The hour has arrived. A destructive flood is upon us. Bolt the door, and bring pitch, and seal it shut. Securely lash everything to the beams, including all the animals." Shocked by the announcement, the dreamy-eyed people, all wearing their colorful robes, and symbol necklaces, are now trembling with fear, and after some attempt a hasty partial bow, they rush about as ordered, with the long 1/2 inch wide leather lashes. They begin to relight the clay oil lamps, securely lashed to the vertical reed beams, and the intensity of the lighting begins to increase, and again softly illuminates the interior, and we hear them repeating both loudly to others, and whispering to themselves, as they rub the sleep from their eyes, and check and securely lash whatever is not already lashed to the vertical, horizontal, and diagonal truss beams, "Flood . . . . destructive flood . . . . upon us?"

Four boatmen position themselves at the nearly square reed bundle door, which already has been coated with pitch, inside and out, and is next to the doorway opening, and has inner protruding reed fastening loops, as does the inner doorway, and also has an inward tapering perimeter. Then, they grasp and hold the door, ready to place it in the doorway. As the boatmen hold the door, Ziusudra takes a last look outside. The south winds are stronger and warmer now, and he sees the thick, now very dark and almost black, solid line of clouds, quickly approaching from the south, turning the new dawn light back into darkness, and hears the hissing, whistling, and crashing sound of the lead wave, as it grows louder; in addition to seeing the fierce lightning, and hearing the thunder,[660] which accompany the now increasing moderate rain and strong south winds. With the rain drumming upon the roof of the boat, and as the rain is now being blown into the boat by the strong south winds, Ziusudra despairingly describes the conditions to the pilot, who stands beside him, "The winds have become more savage now.[661] The weather is frightful

[660] M. Page 93, line 50 As soon as he heard Adad's voice (thunder).

[661] M. Page 93, line 54 The winds became savage as he (God Adad) arose.

to behold,[662] turning into darkness, all that was light."[663]

Other boatmen now push the reed bundle ramp aside, and it falls into the now very choppy river water, which is about 2 feet below the bottom of the doorway. Then the door, which has a 1 inch thick layer of thick pitch around its inward tapering perimeter, is set in place by the four boatmen, who first pass the door to the outside of the doorway, diagonally through the doorway, and then pull it in tight to seal. As they pull the door in to seal, we notice the thick pitch around the perimeter ooze in. Then, the boatmen slide reed bundle bracing beams horizontally through the doorway loops and across the door, through the door loops; and then they drive small wedge shaped reed expanders into place, between the door and the bracing beams, which closes and forces the door even tighter; and they proceed to apply liquid pitch, from clay pots near the door, to the perimeter of the door, further sealing it shut.[664] Then Ziusudra takes a 16 inch long saw-like flint knife from a leather sheath, which is hanging from a reed beam near the door, among several other leather sheaths, one of which is 12 inches long, and personally saws through the 6 inch diameter braided reed hawser,[665] which is pulled out of the boat by its own weight. Then, the boatmen insert a cone shaped reed plug, which is already coated with thick pitch, into the 6-1/2 inch diameter outward tapering hawser hole, and pound it in to seal the hole, and then very quickly apply liquid pitch to it, as they did to the door, to further seal it.

From the river bank, we notice the boat, as it begins to be driven up the very choppy river to the north, by the increasing strong south winds, amidst the rain, fierce lightning and thunder, and the hissing, whistling, and crashing sound of the lead wave, all of which are also increasing.

Inside the boat, Ziusudra turns to his pilot and confides, "For the navigation

---

[662]F. Page 84, line 92 The weather was frightful to behold.

[663]F. Page 85, line 106 (And) turned into darkness all that was light (this is an accurate description; as I further explain under 'The Beginning Of The Flood' on page 346 in my epilogue).

[664]M. Page 93, line 51 Pitch was brought for him to close his door, and line 52 After he had bolted his door; and F. page 84, line 93 I entered the ship and closed my door (these refer to the closing of a single door, without any reference to closing of windows, for there were no windows in the boat when the flood began); and M. page 123, lower line 3 [He] entered and shut up the [boat]; and V. Genesis 7:16 and the LORD (Nibiru; caused effects that) shut him in (similar to this).

[665]M. Page 93, line 55 He severed the hawser and set the boat adrift (implies the boat was already in the river water, and now it begins to drift).

of this boat, to you, I entrust the mighty structure with its goods,"[609] and the pilot responds, as he shakes with fear, "Do not fear, my King."

Everyone is now fully awake, including the teenagers, and the clay oil lamps, lashed to the vertical reed beams are again all relighted, and again softly illuminate the interior of the boat, as everyone is trembling and fearfully excited, while busy securely lashing the few remaining unlashed smaller personal possessions, which rest upon the floor, to the 6 inch diameter vertical and diagonal truss reed beams. They use the long 1/2 inch wide leather strips as lashes, everywhere.

Then, we see the frightened and trembling large and small animals, including their pet dogs, with all beginning to howl and simultaneously paw at the reed bundle floor, being lashed to the vertical, horizontal, and diagonal truss beams within their compartment-like areas, with double lashes for larger animals; and then we notice their pet cats, who with wide open eyes and arched backs and raised tails, have already dug their claws into and cling to the vertical reed beams, as though they were tree trunks, as they all listen to the strange combination of sounds.

Then, we see the birds, all of which are not lashed, with all beginning to squawk and simultaneously vigorously flutter their wings on their horizontal and diagonal beam perches, as they are also frightened by the unfamiliar mixture of sounds.

We observe Ziusudra, as he looks at the water clock, and removes the marker peg, which was on the 17th day. The clock indicates the 17th day, and the spool is on the 2nd month, and we see a close up view of the water clock, securely lashed to the vertical reed beam near the now sealed door. The face of Ziusudra is filled with mixed expressions of fear and anxiety, as he listens to the increasing combined sounds; rain drumming upon the roof; fierce thunder and the strong winds; and the hissing, whistling, and crashing sound of the huge lead wave: as they all become more intense.

In the city, the increasing rain pours down upon the homes, amidst the fierce lightning and thunder, the strong warm south winds, and the hissing, whistling, and crashing sound of the lead wave, all of which are also increasing. All the frightened animals now run aimlessly. A few Sun rays, emitting from the rising Sun to the east, serenely shine upon the city, barely penetrating under the solid line of very dark clouds.

Inside a home, the people are now awake, and are becoming frightened and very concerned, as they listen to the strange combination of sounds. A man comforts his pregnant wife, "Greater wealth and new prosperity, all the gift of God Enlil, now that Ziusudra is departing." Then he looks up at the roof, and exclaims, "God Enlil, we welcome your rain," and then he pauses, and they look at each other with growing fear, and he rationalizes, "listen to the rain storm," as he tries to convince her, and himself, that all the sounds are from the rain storm alone. They try to comfort each other, but their anxiety continues to grow, as the intensity of all the sounds increases.

From space, we observe the planet Nibiru, approaching in all its glory, with its two rings of clouds, within which orbit its four moons, appearing as a radiating and glowing cross, bathed in a sunset red glow, and with star-filled space surrounding the planet.

Then we see the Earth from space, and we zoom in for a close up view, and from the air, water, land, and first wave crest level, we see the 20 huge 300 feet high, rolling lead waves, followed by the 21 smaller 100 feet high, sweeping waves, and hear the hissing and whistling sound of the 20 lead waves, and the crashing sound of the lead wave, and the whooshing sound of the other 19 lead waves, and see the very dense darkness and fierce lightning, and hear the thunder,[666] and see and hear the violently howling wind driven torrents of heavy rain, in the very dark clouds with violent storms, with a raging storm preceding them, as they sweep across the land. The first of the 20 huge 300 feet high rolling waves, all traveling 200 miles per hour and racing toward the north, has now reached 50 miles south of Shuruppak, only 15 minutes away.

Then, from the air, land, and first wave crest level, we see the first wave, the lead wave, a 300 feet high wave, a sloping wall of purple water, as it reaches the southern most tip of Sumer, and quickly begins to cover the trembling and fissuring land, as the waves have already done with most of the Arabian peninsula. All the sounds are very loud, especially of the water crashing down upon the wave, beneath the crest of the lead wave.

From the air, we see a craft, with its brilliant glowing firefly-like light emitting from the underside, which is slightly beneath the solid line of very dark clouds, in the preceding comparatively lighter rain and wind, and traveling

---

[666]M. Page 93, line 53 Adad was roaring in the clouds, and line 5 Into the heavens reached the storm; and F. page 85, line 105 The raging of Adad reached unto heaven (high into the sky).

just ahead of the lead wave and destruction. The craft is in front of the violent storms, which contain very dense darkness, and fierce lightning and thunder, and violently howling wind driven torrents of heavy rain, but beneath the level of the very dark clouds. Fierce lightning, which illuminates the scene below, is flashing near the craft, and the destructive sounds are very loud.

Inside the craft, which is in open communication, it is very quiet, and the destructive sounds are barely audible, for the audio is off. We observe God Enki at a porthole, looking down at the lead wave, in the near distance, and to the south. Because of poor visibility within the violent storms, he can barely see some of the other huge 300 feet high waves, following the lead wave. Behind God Enki is a small viewing monitor, which shows a close up view of the scene below. Even within the flashing lightning, he can now scarcely see the huge waves and destruction through the darkness and raging storm that precedes them below his craft.[667] The craft of God Enki is over the far south of Shuruppak, and he is especially looking at, but can barely see the boat, which is being blown to the north and away from the city, and only barely sees some people, far south of the city, who are beginning to hysterically run toward the north. He notices the huge 300 feet high rolling lead wave, a sloping wall of purple water, has almost reached Shuruppak. God Enki alternates between the porthole, for a better perspective of the destruction[668] below, and the viewing monitor, which shows a close up view of the scene below. There are 2 Gods and 2 Goddesses, 18 Anunnaki Gods, 6 of whom are officers, and 4 Igigi Gods in the craft with God Enki, and all are in their flight attire, except the Igigi Gods. They are all horrified as they look down at the scene below.

God Enki is grief stricken, and trembles with fear, as he astonishingly exclaims to an Anunnaki God officer sitting at a computer console, "The first twenty waves are each three hundred feet high, and the following twenty-one smaller waves are each one hundred feet high." He then looks down, at the violently howling wind driven torrents of heavy rain, quickly approaching the city from the south. Then God Enki runs to a computer console nearby, and touches a heat sensitive audio symbol on the console. Instantaneously, the craft is filled with the loud destructive sounds,[669] and all in the craft become

---

[667]M. Page 125, line 12 Below the chariot (craft) of the gods the storm raged.

[668]M. Page 125, line 13 [It] sweeps forward, it kills, it threshes [all] (violently and abruptly cuts down everything in its path).

startled, and gasp, and are visibly frightened, and begin to tremble.[670] They all listen to the sounds, as they look down at the destruction, alternating between the viewing monitor, and the portholes, with their faces filled with expressions of grief and fear. Then God Enki again touches the audio symbol, and the loud sounds cease in the craft, and all that can now be heard are the barely audible destructive sounds, and the gasps of all those in the craft.

In the city of Shuruppak, the sky overhead and to the south is very dark, with just a few last Sun rays shooting in from the rising Sun to the east, under the solid line of very dark clouds. Then, the very dark clouds cause very dense darkness to prevail, as though it were night, as the conditions and sounds increasingly become more intense, with the south winds becoming howling winds, and the rain becoming torrents of heavy rain,[671] which is driven in sheets, like in a hurricane.

From inside their homes, the people can now plainly hear all the loud destructive sounds, especially of the lead wave; the hissing, whistling, and crashing sound, as well as see the very dense darkness and fierce lightning, and hear the thunder, and see and hear the now torrents of heavy rain, driven by the now howling winds. Strong Earth tremors begin to intermittently shake the land, and then quickly intensify, and become severe, causing the ground to fissure and some of the clay brick houses and courtyard walls to crack and begin to fall apart, with some of the clay bricks breaking loose and falling to the rain soaked trembling and fissuring ground. The people are now panicking, and running out of their homes, into the howling wind driven torrents of heavy rain, accompanied by the very dense darkness, extremely warm temperatures, and fierce lightning and thunder, as they all bear down upon them from the south. With clinging rain soaked colorful robes, and wearing their symbol necklaces, they are hysterically screaming, and as the flashing lightning

---

[669]M. Page 95, line 23 In the heaven of Anu they heard the noise of the [flood], and page 125, line 20 In the heaven of Anu [they heard] the noise of the flood (within the craft in the sky, the Gods heard all the destructive sounds. This could also imply that the sounds were literally transmitted by the craft to the planet Nibiru, where God Anu and all the other Gods also heard them).

[670]M. Page 95, line 24 It was trying [even] of the gods, and page 125, line 21 The loud noise set the [gods] atremble.

[671]M. Page 93, line 7, and page 125, line 16 [Zû] with his talons [rent] the heavens (a mischievous God ripped open the clouds, increasing the rain).

illuminates their way, for in the absence of the flashing lightning, one person can not see another[672] several feet away, some are running toward where their high Priest and King, Ziusudra, and his hexagon shaped reed boat had been, toward the northeast edge of the city, as they quickly realize its purpose and function, with very sobering thoughts; but the reed boat has now been driven away from the river bank, and up the now extremely choppy river to the north, by the howling south winds. The people now notice the increased level of the water,[673] and after not seeing the hexagon shaped reed boat at the river bank, the people look up the river, and within the flashing lightning, they see the reed boat, barely visible in the river to the north. Then, the people desperately turn toward the large and small, rectangular shaped, reed boats nearby, which have also been blown away from the bank, with some still tied to it, but they are all now too far out in the extremely choppy river, and the people are unable to get to them.

Then, the people turn toward the loud destructive sounds, to the south, which they now hear with full, almost deafening intensity, and see the lead wave in the near distance. Illuminated by the flashing lighting, it appears as a giant sloping wall of purple water, 300 feet high, rolling toward them from the south. Some people run, while others are petrified at its sight, and their fear causes them to hopelessly stare at it, as some grasp their symbol necklace pendant. Mothers hold their teenage children, adults hold each other, as they gaze with awe at the lead wave. All the frightened animals have stopped running aimlessly, and are now howling and simultaneously pawing at the ground, and all the birds are squawking as they are being forcefully blown through the air toward the north. The loud destructive sound of the water crashing down upon the lead wave, beneath its crest, is now almost unbearable, and some people are covering their ears with their hands, while others are falling to the ground, and hysterically pounding the rain soaked ground, as the howling wind driven torrents of heavy rain rakes across them.

Inside the boat, which is now being driven further north by the howling south winds, the severe Earth tremors are only slightly noticed, due to the cushioning effect of the water. Ziusudra and the people are finished with securely lashing their smaller personal possessions, which rest upon the floor,

---

[672]M. Page 95, line 13 One person did [not] see another; and F. page 85, line 111 No man could see his fellow (man).

[673]M. Page 125, line 14 Ninurta went on and [made] the dykes [overflow]; and F. page 84, line 102 Ninurta comes along (and) causes the dikes to give way (see footnote 494).

to the vertical and diagonal truss reed beams, and they stand still for a moment on the main deck, within the soft illumination, listening. They can now plainly hear all the destructive sounds, as they listen to the fierce thunder, and the howling wind driven torrents of heavy rain, now heavily pounding upon the roof of the boat. They are frightened, and have a puzzled look upon their faces, as they listen to an unfamiliar sound: the hissing, whistling, and crashing sound of the lead wave, as it grows steadily louder above the other sounds, including the sound of the howling wind driven torrents of rain heavily pounding upon the roof. One person anxiously asks of another, "What is it?" The other person fearfully responds, "We have never heard sounds like this before!" Suddenly Ziusudra shouts, "Quickly, lash yourselves to the vertical beams! Hurry!" All the people immediately proceed to follow his instructions, with no time for bowing, as they pass his words along, and begin securely lashing themselves to the vertical truss reed beams with the long 1/2 inch wide leather strips, while others help those who are not able, especially pregnant women and older people. Men, women, and teenage children are securely lashing themselves to the vertical beams, all as quickly as they can.

Inside the craft of God Enki, over the near south of Shuruppak, the destructive sounds are barely audible, as he again looks down through a porthole at the approaching lead wave, a sloping wall of purple water, which even within the flashing lightning, he can scarcely see through the darkness and raging storm below his craft, as it sweeps over the land and small villages, just south of Shuruppak. Behind him is the small viewing monitor, which shows a close up view of the scene below. God Enki quietly and sorrowfully remarks to an Anunnaki God officer next to him, "The water hits the land, and shatters everything, and like a breaking pot, scatters its counsel.[674] Quickly the flood has set out.[675] Its might comes upon the people like a battle array,"[676] and the Anunnaki God officer nods in agreement.

---

[674]M. Page 93, line 9 [He hit] the land, and line 10 And shattered its noise (noise = mankind = achievements of mankind) [like a pot], and page 125, line 17 [He hit] the land like a pot, he scattered its counsel (components); and F. page 85, line 107 [Hitting] the land he broke it like a po[t] (totally destroying everything and rendering useless: see footnote 326).

[675]M. Page 95, line 11 [Quickly] the flood [set out], and page 125, line 18 [Quickly] the flood set out; and F. page 85, line 109 Fast it blew and [furious].

[676]M. Page 95, line 12 Its might came upon the peoples [like a battle array], and page 125, line 19; and F. page 85, line 110 Like battle [it ca]me over the p[eople] ('like a battle array'/'like battle' implies a rank, formation, rows, or series of waves, as in a military formation).

In the city, as the flashing lightning severs the very dense darkness, we observe the people of Shuruppak, as the violently howling wind driven torrents of heavy rain pours down and rakes across them, and the severe Earth tremors continue to intermittently shake the land, amidst all the other loud destructive sounds. Some people are petrified with fear, as they cringe and brace themselves for the inevitable, and some are covering their ears and eyes with their hands, trying to hide within themselves. Others are hysterically screaming as they run from the lead wave, as if they could escape its wrath, and others continue pounding and now clawing the rain soaked trembling and fissuring ground, like the animals, as if seeking entry through some earthly door that might open to conceal and protect them.

We observe Goddess Ninhursag in her craft, with a small viewing monitor behind her, which shows a close up view of the scene below. The destructive sounds are very loud, for the audio is on. She is also looking down out a porthole. Her craft, like the craft of God Enki, is in open communication, and is slightly beneath the solid line of very dark clouds, in the preceding comparatively lighter rain and wind, and traveling just ahead of the lead wave and destruction. Her craft is also in front of the violent storms, but beneath the level of the very dark clouds. Fierce lightning, which illuminates the scene below, is flashing near her craft. Even within the flashing lightning, she can also scarcely see the huge waves and destruction through the darkness and raging storm that precedes them below her craft. From her location, to the near south of Shuruppak, she does not see the boat, which has been blown further to the north, but only barely sees the hysterically running people, just south of the city. Accompanying her are 2 Gods and 2 Goddesses, 18 Anunnaki Gods, 6 of whom are officers, and 4 Igigi Gods, and all are also in their flight attire, except the Igigi Gods. They are also all horrified at the scene below, and are also all looking down at the destruction, alternating between the viewing monitor, and the portholes, and as they all look at the destruction, their faces are filled with expressions of grief and fear. Then, she touches a heat sensitive audio symbol on a computer console, and the loud destructive sounds are instantaneously reduced to a very low level. She resumes looking down, and astonishingly exclaims to an Anunnaki God officer next to her, "The flood is bellowing like a bull, like a whinnying wild ass the winds howl.[677] All the destructive winds are present.[678] The darkness is very dense,"[679] and the Anunnaki God officer nods in agreement.

We observe God Enlil in his craft, looking down out a porthole, with a small viewing monitor behind him, which also shows a close up view of the scene below. The destructive sounds are barely audible, for the audio is off. His craft, like the craft of God Enki and Goddess Ninhursag, is in open communication, and is also slightly beneath the solid line of very dark clouds, in the preceding comparatively lighter rain and wind, and traveling just ahead of the lead wave and destruction. His craft is also in front of the violent storms, but beneath the level of the very dark clouds. Fierce lightning, which illuminates the scene below, is also flashing near his craft. Even within the flashing lightning, he can also scarcely see the huge waves and destruction through the darkness and raging storm that precedes them below his craft. From his location, to the near south of Shuruppak, he does not see the boat, which has now been blown even further to the north, but also only barely sees the hysterically running people, just south of the city. With him are 2 Gods and 2 Goddesses, 18 Anunnaki Gods, 6 of whom are officers, and 4 Igigi Gods, and all are also in their flight attire, except the Igigi Gods. They are also all horrified at the scene below, and are also all looking down at the destruction, alternating between the viewing monitor, and the portholes, and as they all look at the destruction, their faces are filled with expressions of grief and fear; except God Enlil, who smugly smiles as he looks upon the destruction.

At the southern most edge of the city of Shuruppak, with the flashing lightning continuing to sever the very dense darkness, at the moment the lead wave hits[674] the city, a 300 feet high sloping wall of purple water, accompanied by the violently howling winds, and the torrents of wind driven heavy rain, and the hissing, whistling, and crashing sound of the lead wave, amidst the very dense darkness, and fierce lightning and thunder, and the severe Earth

---

[677]M. Page 95, line 15 [The flood] bellowed like a bull (the sounds were very loud), and line 16 [Like] a whinnying wild ass the winds [howled].

[678]M. Page 93, line 6 All the destructive winds were yoked (working together), and page 143, line 201 All the destructive winds (and) gales were present (see footnote 645).

[679]M. Page 95, line 18 The darkness [was dense], there was no sun.

tremors and fissures: the devastation is total. The wave overwhelms[680] everything in its path, as the hissing, whistling, and foam filled lower wave first pushes everything down in front of it; then lifts and pulls everything that floats, up and into the sloping lower part of the wave; then carries everything up the sloping side of the wave; then sweeps everything up to its crest; then throws everything down and into the wave, beneath the crashing, foam filled crest; and then buries everything beneath the water of the main part of the wave, in its wake.[681] The clay brick houses and courtyard walls, outside oval wall, and colorful central ziggurat compound with its eight temples near the base, all shatter beneath the lower part of the wave, as they are pushed down before the main part of the wave reaches them, as if they were merely sand castles on a beach.

Teenagers are being ripped from the clutching arms of their mothers, and adults are being torn from the grappling arms of other adults. Both mankind and animals[682] are knocked off their feet and swept[683] into the water, as they are pulled up and into the foam and debris filled sloping side of the wave. The reed huts are pulled apart into sections, and together with the contorted woven reed mats, they are swept up and into the water. The large and small, rectangular shaped, reed boats are also immediately pulled apart into sections,[684] and swept up and into the water, because of their rectangular shape and their normal structural design.

Most of the people are immediately drowned, but some grasp for and cling onto floating debris, consisting of: reed bundles, left over from the boat construction; reed bundle floors, ramps, and exterior hull sections, from the severed rectangular shaped reed boats; woven reed mats, from the farewell banquet; reed bundle wall sections, from the reed huts; small tree branches, and palm leaves, all of which are not being broken by the wave, as they bend

---

[680]F. Page 85, line 128 The wind blew, the downpour, the tempest, (and) the flo[od] overwhelmed the land.

[681]M. Page 143, line 202 The storm swept over the capitals (plural form of the word implies it was not a local flood).

[682]V. Genesis 7:21, and 7:22 All in whose nostrils was the breath of life, of all that was in the dry land, died (similar to these).

[683]M. Page 97, line 26 Where it swept, mankind, and line 27 Was thrown down before it; and H. page 77, line 39 The Flood swept thereover.

[684]M. Page 125, line 15 Errakal tore up [the mooring poles]; and F. page 84, line 101 Irragal pulls out the masts (of boats. Irragal is a differently spelled identical epithet for God Nergal, a son of God Enki: see footnote 494).

and bow, but spring back; or cling onto pieces of the wood tables and benches from the banquet area, and tree trunks, and large tree branches, all of which are being broken and splintered by the force of the wave. Together, they are swept away and buried in the foaming water. The foam of the wave is very thick, and of the few people who appear in it, one person can not see another.[672]

A few people are able to swim several feet to debris, as they are being swept up in the foaming water, and are able to maintain a grasp onto the debris. But most people and debris are quickly separated, as they are unable to hold on for more than a few seconds, as the strong and violent upward movement of the wave tears the debris out of their clutching hands, and the people disappear into the foam filled water, as their outstretched arms slowly sink into the side of the wave: all as they are being pulled up the sloping side of the wave by its force, amidst the violently howling wind driven torrents of heavy rain, which rakes across the water.

From outside the boat, as the flashing lightning illuminates the scene within the very dense darkness, we see the shiny black pitch coated hexagon shaped reed boat, which has now been blown even further north of Shuruppak, being driven up the now exceedingly choppy Euphrates River by the violently howling winds, riding 10 feet above the water line, amidst all the loud destructive forces and sounds; fierce lightning and thunder; violently howling winds; torrents of wind driven heavy rain; and the hissing, whistling, and crashing sound of the lead wave. We see the huge 300 feet high foam and debris filled lead wave, a sloping wall of purple water, as it hits the boat. The sloping lower part of the rolling wave lifts and pulls the boat out of the river,[685] as it seizes the boat, and begins to carry and sweep the boat up to its crest. The boat is now almost sideways[686] in the sloping side of the wave, surrounded by foam and debris filled water, and quickly moving up toward the crest, as it is now simultaneously being driven north, by the predominant driving force of the wave.

Inside the boat, within the soft illumination, Ziusudra now hears all the loud destructive sounds with full, almost deafening intensity, as do all the people, and they all shake with fear, and some begin to weep, and

---

[685]V. Genesis 7:17 and the waters increased, and bare up the ark, and it was lift up above the earth (similar to this).

[686]M. Page 145, line 205 And the destructive wind had rocked the huge boat in the high water.

others begin to cry out for help, as they passionately and pleadingly shout, "Help . . . . please God Enki . . . . help us . . . . please help us." The people, all securely lashed to the vertical reed beams, are now becoming nearly horizontal, and are horrified, and some manage to grasp their symbol necklace pendant, and others are burying their ears in their shoulders, while holding on tight to the beams, to which they are lashed. As the boat rolls, small personal possessions are thrown out of their securely lashed containers, and onto the floor, where they slide against the now horizontal hull, and the unsealed securely lashed large clay vases of drinking water and wine spill their contents, while the sealed securely lashed large clay vases of drinking water and wine slightly leak, and the securely lashed large reed baskets of grain also spill their contents, and most of the securely lashed clay oil lamps are extinguished, as they spill their oil and snuff out their own flames, and the interior of the boat is thrust into a dimly illuminated shadow filled eerie atmosphere, adding a new dimension to the people's fear.

We see the trembling animals within their compartment-like areas, the third and fourth decks, howling and simultaneously pawing at the now vertical reed bundle floor, and cats clinging onto the now horizontal reed beams, as though they were glued onto them, with a few small animals slipping out of their lashes, and being thrown against other animals, the vertical and diagonal truss beams, or the sides of the hull, and their large bundles of grass are strewn about, and their large flat clay bowls of drinking water overturn and slide against the hull.

We see the birds, among which are doves, swallows, and ravens, within their compartment-like areas, the first and second decks, squawking and simultaneously vigorously fluttering their wings to keep their balance upon their former horizontal, but now vertical perches, upon the 6 inch diameter reed beams bracing the hull, and their large flat reed baskets of seeds, and their large flat clay bowls of drinking water are both overturned and slide against the hull, and the seeds are also strewn about, except on the first deck, where both are somewhat contained within the bottom exposed bracing beams.

Meanwhile, back among the people, we notice the inner side of the boat hull, which is coated with shiny black pitch. As the boat continues to roll, it makes loud creaking and cracking sounds, as if it were breaking apart, which also greatly frightens the people. Suddenly the hull momentarily slightly bulges in from the force of the wave that hits the boat. The bulge knocks small personal belongings loose from their lashes, and they fly across the slightly beyond vertical reed floor, and slam against the pitch coated hull on the opposite side, but the boat springs back and holds tight, due to the cushioning effect of the water, and the flexibility and hexagon design of the reed boat, and water does not leak in.

From outside the boat, with the flashing lightning illuminating the scene within the very dense darkness, the huge sloping, rolling lead wave is now cresting, with the boat and debris just beneath its crest, and the lifted boat is now being tossed forward. The boat almost rolls over, as the crashing and foaming crest collapses and crashes down upon the boat, and throws it down into the wave, and completely covers it under the foam and debris filled water of the wave. Then, the main part of the wave buries the boat beneath the water. The boat has disappeared. Everywhere, all we see are foam and debris in the wake of the wave, amidst the violently howling wind driven torrents of heavy rain, which rakes across the water.

We scan over the water, among the foam and debris, searching for the boat, and we see debris popping up, out of the tossing, foaming water. There is a large amount of debris, including some broken and splintered wood tables and benches, tree trunks, and large tree branches; and some intact small tree branches, and many palm leaves; and other intact debris, among which are reed bundles; reed bundle boat floors, ramps, and exterior hull sections; woven reed mats; and reed bundle hut wall sections. The reed bundles; reed bundle boat floors, ramps, and exterior hull sections; woven reed mats; reed bundle hut wall sections, and some of the other debris looks familiar, and we do not know if they are left over construction bundles, or floors, ramps, and exterior hull sections from the severed rectangular shaped reed boats, or farewell banquet woven reed mats, or wall sections from the reed huts: or all parts and sections from the destroyed hexagon shaped reed bundle boat.

Then, after approximately 30 seconds of searching, out of the tossing foam and debris filled wake of the wave, amidst all the destructive forces and sounds, including the violently howling wind driven torrents of heavy rain, the boat bobs up, slightly less shiny, but intact and right side up. It is now being swept away in the foam and debris filled water, toward the north, now riding slightly less than 10 feet above the water line, along with a large amount of debris, some of which is laying on and some hanging off the roof of the boat.

From the air, amidst all of the destructive forces and sounds, we can barely see the hissing, whistling, and whooshing second, third, and fourth, of the remaining 19 huge 300 feet high sloping, rolling waves, spaced at short intervals, as they begin to sweep through the former city of Shuruppak area. These waves do not have as severe a destructive and crashing crest, as the first wave had. They have a whooshing crest, and except for the crest, they are like the first wave; sloping and rolling, as they also lift and pull everything that floats, up their sloping side, and up to their crest, and then throw everything down and into the wave, beneath their collapsing and whooshing, foam filled crest.

From the level of the water, as the flashing lightning illuminates the scene within the very dense darkness, amidst all the destructive forces and sounds, we see a close up view of the former city of Shuruppak area, now covered with debris filled deep water. Amidst the violently howling wind driven torrents of heavy rain, which rakes across the water, we notice a few tired and weak people, who have miraculously survived the first wave by clinging onto debris, which they have now draped themselves over. Because of their weakened and exhausted condition, they are being torn from the debris, as they have now expended all their energy, and can no longer hang on. We observe them being drowned, sinking beneath the foam and debris filled water, with arms outstretched for help, as they are pulled up the sloping side, and toward the whooshing crest of the second huge 300 feet high sloping, rolling wave.

Inside the boat, as it is again being lifted and pulled up, it rolls and bobs, and within the dimly illuminated shadow filled eerie atmosphere, all the people are now white with fear, as they again weep or cry out for help, while still securely lashed to the vertical reed beams, which again begin to tilt toward the horizontal. We observe Ziusudra, with tears streaking down his face, and flowing into his beard, praying and exhorting as he shouts, "God Enki. May God Enki help us find our way through."

Again we notice the inner side of the boat hull, as the boat continues to roll, and again it makes loud creaking and cracking sounds, and again suddenly the hull momentarily slightly bulges in, as the boat is tossed in the water. But the boat again springs back and continues to hold fast, and does not leak water, because the water is cushioning the full destructive force of the wave, and because of the flexible reed construction and hexagon shaped design of the boat.

Outside the boat, with the flashing lightning continuing to illuminate the scene within the very dense darkness, amidst all the destructive forces and sounds; fierce lightning and thunder; violently howling winds; torrents of wind driven heavy rain; and the hissing, whistling, and whooshing sound of the remaining 19 sloping, rolling waves, we see the boat as it is continued to be tossed and rolled about, and swept over by the whooshing crests, and covered with the foam and debris filled water of the second and third huge 300 feet high sloping, rolling waves. Two more times, with each wave, the boat is momentarily lost beneath the foam and debris filled water of the wake, and the submerged time increases with each wave, as the boat is covered with the foam and debris filled water.

Miraculously, it bobs up faithfully the second and third time, amidst the foam and debris filled wake, now riding considerably less than 10 feet above the water line, with an exceedingly large amount of debris laying on and hanging off the roof, and with the boat slightly less shiny each time, as it is constantly being driven further north by each wave,[687] for the waves are now the dominant driving force, aided by the violently howling winds. But the fourth time, as the fourth huge 300 feet high sloping, rolling wave sweeps the boat, and the wave throws the boat down and into the wave, beneath its collapsing, whooshing foam filled crest, the boat submerges in the foam and debris filled wake, and we wait longer than we had on any of the previous three times for it to bob up and reappear, but it does not. Amidst the violently howling wind driven torrents of heavy rain, which rakes across the foaming water, we only see large amounts of debris popping up to the surface, some of which looks familiar, especially the reed bundles and woven reed mats. We have now lost the boat, and do not see it reappear, and it impressibly seems as though the waves have become triumphant.

*

From the air, amidst all the destructive forces and sounds, we can barely see the remaining 16 huge 300 feet high sloping, rolling waves, as they sweep

---

[687]V. Genesis 7:18 And the waters prevailed, and were increased greatly upon the earth; and the ark went upon the face of the waters (rode upon the upper surface of the waves, and/or near the crest of the waves, similar to this).

through the former city of Shuruppak area, and then the 21 smaller 100 feet high rippling waves begin to sweep through the area.

Several Hours Later

We observe God Enki in his craft, which is now being very slightly buffeted, for it is now within the violent storms. His craft is in open communication with the other craft, including the craft of Goddess Ninhursag. All the destructive sounds are barely audible within the craft, except the hissing, whistling, and whooshing sound of the sloping, rolling waves, which have now passed through the area. God Enki is alternately looking down through a porthole, and at the small viewing monitor, amidst the very dense darkness, and fierce lightning and thunder, through the violently howling wind driven torrents of heavy rain, at the violently howling wind whipped foam which stretches over the debris filled water, and at the torrents of heavy rain, which rakes across the foam and debris. Because of the increased poor visibility within the violent storms, and the large amount of debris, he has now lost sight of the boat. Even within the flashing lightning, he can only barely see a few of the smaller 100 feet high waves, all rippling and without crests, as they sweep across the foam and debris filled water, as he looks at the former city of Shuruppak area, and especially to the north, where he last saw the boat.

God Enki softly and despairingly[688] remarks to himself, as he looks down from a porthole, "One person could not see another.[689] The boat is not recognizable to me in the destruction,"[690] and he now turns away from the porthole, and empathetically looks at the Gods in his craft, and slowly scans over them, and we see a close up view of them. All the Igigi Gods, and all the Gods and Goddesses, and many of the Anunnaki Gods are weeping, and most

---

[688]M. Page 95, line 25 [Enki] was beside himself.

[689]M. Page 95, line 13 One person did [not] see another; and F. page 85, line 111 No man could see his fellow (man. The people on the ground, and then the people within the water. Aside from the obvious actual words, this is a reference to visibility, and means that prevailing conditions caused extremely poor visibility: see footnote 672).

[690]M. Page 95, line 14 They (probably means the people on the ground and then within the water, but it could imply the people within the boat, which would mean the boat itself) were [not] recognizable (observable) in the destruction (amidst all the forces, including the waves); and F. page 85, line 112 The people (within the boat, therefore the boat) could not be recognized (observed) from heaven (in this context, from within the craft in the sky over the destruction).

of them are crouched in fear beneath the portholes, painfully looking at the viewing monitor, but afraid to look down through the portholes, directly at the total devastation. God Enki concedes in a loud tone, to an Anunnaki God officer seated at a computer console, who is also grief stricken, "The Gods are terror stricken by the deluge."[691] Then, God Enki again looks down through a porthole, and then quickly turns away, and loudly exclaims to Goddess Ninhursag, through the open communication system, "Goddess Ninhursag, my sons were thrown down before me!"[692]

We observe Goddess Ninhursag in her craft, which is also now being very slightly buffeted, for it is now within the violent storms. Her craft is in open communication with the other craft, including the craft of God Enki. All the destructive sounds are barely audible within the craft. Goddess Ninhursag is also alternately looking down through a porthole, and at the small viewing monitor, amidst the very dense darkness, and fierce lightning and thunder, through the violently howling wind driven torrents of heavy rain. She has just heard the words of God Enki, and after hearing them, she is openly weeping,[693] as she looks down at the destruction, and even within the flashing lightning, she also only barely sees the smaller 100 feet high rippling waves, as they sweep across the foam and debris filled water. She is also looking at the former city of Shuruppak area, but away from the north, where the boat was last seen by God Enki, for she has not seen the boat, or know that it had existed.

Goddess Ninhursag responds to God Enki, through the open communication system, as she emotionally wails, "God Enki, mankind was thrown down at my own command,"[694] as she tries to place the blame upon herself, for taking the oath of secrecy: to withhold information of the impending flood, and keep it a secret from mankind. She then turns, and affectionately looks at the Gods in her craft, and slowly scans over them, and we see a close up view of them. All the Igigi Gods, and all the Gods and Goddesses, and many of the Anunnaki Gods are also weeping, as in the craft of God Enki, and most of them are also cowering like dogs, and crouched in terror beneath the

---

691F. Page 85, line 113 (Even) the gods were terror-stricken at the deluge.

692M. Page 95, line 26 [Seeing that] his sons (his creation = mankind) were thrown down before him.

693M. Page 125, line 23 Ninhursag spent her emotion.

694M. Page 125, line 22 Seeing that her sons (her creation = mankind) were thrown down at her own command.

portholes, as they painfully look at the viewing monitor. She then points to them, and affirms and describes in a loud tone to God Enki, through the open communication system, "The Gods are cowering like dogs, and crouched in distress,[695] for their legs tremble with fear, and they cannot stand up."

From the level of the water, amidst the very dense darkness, and fierce lightning and thunder, and through the violently howling wind driven torrents of heavy rain, which rakes across the water, and with poor visibility, we see a close up view of the former city of Shuruppak area, as we scan over the foam and debris filled water. We witness a total water scene, where the city of Shuruppak was formerly located. After being inundated by the first huge 300 feet high hissing, whistling, and crashing wave, and then the 19 huge 300 feet high hissing, whistling, and whooshing waves, and now swept by the smaller 100 feet high rippling waves without crests, everything is totally covered with very deep water. We do not see the hexagon shaped reed bundle boat of Ziusudra, but again, we do see some familiar floating debris, among all the other debris. There are: a few reed bundles; some reed bundle boat floors, ramps, and exterior hull sections; a few woven reed mats; some reed bundle hut wall sections; a few small tree branches; and many palm leaves, all of which are intact, and not broken, because they are flexible. Additionally, there are: a few wood tables and benches; some tree trunks; and a few large tree branches, all of which are broken and splintered from the force of the waves, because they are rigid.

From high in the air, amidst the very dense darkness, and fierce lightning and thunder, through the violently howling wind driven torrents of heavy rain, and with obscured vision and difficulty, we continue to see the many smaller 100 feet high rippling waves sweeping across the area, and from this height, they look like ripples upon a pond of water. Spaced at short intervals, they are all rippling, sweeping waves, without crests, and each of them is rapidly increasing the already high water level. Also, from this height, if the boat survived, we cannot distinguish it from the large amount of debris. There are no visible signs of mankind, or of the cities[696] of mankind, and the foot hills and lower Zagros Mountains,[697] far to the east and northeast of Shuruppak, are quickly being covered by the rapidly rising water level. The water now

---

[695]F. Page 85, line 115 The gods cowered like dogs (and) crouched in distress.

[696]M. Page 143, line 202 The storm swept over the capitals.

[697]V. Genesis 7:19 and all the high hills, that were under the whole heaven, were covered (similar to this).

sloshes back and forth, after slapping and glancing off the higher Zagros Mountains.[698] The very dense darkness, and fierce lightning and thunder, and the violently howling winds, as well as the wind driven torrents of heavy rain, all continue unabated.

We observe God Enlil in his craft, which is also now being very slightly buffeted, for it is now within the violent storms. His craft is in open communication with the other craft, including the craft of God Enki and Goddess Ninhursag. All the destructive sounds are barely audible within the craft. God Enlil is also alternately looking down through a porthole, and at the small viewing monitor, amidst the very dense darkness, and fierce lightning and thunder, through the violently howling wind driven torrents of heavy rain. He smiles, after having just heard the words of God Enki and Goddess Ninhursag. As he looks down at the smaller 100 feet high rippling waves, which he can also only barely see, even within the flashing lightning, as they sweep across the foam and debris filled water, he surveys the destruction. He is also looking at the former city of Shuruppak area, but away from the north, for he also has not seen the boat, or know that it had existed.

As he views the destruction and smiles, he smugly responds to God Enki and Goddess Ninhursag, through the open communication system, "Total destruction!"[699] He then turns, and casts admonishing looks at the Igigi Gods in his craft, and quickly scans over them, and we see a close up view of them. They are also all weeping, as are all the Gods and Goddesses, as are many of the Anunnaki Gods, as in the craft of God Enki and Goddess Ninhursag, and most of them are also cowering like dogs, crouched in terror beneath the portholes, and painfully looking at the viewing monitor.

We see time passing through a series of alternating cameo views of the rippling, sweeping waves, the solid very dark clouds, the very dense darkness, the fierce lightning and thunder, the violently howling winds, and the wind driven torrents of heavy rain, all of which, except the solid very dark clouds and the heavy rain, slowly diminish and finally cease, leaving a slightly lighter scene, with the solid very dark clouds and the heavy rain. Then, the solid very dark clouds begin to moderate to dark gray, as the sky begins to become slightly

---

[698]V. Genesis 7:20 and the mountains were covered (similar to this); and X. lower page 114 through upper page 115 (according to ancient Chinese annals) "The water was well up on the high mountains, and the foothills could not be seen at all." (etc.), "threatening the heavens with their floods."

[699]M. Page 99, line 44 You decided on total destruction.

lighter in color, and again we see daylight, and the heavy rain diminishes and becomes moderate rain, which continues at a steady pace, falling almost straight down, being driven by the now gentle wind, which is causing gentle waves. No Sun rays have yet pierced the clouds.

*

From high in the air, we witness a calm destruction scene of total water, with a moderate rain falling almost straight down from a dark gray colored overcast daylight sky, completely filled with dark gray clouds. There are no Sun rays penetrating the clouds.

7 Days After The Flood Began[700]

We zoom down to the level of the water, and then scan the area, and we see a close up view of the calm destruction scene, the former Shuruppak area, with the total high water filled with scattered debris, containing bloated animal carcasses, and including bloated human bodies, all serenely floating upon the water, some of which is rafting together, all being washed by the rain, and gently rocked by the gentle waves, but we do not see the boat of Ziusudra. The high waters, which have stopped rising, after reaching a height of approximately 8,100 feet, are almost calm, with gentle waves upon them. There is a moderate rain falling almost straight down, and a gentle wind, as all the other destructive forces have now ceased, and the overcast sky, with dark gray clouds, prevails.

---

[700]M. Page 97, line 24 For seven days and seven nights, 25 Came the deluge, the storm, [the flood], 26 Where it swept, mankind, and line 27 Was thrown down before it, and page 143, line 203 After the storm had swept the country for seven days and seven nights; and X. page 74 through upper page 78 An exceedingly strong wind endured seven days. All the time the land was shrouded in darkness (similar to this, but at an earlier time); and H. page 77, line 40 After the Flood had swept thereover; and M. page 145, line 205 And the destructive wind had rocked the huge boat in the high water; and F. page 85, line 127 Six days and [six] nights, 128 The wind blew, the downpour (rain), the tempest (winds), (and) the flo[od] (waves) overwhelmed the land, 129 When the seventh day arrived, the tempest (winds), the flood (waves), and line 130 Which had fought (wrought destruction) like an army, subsided in (its) onslaught (only these two, the winds and waves, subsided; but some rain still continued), and page 86, line 131 The sea grew quiet, the storm abated (winds subsided, but some rain still continued), the flood ceased (waves stopped, water stopped rising).

We observe Goddess Ninhursag in her craft, which is still in open communication, and throughout, we see cameo views of God Enki in his craft, listening to her words. Both are still in their flight attire. Her craft is above the former Shuruppak area, amidst the moderate rain and gentle wind. It is very quiet in the craft, for the audio is off, and the moderate rain and gentle wind sounds are not audible within the craft. She is searching for signs of life among the scattered debris, bloated animal carcasses, and bloated human bodies, all being washed by the rain, and gently rocked by the gentle waves. She is tired and emotionally spent. Her eyes are dark, her face shows anguish and stress, and her lips are parched.[701] She is still in grief, and is softly weeping, as she alternately looks down through a porthole, and at the small viewing monitor, for a close up view of the devastation below.[702] We scan within her craft, and observe the Gods, and notice all of them are seated, and all of them have parched lips, with some still weeping, and some still weeping while crouched[703] beneath the portholes, fearfully taking glancing looks at the viewing monitor.

One of the Anunnaki Gods sitting beneath the portholes, near the standing Goddess Ninhursag, consoles her, "Goddess Ninhursag, great lady, midwife of the Gods.[704] For mankind you cry. Like a woman in travail,[705] you wail and spend your emotions.[706] The wind, the downpour, the tempest, and the flood have overwhelmed the land for seven days and seven nights. The destruction is subsiding in its onslaught."[707] Then another Anunnaki God sympathetically relates to her, "We are all sitting, suffering cramps from thirst and hunger.[708] Our lips are feverishly

---

[701]M. Page 95, line 29 Her lips were covered with feverishness.

[702]M. Page 95, line 32 The goddess saw it as she wept, and page 97, line 4 Nintu (an epithet for Goddess Ninhursag) was wailing for mankind.

[703]F. Page 85, line 115 The gods cowered like dogs (and) crouched in distress.

[704]M. Page 95, line 28 Nintu, the great lady, and line 33 The midwife of the gods, the wise Mämi (see footnote 233).

[705]F. Page 85, line 116 Ishtar cried out like a woman in travail (suffering pains of childbirth. I believe Ishtar is a scribal error, intentional or accidental, and the name should be Ninmah/Goddess Ninhursag).

[706]M. Page 97, line 13 Nintu wailed and spent her emotion.

[707]F. Page 85, line 128 The wind blew, the downpour, the tempest, (and) the flo[od] overwhelmed the land, 129 When the seventh day arrived, the tempest, the flood, and line 130 Which had fought like an army, subsided in (its) onslaught (see footnote 700).

[708]M. Page 95, line 30 The Anunnaki, the great gods, and line 31 Were sitting in thirst and hunger, and page 97, line 22 They were suffering cramp from hunger.

athirst.[709] We bow, and weep with you for the people,[710] for we miss their labor, and their food and drink," and we see a close up view of the weeping Gods, those who are crouched beneath the portholes. Then the Anunnaki God continues, "Like sheep, we fill the trough,[711] and wait for mankind to serve us. But mankind is destroyed, and we must again provide for ourselves."

Goddess Ninhursag softly and emotionally reflects[712] to the Anunnaki Gods, as she weeps and looks down through a porthole, "As I weep, I ease my feelings.[713] Let the day become dark. Let it become gloom again,[714] for without mankind to provide sustenance for the Gods, our days shall again be dark and gloomy. My lips already thirst for their beer.[715] In truth, the olden time has turned to clay."[716] She now pauses, turns and looks beneath the portholes, at the seated Gods, and admonishing and questioning herself in a louder tone, she continues, "The oath of secrecy! How did I? How could I, with them, the assembly, command total destruction?[717] How could I command

---

[709]M. Page 97, line 21 Their lips were feverishly athirst; and F. page 85, line 126 Covered (parched and cracked) were their lips athirst.

[710]M. Page 97, line 15 The gods wept with her for the land (people), and line 18 Where she sat, they sat weeping; and F. page 85, line 124 The Anunnaki-gods wept with her, and line 125 The gods sat bowed (and) weeping.

[711]M. Page 97, line 20 Like sheep, they filled the trough.

[712]F. Page 85, line 117 The lovely-voiced Lady of the g[ods] lamented (lovely-voiced = advocate of mankind).

[713]M. Page 97, line 12 She wept and eased her feelings.

[714]M. Page 95, line 34 (She spoke) Let the day become dark, and line 35 Let it become gloom again (as it was for the Gods before mankind was created).

715M. Page 97, line 16 She was surfeited with grief and thirsted for beer (which I equate and refer to as liquid bread. From the creation or earliest days of mankind, the Gods probably imparted the art of making low alcohol beer and wine to mankind as a means of purifying water and maintaining it, preparing disease free water for drinking, probably at first for the exclusive use of the Gods; and these beverages may be the 'pure water', as opposed to the ordinary/normal natural drinking water of mankind, referred to in the ancient texts, for the use of the term 'pure water' implies a special quality through treatment of ordinary water, such as can be obtained through distillation (see footnote 38 and 295) or fermentation. Subsequently, the Gods had developed a liking/taste for the purified beverages of mankind, in this instance, beer, and closely associated both of them. Therefore, when the God's lips thirst for beer, they were metaphorically and possibly euphemistically longing for mankind: some of the Gods missed mankind).

[716]F. Page 85, line 118 In truth, the olden time has turned to clay (clay = inanimate/lifeless/death).

[717]M. Page 95, line 36 In the assembly of the gods, and line 37 How did I, with them, command total destruction?; and F. page 85, line 119 Because I commanded evil in the assembly of the gods, and line 120 How could I command (such) evil in the assembly of the gods (line 37 could be linked with footnote 491 and 634).

war to destroy my people?[718] For it is I who brought forth these people!"[719] Again she pauses, and stops weeping, as do the Gods, and then she continues in an angry tone, "God Enlil has had enough of bringing about an evil command![720] He uttered abominable evil!"[721] Again she pauses, and then remorsefully continues in a soft tone, "As a result of my own choice,[722] and to my own hurt, I have listened to the assembly's nonsense,"[723] and again she pauses, and looks to the far distance through a porthole, and views a mountain peak, barely sticking out of the water, to the far north of the former Shuruppak area.

---

[718]F. Page 85, line 121 (How) could I command war to destroy my people (war = evil).

[719]F. Page 85, line 122 (For) it is I who bring forth (these) my people! (If, according to Plato in his circa 400 B.C. 'Timaeus' and 'Critias' (see footnote 57), the Egyptian priests were correct when they revealed to his predecessor Solon in circa 600 B.C., that according to) "our sacred registers preserved in our temples," (the original Hellenic people/nation) "race/city" (was established) "founded" (by Athena (Ninhursag) about 9,600 B.C., 1,000 years before the Egyptian counterpart in 8,600 B.C., 200 years after the flood, was established by the same Goddess) "she is called in the Egyptian tongue Neith, which is asserted by them to be the same whom the Hellenes called Athene," (and Atlantis predated and later was contemporaneous to the original Hellenic nation, then the Sumerian flood story could very well be an account of the submergence of Atlantis; as I further explain under 'Ice' on page 366 in my epilogue. Therefore, because the names of the principal players in the later Hellenic/Greek accounts through Berossus are variants of the Sumerian story, we now presume the later Hellenic/Greek accounts and names followed the Sumerian, but it may very well be that the Sumerian account and names followed the earlier original Hellenic story of Deucalion. The Deucalion story is probably older than and not a variant of the Xisuthros/Sisithros (Ziusudra) flood story rendered by Berossus in circa 290 B.C., because in the report rendered by Solon in circa 600 B.C. through Plato in circa 400 B.C.), "the most ancient things in our part of the world — about Phoroneus, who is called 'the first,' and about Niobe; and, after the Deluge, to tell of the lives of Deucalion and Pyrrha," (the descendants of Hellen, son of Deucalion, were already known as Hellenes: see footnote 621. Additionally , and of considerable interest, further quoting Plato from the above source), "O Solon, Solon, you Hellenes are but children, and there is never an old man who is an Hellene," (which is understood and accepted to mean that the Hellenes had no actual written record, only a story of) "Phäethon, the son of Helios," (either a planet/Nibiru or lesser/dark companion star which orbits our Sun, preserved in) "the form of a myth, but really signifies a declination of the bodies moving around the earth and in the heavens," (of ancient devastating events) "upon the earth recurring at long intervals of time: (etc.), at the usual period, (etc.), and thus you (Hellenes) have to begin (learning) all over again as children, and know nothing of what happened in ancient times, either among us (Egyptians) or among yourselves (Hellenes)." (Thus, in this sense, due to our lack of knowledge of ancient devastating events, we earthlings have all become Hellenes).

[720]M. Page 95, line 39 Enlil has had enough of bringing about an evil command.

[721]M. Page 95, line 40 Like that Tiruru (a derogatory epithet for Goddess Ishtar), he uttered abominable evil.

[722]M. Page 95, line 42 As a result of my own choice.

[723]M. Page 95, line 43 And to my own hurt I have listened to their noise (the evil conduct/words/decision of the assembly: see footnote 325).

From the air, amidst the moderate rain and gentle wind, we see the craft of Goddess Ninhursag, with its brilliant glowing firefly-like light emitting from the underside, zoom away, toward the mountain peak, and quickly arrive and hover next to it. Then we look down from the air, and notice 30 bloated human bodies, floating upon the water, like a raft, clustered next to the mountain peak, among the other debris and bloated animal carcasses, all being washed by the rain, and gently rocked by the gentle waves, amidst the moderate rain, which is falling almost straight down, and the gentle wind.

Again we observe Goddess Ninhursag in her craft, as she again openly weeps, as she looks away from a porthole, after looking down and being appalled by the 30 bloated human bodies, rafting next to the mountain peak. She loudly and emotionally reflects to the seated Gods, as she turns toward them, "For seven days and seven nights came the deluge, the storm, the flood. Where it swept, mankind was thrown down before it."[724] She now turns, and again looks down at the 30 bloated human bodies, which look like a raft, among the other debris and the bloated animal carcasses, all floating upon the water, clustered next to the mountain peak, amidst the moderate rain, and all being washed by the rain, and gently rocked by the gentle waves. We see a close up view of the scene below on the small viewing monitor, behind Goddess Ninhursag, as she emotionally declares in a quiet tone, "My offspring, cut off from me, have become like flies![725] They have filled the sea like dragon flies.[726] What? Were they given birth by the rolling sea?[727] Like the spawn of fish, they fill the sea.[728] Like a raft, they have put in to the edge.[729] Like a raft, together they have put in to the bank,"[730] and some of the Gods resume their weeping with her, as they bow their heads, after looking at the close up

---

[724]M. Page 97, line 24 For seven days and seven nights, 25 Came the deluge (rain), the storm (winds), [the flood] (waves), 26 Where it swept, mankind, and line 27 Was thrown down before it.

[725]M. Page 95, line 44 My offspring—cut off from me—have become like flies! (Flies metaphorically refers to a large number, and in this circumstance, numerous floating bloated human bodies rafting together upon the surface of the water).

[726]M. Page 97, line 6 They have filled the river (former river, now a sea) like dragon flies! (The outstretched arms and legs of the floating bloated human bodies appeared like the wings of drowned dragonflies spread out upon the water), and page 162, note iv. 5-9.

[727]M. Page 97, line 5 What? Have they given birth to the [rolling] sea? (Could be: What? Were they given birth (brought forth/created) by the rolling sea? In this context, they would be a reference to the bodies: see footnote 634).

[728]F. Page 85, line 123 Like the spawn of fish they fill the sea.

[729]M. Page 97, line 8 Like a raft they have put in to the edge.

view of the bloated human bodies on the viewing monitor.

Then, after a long moment passes, Goddess Ninhursag recovers her composure, stops weeping, and turns away from the porthole, and now announces to the seated Gods in a firm and moderate tone, "And as for me, I have seen and wept over them.[731] I have ended my lamentation for them.[732] My cry has died away,"[733] and she pauses for a brief moment, and then she continues, as again she questions herself in an increasing tone, "Now that we have lost all mankind, how can I stay on this planet, Earth? Where shall I go? Shall I go up to our heavenly[734] planet, Nibiru, as if I were to live in a treasure house,[735] and ignore this loss? How can I forget mankind?" She now raises her tone, and becomes angry, as she loudly questions and denounces,"Where has God Anu, the president of the assembly gone,[736] whose divine sons obeyed his command,[737] and took the oath of secrecy. God Enlil did not consider the people individually, but consigned all the people to destruction."[738]

We observe God Enki in his craft, which is still in open communication, and throughout, we see cameo views of Goddess Ninhursag in her craft, listening to his words. His craft is above the former Shuruppak area, but to the north, amidst the moderate rain and gentle wind. It is also very quiet in his craft, for the audio is off, and the moderate rain and gentle wind sounds are not audible within the craft. He is deeply saddened, and is searching the scattered debris, bloated animal carcasses, and bloated human bodies, all strewn upon the water, and all being washed by the rain, and gently rocked by the gentle waves, and alternately looking down through a porthole, and at the

---

[730]M. Page 97, line 9 Like a raft together they have put in to the bank.

[731]M. Page 97, line 10 I have seen and wept over them.

[732]M. Page 97, line 11 I have ended my lamentation for them.

[733]M. Page 95, line 46 And as for me, like the occupant of a house of lamentation (a funeral home), My cry has died away.

[734]M. Page 95, line 48 Shall I go up to heaven (in this context, Nibiru).

[735]M. Page 95, line 49 As if I were to live in a treasure house? (Surrounded by abundance on Nibiru).

[736]M. Page 97, line 51 Where has Anu the president gone? (Possibly implies; how could God Anu, the president of the assembly, acquiesce and allow this to occur).

[737]M. Page 97, line 52 Whose divine sons obeyed his command (in this context, sons could be a reference to the assembly and the oath of secrecy, or actually to his sons involved in the destruction, or all the forces of destruction, whom God Enlil was the creator/parent of).

[738]M. Page 97, line 53 He who did not consider (the people individually) but brought about a flood, and line 54 And consigned (all) the peoples to destruction.

small viewing monitor, for a close up view of the devastation below; looking for the hexagon shaped reed boat, or debris he can identify as part of the boat. Some debris resembles the reed construction of the boat; reed bundle beams, floors, ramps, and exterior hull sections, but could also be from the rectangular shaped boats, or from the reed huts. He is not sure, as he envisions the boat in his mind, and tries to match from what part of the boat the debris has come.

God Enki declares and whispers to himself, as he looks down and scans over the water through a porthole, and the other Gods within his craft, or of the other craft through the open communication system, do not hear his words, "How could they have survived those seven days? It was more frightening than I had envisioned. If my plan was successful, and Ziusudra[739] survived the deluge, he shall emerge and announce his presence to me, for I can not find or identify the boat among all the debris."[740] Then we see the water below the craft of God Enki, which includes bloated human bodies, floating upon the water, among the scattered debris and bloated animal carcasses, some of which is rafting together, all being washed by the rain, and gently rocked by the gentle waves, amidst the moderate rain, which is falling almost straight down, and the gentle wind.

God Enki turns from the porthole, and looks at the viewing monitor, and then in a loud tone with disgust, exclaims to the Gods in his craft, all seated and all with parched lips, with some still weeping, and some still weeping while crouched beneath the portholes, fearfully taking glancing looks at the viewing monitor, "Look at the Earth! Who could have thought it would be like this?" Then God Enki walks to a computer console, and touches the heat sensitive numbers, and his words are instantly displayed, in hologram-like pictures of the past, on the small viewing monitor, and now in a normal tone, without disgust, he reflects, "Coming from deep space, and orbiting toward the star we call the Sun, our heavenly planet, Nibiru, which crosses through the solar system of planets, between the planets Mars and Jupiter, and whose symbol is the cross: normally, during previous orbits near this planet, Earth, every three thousand six hundred Earth years, as it approaches the region near the Earth, with the north axis of the Earth being nearest to Nibiru, coming

---

[739]V. Genesis 8:1 And God (God Enki) remembered Nō´-ăh (Ziusudra), and every living thing, and all the cattle that was with him in the ark (similar to this).

[740]M. Page 95, line 14 They were [not] recognizable in the destruction; and F. page 85, line 112 The people could not be recognized from heaven (now the boat cannot be found or identified among all the debris: see footnote 690).

closer to the perihelion of its orbit around the Sun, its closest distance to the Sun, and simultaneously closer to the perigee of its orbit, its closest distance to the Earth; the gravitational and electromagnetic force fields of Nibiru affect the Earth, causing instability and vibration of the Earth, as the Earth is quickly seized by the force fields of Nibiru."

"When the sequence begins, the fluid components of the Earth, the molten core, the waters, and the air, are affected, as they are all drawn toward Nibiru: for a period lasting approximately fourteen Earth days, as Nibiru begins its secondary seizure of the Earth, very few light Earth tremors begin, as the molten core of the Earth is attracted toward Nibiru. Then, for a period lasting seven Earth days, as Nibiru begins its primary seizure of the Earth, the very few light Earth tremors increase, and become moderate Earth tremors with fissures, and volcanic eruptions with lava and thick clouds of smoke, ash, and steam begin, and temperatures of both the surface of the Earth and air begin to increase, as the molten core of the Earth is drawn toward the surface, away from the interior, toward Nibiru. Also, the rapid displacement of the underground waters and of the seas begins, in the form of flooding, with mild flooding in the central regions of the Earth, and very severe flooding at the north and south axis regions, as the waters of the Earth are drawn across the Earth, primarily toward Nibiru. This is due to the gravitational and electromagnetic effect of Nibiru upon the waters of the Earth, especially the equatorial waters, and of the Earth itself, like the tidal effect upon the Earth caused by the Moon, except far more severe. For the equatorial waters, along with the Earth at the equator, are thirteen Earth miles higher than at the axis, therefore, as the forces of Nibiru overcome, neutralize, and transcend the forces of the Earth, the equatorial waters are forced to recede downward, toward both axis areas of the Earth."

"At the north axis area, the waters are pulled away from the Earth, causing them to bulge, and this is the area of the strongest pulling effect, because it is nearest to Nibiru, while at the opposite side of the Earth, at the south axis area, the Earth itself is pulled away from the waters of the Earth, toward Nibiru, and concurrently the waters float away from the Earth, also causing a bulge of the waters. Also, the air is disturbed in a similar manner, as it is drawn across the Earth toward Nibiru, causing mildly blowing winds to begin. Concurrently, the surface waters of the Earth are excessively evaporated, due to the increased frictional heat of the molten core acting upon the waters, as the molten core is pulled toward the exterior, and the resulting pure water mist is drawn up from the Earth and appears as a rainbow, and then the mist condenses, causing a profuse moderate rain to fall upon the Earth. Within this seven Earth day period, the moderate Earth tremors with fissures, the volcanic eruptions, the temperatures, the rain, the flooding, and the mild winds are at their maximum levels. At the end of these seven Earth days, as Nibiru concludes its primary seizure of the Earth, the moderate Earth tremors diminish, and again become very few light Earth tremors, and the volcanic eruptions cease, and the temperatures slightly decrease and remain near that level, and the rain ceases, and the mild winds cease, and also, the rapid displacement of the underground waters and of the seas, in the form of flooding, ceases, but the flood waters are held at their highest levels. Then, after a period lasting approximately fourteen Earth days, as Nibiru concludes its secondary seizure of the Earth, the very few light Earth tremors cease."

"At the end of this entire quick seizure period, after which Nibiru has completely seized the Earth within its gravitational and electromagnetic force fields, holding the Earth captive, the flood waters are held at their highest levels for a period of five Earth months, as Nibiru reaches the perigee of its orbit, its closest distance to the Earth. Then, after Nibiru passes through the perigee of its orbit, and for the following five Earth months, as Nibiru orbits away from the perigee of its orbit, the gravitational and electromagnetic force fields of Nibiru begin to very slowly diminish, and simultaneously very slowly release the Earth, and the flood waters are released from their highest levels, and begin to very slowly recede. This normal water displacement period totals ten Earth months."

"Then, normally, after these ten Earth months, the last five of which the Earth is being very slowly released from the force fields of Nibiru, for a period lasting approximately fourteen Earth days, as Nibiru begins its secondary release of the Earth, very few light Earth tremors again begin, as the molten core of the Earth retreats from Nibiru. Then, for a period lasting seven Earth days, as Nibiru begins its primary release of the Earth, the very few light Earth tremors increase, and become moderate Earth tremors with fissures, and volcanic activity with thin clouds of smoke, ash, and steam again begin, and the temperatures of both the surface of the Earth and air begin to slightly increase, as the molten core of the Earth withdraws from the surface, toward the interior. Also, the rapid replacement of the underground waters and of the seas begins, and also, the mildly blowing winds begin. Within this seven Earth day period, the moderate Earth tremors with fissures, the volcanic activity, the rapid replacement of the underground waters and of the seas, and the mild winds are at their maximum levels. At the end of these seven Earth days, as Nibiru concludes its primary release of the Earth, the moderate Earth tremors diminish, and again become very few light Earth tremors, and the volcanic activity ceases, and the temperatures decrease and approach their normal level, and the mild winds cease, and also, the rapid replacement of the underground waters and of the seas ceases, as they now have completely returned to their former place. Then, after a period lasting approximately fourteen Earth days, as Nibiru concludes its secondary release of the Earth, the very few light Earth tremors cease, signaling the end of this entire slow release period, and the end of the passing period of Nibiru, and the Earth is completely released from the gravitational and electromagnetic force fields and effects of Nibiru, and again, the freed Earth is quiet and stable, as our heavenly planet, Nibiru, continues toward the perihelion of its orbit around and then away from the Sun, and begins its return to deep space." Then God Enki pauses, and scans within the craft, and we observe the faces of the Gods within his craft, all closely listening to the words of God Enki, as they simultaneously watch the viewing monitor.

God Enki then emotionally and empathetically continues, "But with this passing of Nibiru, the normal gravitational and electromagnetic effects were drastically increased. Because of the acute proximity of Earth to Nibiru during this orbit,[741] and the location of the other nearby planets, the solid Earth itself was disturbed, as well as the fluid components, with the degree of disturbance being unusually extreme. The Earth was temporarily pulled out of its normal orbital path, as were the other nearby planets,[742] and the terrestrial axis of the

[741]X. Middle page 274 the catastrophe was caused by a planet (Nibiru) in contact with the earth (similar to this, but at an earlier time).

[742]X. Lower page 209 (quoting Job 38:32) Canst thou bring forth Măzz´-ă-rōth in his season? or canst thou guide Ärc-tū´-rŭs with his sons? The (Greek) translation of the Seventy (Septuagint) reads: Canst thou bring forth Mazzaroth in his season and guide the Evening Star by his long hair? (Mazzaroth = Marduk = Nibiru, and season = orbit duration, and guide = influence, and evening star = Venus, and long hair = long reaching or powerful radiating forces. Interpreted, it could read: can you (Job) produce Nibiru during its orbit, and influence Venus by its powerful radiating forces? Or could the word Ärc-tū´-rŭs/Arctures, which is a very bright star in the northern sky, be a symbolic word, and actually be a reference to and represent the arctic/north passing/orbiting bright Nibiru, and his sons be a reference to the 4 moons/Winds of Nibiru), and lower page 250 The common cause may have been some comet (or the planet Nibiru) which changed the orbits of Venus and Mars.

Earth was temporarily tilted,[743] compounding the interim displacement of the waters, and especially the equatorial waters, toward the axis areas.[744] All these additional factors caused the huge waves and the violent storms, with the waters of the huge waves from the equatorial bulge regions being primarily pulled toward the north regions of the Earth, and held at their highest levels. The severity and destruction of these flood waters was overwhelming: unprecedented. Instead of the normally mild flooding, this has been an almost complete covering of all the lands, except for the highest mountains." Again, God Enki pauses, and scans within the craft, as the viewing monitor returns to the present, and again, we observe the faces of the Gods within his craft, all closely listening and empathizing with God Enki, as they now look at God Enki.

God Enki resumes, as he optimistically and hopefully exclaims, "Some people may have survived on the high mountains!" Then, he looks down through a porthole, and declares, "There are no signs of life here, let us continue our search in the mountainous areas, to the far north of the former Shuruppak area," as he looks down through the porthole, amidst the moderate rain and

---

[743]X. Middle page 317 After a cosmic upheaval shifted the sky (similar to this, but at an earlier time); and Z. upper page 128 through upper page 130 The largest interplanetary bodies probably carried so much energy and momentum that, depending on the direction they approached a planet, they could have tilted it (the Earth), speeded up its spin, slowed down its spin.

[744]X. Middle page 332 When the earth's rotation was disturbed (retarded/slowed, but evidence during the flood orbit for stasis is lacking, but it is very possible), waves of translation moved eastward, because of inertia, and poleward, because of the recession of the waters from the equatorial bulge where they are held by the rotation of the earth (but the pulling forces of Nibiru upon the waters could have been greater than the rotational forces of the Earth upon the waters, and overcame them, without slowing the Earth's rotation, thus producing the same effect: a recession of the equatorial bulge waters toward the axis, especially the north), and lower page 85 but the celestial body (Nibiru) nearby would disturb (overcome) this poleward recession, drawing the water toward itself (as I further explain under 'Catalyst' on page 322 in my epilogue. Incidentally, the word catastrophe in Hellenic means down turn, a sudden down turn/slowing of the Earth's rotation; which would cause destructive earthquakes, for the plates of the Earth's crust, because of their inertia, would continue to turn at or near their former speed, until they were piled up or driven down; as I further explain under 'Continental Drift' on page 368 in my epilogue: see footnote 630. And since this is a specific use word, describing a very unique event, involving rotation, it may have been contemporaneously composed and initially used to describe the effects upon the Earth of a passing/orbit of Nibiru; and today used to describe a destructive event, which would be synonymous. Also, another word comes to mind, cataclysm, I believe of similar origin, which means a down wash, such as a deluge, or an overflowing of waters, a sudden violent change involving water).

gentle wind, at the scattered debris, bloated animal carcasses, and bloated human bodies floating upon the water, all being washed by the rain, and gently rocked by the gentle waves.

From the air, amidst the moderate rain and gentle wind, we see the craft of God Enki, with its brilliant glowing firefly-like light emitting from the underside, zoom away, toward the far north mountainous areas.

*

From the air, amidst the moderate rain, which is falling almost straight down, and the gentle wind, we see a large area, much further north of the far north mountainous areas. The daylight sky in this area is also filled with dark gray clouds, and there are no Sun rays piercing the clouds.

Several Hours Later

The debris strewn water stretches in all directions, with no mountain peaks visible. Most of the debris is scattered, but some of it is rafting together, with a few very large piles among the rafting debris, and we zoom in for a close up view. The debris is composed of intact: reed bundles; reed bundle boat floors, ramps, and exterior hull sections; woven reed mats; reed bundle hut wall sections; small tree branches; and palm leaves, in addition to broken and splintered: wood tables and benches; tree trunks; and large tree branches. Among the debris, there are bloated animal carcasses, and bloated human bodies, all being washed by the moderate rain, and gently rocked by the gentle waves. As we scan over the debris, stopping and examining it for possible similarities to the boat of Ziusudra, we suddenly stop, and scan back over an extremely large pile of debris.

Then, we zoom in further and take a closer look, and see the debris cluttered, now dull black pitch coated weathered and wet roof of the reed boat. As we now slowly scan, we see the entire perimeter of the hexagon shaped reed boat. The boat is not easily identifiable, for it is disguised by the very large amount of debris that surrounds it, as well as the exceedingly large amount of debris laying on and hanging off the roof. It is floating in the high, but almost calm water, among the gentle waves, still riding considerably less than 10 feet above the water line. The boat is now being driven by the gentle wind and gentle waves, as the moderate rain continues in its uninterrupted fall upon the silent boat. There are no signs of life, and we hear no sounds coming from the slightly deteriorated and battered boat, except the drumming sound of the moderate rain upon the cluttered roof of the boat. The boat is still tightly sealed, as it had been when the flood waves first swept it away, 7 days earlier, although some of the pitch coating has been worn off, and the appearance of the remaining pitch has changed from shiny black to dull black, even though it is wet. It appears that Ziusudra and all his people have perished.

Inside the boat, we see a shadowy close up view of the water clock, still securely lashed to the vertical reed beam near the sealed door, and it shows the 24th day, and the spool is still on the 2nd month. We slowly scan within the sixth deck, the uppermost, or main deck of the boat, which has a very dimly illuminated shadow filled eerie atmosphere, for nearly all of the clay oil lamps, still securely lashed to the vertical reed beams, are now out, and we notice the coating of shiny black pitch on the inner side of the reed hull, still as shiny as it had previously appeared. It is a very quiet scene. We observe emotionally and physically exhausted people,[745] with a very unkempt look, who are tense, tired, weak, and hungry, within their compartment-like areas. They listen to the moderate rain drumming upon the roof, the predominant sound, and frequently look up at the roof, as they are untying the long 1/2 inch wide leather lashes from their drooping bodies, and releasing themselves from the vertical reed beams. The lashes are being carefully hung on the inner horizontal reed beams, for future use, if necessary, for the people are not

---

[745]V. Genesis 7:23 and Nō´-ăh (Ziusudra) only remained alive (from this area), and they that were with him in the ark (similar to this. I believe other human beings and animals, on high elevations or mountains in other areas of the Earth, must have also survived the flood; as I further explain under 'Continental Drift' on page 368 in my epilogue, for how else can we account for the widely varying skin pigmentation/color and physical attributes of mankind upon the Earth today. If others did not survive, we would all share the same characteristics as the children of Ziusudra/Noah or the other occupants of the boat, his relatives and friends, presumably one skin color and similar physical attributes, for they were all from the same local area; or the characteristics of the children of Ziusudra/Noah and his relatives and friends covered the entire spectrum, which would be highly unlikely, for he, and it must be presumed his relatives and friends, was selected and saved because he was free from sin and transgression, which can also be interpreted as possessing pureness; or indeed there was further genetic engineering after the flood, as is referred to later, of new categories for the people, which could have included different races, and therefore different characteristics, including different skin pigmentation/color and physical attributes).

convinced the worst is over, and they are proceeding cautiously, with many huddling together for comfort and strength. The interior of the boat is dry, and is quickly again becoming softly illuminated, as the clay oil lamps are now being refilled with oil and relighted; some having exhausted their supply of oil by burning, and others having it spilled 7 days earlier. The expedient soft illumination does not hesitate at crowding the shadow filled eerie atmosphere back into seclusion. No daylight is penetrating the boat. Some small personal possessions and noticeable amounts of grain are strewn about the floor, where they were thrown by the force of the waves, in addition to evidence of spilled water and wine, all from their still securely lashed containers, baskets, and vases. Most of the people are shaky on their feet, as they attempt to regain their equilibrium. A few people search for and retrieve their strewn personal possessions, while others collect the spilled grain, and still others embrace and comfort each other. A mother cautiously instructs her teenage daughter, "Hang the lashes carefully, as we may need them again," and the daughter respectfully obeys.

Then, we observe the people untying the animals within their compartment-like areas, the third and fourth decks, and they are also a little shaky, like the people, and they quickly begin to eat the large bundles of grass, which like the still overturned large flat clay drinking bowls, are strewn about upon the reed bundle floor from the force of the waves. Then we notice the teenagers untying their pet dogs and retrieving their pet cats from the reed beams, and then petting and hugging them with affection.

Then we see the birds, within their compartment-like areas, the first and second decks, with the first deck being the lowest deck of the boat. They are mostly settled, and again peacefully sit upon the horizontal reed beams bracing the hull, and the horizontal and diagonal truss reed beams, using them as a perch. Their large flat clay drinking bowls are still overturned, as are their large flat reed baskets, with seeds for the birds also strewn about upon the reed bundle floor, except on the first deck, where both are somewhat contained within the bottom exposed bracing beams, as the people begin to collect the seeds, and place them in the large flat reed baskets. But unlike the animals, which had been lashed to the beams and unable to eat, the unrestrained birds have already been eating, and do not quickly rush to the seeds.

Then we observe Ziusudra, looking at the water clock, and declaring to the pilot, and requesting of the spent people as they expectantly look on,

"After seven days, the flood waves have ceased, and the sea has grown quiet. Let us pray to God Enki." Ziusudra and all the people sit upon the reed bundle floor, and cross their legs, and bow their heads, and amidst the drumming sound of the moderate rain upon the roof, Ziusudra pleadingly exhorts, "God Enki, please guide us through this flood. Please save this boat. Save mankind."

We see time passing through a series of alternating cameo views, with the advancing water clock superimposed on them, of the exterior of the boat of Ziusudra, and of the moderate rain diminishing and ceasing within the gentle wind, as the dark gray clouds increasingly lighten and begin to scatter. As the sky clears, the first few rays from the Sun break through the now almost normal clouds, and intermittently shine upon the boat, and also upon the debris scattered over the water, of which the bloated animal carcasses and bloated human bodies now begin to decompose, all being gently rocked by the gentle waves, and in the distance a rainbow is forming, with the days on the water clock, and the spool marking the month, both advancing from the 24th day of the 2nd month, and progressing through the 3rd month, but before the days reach the 27th day of the 3rd month, the superimposed advancing water clock fades away.

*

It is middle morning, and amidst the gentle wind, again we see the exterior of the boat, and as we scan, we see the entire perimeter of the dull black pitch coated weathered hexagon shaped reed boat, looking as it had previously, surrounded and cluttered by the large amount of debris, in the high and almost calm debris strewn water, which stretches in all directions, and still riding considerably less than 10 feet above the water line. The mostly scattered debris, including bloated and now decomposing animal carcasses, and bloated and now decomposing human bodies, with some of it rafting together, is being gently rocked by the gentle waves. The sky is clearing further, and now the clouds are normal, and the rays from the Sun,[746] to the east, uninterruptedly fall directly upon the boat, which is still being driven by the gentle wind and gentle waves.

---

[746]M. Page 145, line 206 The Sun came out, illuminating the earth and the sky.

40 Days After The Flood Began

Inside the boat, we see a close up view of the water clock, still securely lashed to the vertical reed beam near the sealed door, and it shows the 27th day, and the spool is on the 3rd month. We scan the main deck, and as the still securely lashed clay oil lamps softly illuminate the interior, we observe Ziusudra, still wearing his special full length purple robe with gold edging, which covers both shoulders, and his gold God Enki symbol necklace, and his gold winged sphere symbol of Nibiru on his robe, above the left breast, and his gold bracelet. He is looking at the water clock, and again declaring to the pilot, and then commanding him, as the people, who are now slightly improved in appearance, but still slightly unkempt looking, and more relaxed, and still wearing their colorful robes, and symbol necklaces, expectantly look on, "Now, after forty days, the rain has abated,[747] and truly the flood has ceased.[748] Take your knife, and make a slit here," as Ziusudra turns and points to a spot high on the shiny black pitch coated inner side hull, 2-1/2 feet below the roof bracing beam, near the top and center of one of the six, 100 feet wide hull sections, between two 6 inch diameter horizontal hull bracing beams, and to the left of one of the 6 inch diameter vertical truss reed beams, and to the right of one of the 6 inch diameter diagonal truss reed beams, and he makes a 2 inch vertical motion with his finger, and then the pilot bows. There is a large storage chest nearby, still securely lashed to an inner vertical and diagonal truss beam.

Ziusudra then turns to and requests of the people, "But first, let us pray to God Enki, my personal God," and Ziusudra and all the people again sit upon the reed bundle floor, and cross their legs, and bow their heads, and Ziusudra imploringly exhorts, "God Enki, I beg of you God Enki. We pray to you everyday, with reverence and humility. We appeal to you, with our full devotions. Mankind is obedient. Hear my plea and save mankind, save us. For it is your duty, as my personal God, to care for us in time of need."

Ziusudra and the pilot then rise, as the others remain seated in prayer, but with their heads now raised, and some people grasp their symbol necklace

---

[747]V. Genesis 7:12 And the rain was upon the earth forty days and forty nights (similar to this).

[748]V. Genesis 7:17 And the flood was forty days upon the earth (similar to this).

pendant. As Ziusudra stands back nearly 10 feet from the hull, the pilot takes a 12 inch long flint knife from its leather sheath, which is hanging from the reed beam near the door, among the other sheaths. He prepares to insert it through the shiny black pitch coated reed hull, and proceeds to work it into the 6 inch thick reed bundle hull, at an angle, and in a poking manner, to pry the reeds apart. Then, he makes a 2 inch vertical slit in the hull, and as he slowly pulls the pitch covered knife out, a ray of sunlight follows the knife out of the slit, and it brightly shines directly upon the reed bundle boat floor, with the ray passing next to where Ziusudra is standing.

All the people see the ray, and are silent, as their faces begin to slightly beam with reserved joy and anxiety at the sight. Then, Ziusudra rushes to the slit, and presses his eye to it, and excitedly turns and commands the pilot, "Make an opening for a window."[749] The pilot bows, and then Ziusudra steps 5 feet away from the hull, directly behind the pilot, and anxiously observes.

As the people watch, the pilot places the knife into the slit, and works the knife up and then down at an angle, a total of 14 inches vertically, to form the right side of the window. Then he carefully does the same, except 29 inches to the left of the first slit, to form the left side of the window, which is also 14 inches in height, all between the 6 inch diameter vertical, diagonal, and horizontal reed beams. Then the pilot sets the knife down, and takes the 16 inch long saw-like flint knife from its leather sheath, which is also hanging from the same reed beam, the same knife Ziusudra used to saw through the hawser, and proceeds to saw at an angle, 29 inches horizontally, joining the top of each of the two slits, and then the bottom of each of the two slits. This is more difficult, and is done in a sawing manner, as he is now cutting through the reeds, whereas before, he was spreading the reeds apart. As he saws, more light eagerly enters the boat through the increasing new slits.

Soon, he has the 14 inch by 29 inch hull section free, an inward tapering perimeter reed bundle plug, 6 inches thick, and he grasps it and first holds it outside, as the Sun now streams in, and then he turns it lengthways, to fit it through the opening, and then he brings it into the boat, and we notice the

---

[749] M. Page 145, line 207 Ziusudra made an opening in the huge boat ('made an opening', implies that no sealed windows existed, and a new opening was cut to make a window), and page 171, note 207 'to open a window' is to be taken in the sense of 'making the opening', not of 'opening a closed window'; and V. Genesis 6:16 A window shalt thou make to the ark, and in a cubit shalt thou finish it above (etc., similar to this: see footnote 664).

contrast in colors of the pitch coated hull section; the outer side being dull black, and the inner side a shiny black. Then he steps aside, allowing Ziusudra, his high Priest and King, the privilege of the first look out the window opening, the bottom of which is 4-1/2 feet from the reed bundle floor. The Sun now brilliantly shines through the window opening, and falls directly upon the face of Ziusudra,[750] who now squints, as it somewhat illuminates the interior area beyond the window.

The squinting pilot is joyous, as he laughingly compliments Ziusudra, "The sunlight falls upon your face, my King. What better place could it have first chosen?" Upon hearing this, Ziusudra broadly smiles. As they also squint, all the people give a sigh of relief, and momentarily forget their plight, as they also break out in smiles, and with some people breaking out in laughter. Then Ziusudra, with his eyes now adjusted to the bright sunlight, and who is still broadly smiling, steps to the window, and slowly and curiously puts his head directly into the opening, and looks out the window, and as he silently stares out, his face suddenly becomes serious, and his broad smile fades into a shocked look of disbelief.

The people also quickly become quiet, and again become serious, as they look at the strained expression upon the face of Ziusudra, for they cannot see out, they are still in a cross legged seated prayer position, upon the reed bundle floor. Total silence now permeates the boat.

As Ziusudra looks out the window, to the east, he sees the scattered debris upon the almost calm water, and he notices the debris which is rafting together in piles, and especially the large amount of debris next to the boat, all of which is being gently rocked by the gentle waves, and other debris hanging off the roof of the boat. It is very quiet, and the high water is everywhere. There are no signs of life. He sees no mountain peaks, as the boat is much further north of the far north mountainous areas, for the mountainous areas which God Enki and Goddess Ninhursag are searching, are much further south of the boat. As he scans over the water, he notices a few bloated and decomposing animal carcasses, and bloated and decomposing human bodies among the debris. Then Ziusudra laments, in a loud but sad toned voice, so all the people can hear, "Upon the sea, all is silence.[751] All mankind has turned

---

[750]M. Page 145, line 208 And the Sun with its rays entered the huge boat; and F. page 86, line 135 I opened (cut open) a window, and light fell upon my face.

[751]F. Page 86, line 132 I looked upon the sea, (all) was silence.

to clay.[752] The sea is as level as a roof,"[753] and then he pauses, and turns around to the people, and solemnly declares, "We must wait until the waters recede."

Ziusudra slowly and very sadly walks away from the window, and sits on the large storage chest nearby and weeps,[754] and tears slowly streak down his face,[755] and flow into his beard. The people curiously now get up from their seated prayer position, and make their way to the window by working themselves around the diagonal truss bracing beams, located every 5 feet on center, and most people timidly and cautiously look out the window, standing a good distance away from the window, with a few people courageous enough to stand very close to the window. They are also shocked and very saddened by the scene, and some sit down and begin to weep with Ziusudra, while others stare out the window, dazed by the total high water and scattered debris view that fills their eyes.

We see time passing through a series of alternating cameo views, with the advancing water clock superimposed on them, of normal clouds passing through the sky, and sunrise to sunset sequences, and sprinkled throughout the passing time, we see the gentle wind and gentle waves diminish and cease, and the boat of Ziusudra become motionless upon the now calm water, with the days on the water clock, and the spool marking the month, both advancing from the 27th day of the 3rd month, and progressing through the 7th month, but before the days reach the 17th day of the 7th month, the superimposed advancing water clock fades away.

*

It is middle morning, shortly before 8:00 a.m., and again we see the exterior of the boat, now totally free of debris, and as we scan, we again see the entire perimeter of the dull black pitch coated weathered hexagon shaped reed boat, now with four identical window openings, one on each opposite side, near the

---

[752]F. Page 86, line 133 And all mankind had turned to clay (clay = inanimate/ lifeless/death: see footnote 716).

[753]F. Page 86, line 134 The sea was as level (flat) as a roof.

[754]F. Page 86, line 136 I bowed, sat down, and wept.

[755]F. Page 86, line 137 My tears (began) running down over my face.

top and center of four of the six, 100 feet wide hull sections, except the side where the sealed door is located, and opposite it. The boat is motionless upon the calm and quiet, waveless water, bathed in sunlight from a cloudless bright blue sky. The debris has now floated free by itself, away from the boat, or has been removed and pushed away, by the boatmen climbing out the windows and onto the roof, and throwing the debris into the water: thus, there is no longer any debris surrounding the boat, or cluttering the roof of the boat, and the boat again rides 10 feet above the water line. The debris is now thinly scattered over the water, and we no longer see bloated and decomposing animal carcasses, or bloated and decomposing human bodies among it.

5 Months After The Flood Began[756]

Inside the boat, we see a close up view of the water clock, still securely lashed to the vertical reed beam near the sealed door, and it shows the 17th day, and the spool is on the 7th month. We scan the main deck, and see the three new window openings, now a total of four, which have been cut into four opposite sides of the main deck hull of the boat. All the new window openings are the same size, 14 inches high by 29 inches wide, and are at the same location as the first window opening, near the top and center of the sides of the boat. The sunlight now moderately illuminates the main deck, where all the still securely lashed clay oil lamps have been extinguished, however, the securely lashed clay oil lamps continue to softly illuminate the five lower decks of the boat. Life is going on as normally as possible, and the people are in a cheerful state of disposition, now completely relaxed, and better groomed, but still slightly unkempt looking, and now with slightly dirty faces, but generally improved over how they had looked almost 4 months earlier. They are still wearing their now slightly soiled colorful robes, and symbol necklaces, and all are still without their head and ear wear, pendant cylinder seals, and bracelets. We notice water being moderately rationed, and being somewhat attentively measured and poured into clay cups from a clay vase.

The animals are being cared for, and freely given additional large bundles of grass to eat, and additional moderately rationed amounts of water in their

---

756 V. Genesis 7:24 And the waters prevailed upon the earth an hundred and fifty days (similar to this).

large flat clay drinking bowls, which are both neatly placed about them, and goats are being milked by the teenage children and young adults. All the animals are peaceful.

The birds are also being cared for, and given additional seeds to eat, as the teenage children and young adults freely add seeds to the large flat reed baskets, and moderately rationed amounts of water to their large flat clay drinking bowls, which are both also neatly placed about them. All the birds are calm.

We observe Ziusudra, appearing in the same clothing and adornments that he had previously worn, including his gold God Enki symbol necklace, except his purple robe is now slightly soiled, his wrinkled face is now slightly dirty, and his graying black tight-curly hair and beard are still slightly unkempt. He is on the main deck, looking out the west window, with the pilot standing next to him. The pilot holds the sextant-like device in his hand, which he has just finished looking through out the window.

From the window, we view the water, including the water directly beneath the window, which is about 6 feet below. There is no movement by the boat in the quiet water, for it is calm, and there is no current. There are several small pieces of thinly scattered debris beneath the window, that sit motionless upon the water.

Ziusudra attests to the pilot, "For nearly four months, I have looked in all directions for the boundaries of the sea.[757] They are not to be found. Look, to the west," and Ziusudra motions out the window with his hand, and he declares, "all is water!" Then the pilot consolingly replies, "There has been little wind and current, my King, for this reason, we have not moved very far in all this time," and they both resume looking out the west window. Suddenly, directly beneath the window, they notice the small pieces of thinly scattered debris moving past the boat, and toward the south, in a slowly developing current. Both are surprised, as they look at each other with an astonished look upon their face, and then quickly, they again look out the window. (After 5 Earth months, Nibiru has now passed through the perigee of its orbit, its closest distance to the Earth, and the waters begin to recede, as its gravitational and

[757]F. Page 86, line 138 I looked in (all) directions for the boundaries of the sea (boundaries = land).

electromagnetic force fields begin to diminish.)[758]

From the window, we again see the small pieces of thinly scattered debris, being carried toward the south, and then the pilot excitedly exclaims to Ziusudra, "We are moving! We have begun to move!"

Everyone on the boat immediately notices the movement. Some people notice the lashes, hanging on the inner horizontal reed beams, as they begin to slightly sway, and others notice the movement by their own very slight imbalance. All the people instantly stop whatever they are doing, and take notice. The people quickly run up the ramps to the main deck, to where Ziusudra is, and most crowd around, while others again prepare to lash themselves to the vertical reed beams, as they begin to tremble, fearing a resumption of the destructive activity of the huge flood waves.

Ziusudra turns away from the window, and enthusiastically announces to the people, some of whom are nearly out of breath from having just run to him from the lowest decks, "The waters are receding! Returning to the south, from where they came, and the boat is being carried in the return flow!" Then Ziusudra turns to the pilot and exclaims, "To the south window! To see where the waters take us," and they both run toward the south window, working their way around the 5 feet on center diagonal truss bracing beams, followed by the now very curious people, for a better view of where the boat and the thinly scattered debris are being carried toward. But some people cannot wait, as they crowd around the west window, some with their head out the window for a better view.

From outside the boat, we observe the people at the west window, their faces filled with mixed expressions, from joy to fear, all crowding to the window, as the boat is being carried to the south, at a rate of 2 miles per hour, and slowly but steadily increasing.

Meanwhile, back inside the boat, Ziusudra and the pilot, with the people following, arrive at the south window, and both resume looking out over the water, and closely watch the slowly increasing current. Then, after a few

---

[758]V. Genesis 8:1 and God made a wind (an actual reference to the stimulated wind felt upon the Earth, caused by the force fields of Nibiru; or Nibiru itself, accompanied by its four satellites called Winds; and possibly a reference to one of these satellites called the North Wind, which initially drove the Earth and Moon into our present orbit) to pass (orbit) over (beyond) the earth (and through its perigee), and the waters asswaged (abated, similar to this; as I further explain under 'Winds' on page 344 in my epilogue).

moments, as Ziusudra scans the distant expanse of water, he spots a speck of land on the far south horizon; the protruding twin peaks of a mountain, which they are being carried toward, and as he points with his excited finger out the window, he shouts, "Look, to the south! A stretch of land!"[759]

We see the speck of land from the south window; it is barely visible, mostly due to the curvature of the Earth. The pilot then holds up the sextant-like device, looks through it, and takes a measurement, and declares to Ziusudra, "The water has carried us three hundred and ten miles north of Shuruppak, our city, and the twin peaks of that mountain top are forty[760] miles south of us. That fixes the high mountain two hundred and seventy miles north of Shuruppak." The pilot pauses for a moment, and again looks out at the twin-peaked mountain in the far distance, and then questioningly asserts, "It appears to be the top of Mount Nisir? We are being slowly carried back toward

---

[759]F. Page 86, line 139 At (a distance of) twelve (double-hours) there emerged a stretch of land (this could be a reference to a small amount of land, however, the use of the term stretch of land could imply an elongated element and/or verify a twin-peaked mountain, in which the lower or connecting land portion of the mountain would appear to be stretched between the two higher points of each peak); and T. lower page 360 Who in the twin-peaked mountain, Arrested the boat (a description of the physical characteristics of the mountain on which the boat landed; to my knowledge, the clearest reference to the appearance of the landing site; as I further explain under 'Location Of The Boat' on page 351 in my epilogue).

[760]F. Page 86, line 139 At (a distance of) twelve (double-hours) there emerged a stretch of land (1 double-hour is equal to 7 linear miles, therefore 12 double-hours equal 84 linear miles. Since 12, like 7, is a mystical number, and there is a variant of 98 linear miles in) F. page 86, note 194 Var. : fourteen (which could imply that it has been changed, and that possibly neither is correct. I prefer to use my own figure of 6 double-hours, 1/2 the distance given, or about 40 nautical miles. Keep in mind, the greater the distance, the greater the probability Ziusudra would have seen an even higher mountain first, for we must also double the total area of visibility; if he can see 40 miles in one direction, he can also see 40 miles in the opposite direction, or a total area of 80 miles. Therefore, at a distance of 84 miles, approximately 9,000 feet high Mount Nisir would have to be protruding 4,699 feet above the water to be barely visible, and at 98 miles, 5,885 feet, for as the visibility distance of the protruding mountain increases, the mountain must also increase in height, or grow, which is improbable, or the depth of the water must decrease, exposing more of the mountain to see, and eventually Mount Nisir would be totally out of the water, and there would not be a flood. But if the distance is truly 84 or 98 miles, this means the water was not as deep as I have given, approximately 8,100 feet deep: it was only either about 4,301 or 3,115 feet deep, thus giving Ziusudra the ability to see the mountain at a much further distance. Therefore, according to the best figures available to me, which allow for the curvature of the Earth, a 900 feet height is visible at a distance of 34.4 nautical miles, and adding the 10 feet above water level height of the boat, a visibility distance of 3.6 nautical miles, I arrive at a total visibility distance of 38 nautical miles, which I round off to 40 nautical miles).

Mount Nisir, to the south, by the receding waters!"

As Ziusudra looks on, the pilot turns and instructs some of the boatmen, "Boatmen, prepare a steering oar!" Six boatmen quickly run toward the 3 inch diameter by 20 feet long wood punting-poles, and begin lashing them together with the long 1/2 inch wide leather strips, to form a steering oar.

From the air and water level, we see the hexagon shaped boat being slowly and gently carried in the now strong current, which has now increased to and remains at the rate of 5 miles per hour, along with thinly scattered debris, toward the protruding twin peaks of Mount Nisir, which are becoming more visible and simultaneously larger in height, and seem to be rising out of the water. This effect is due to the decrease in the curvature of the Earth, as the current continues to lessen the distance between the boat and the mountain.

From the air, within the bright sunlight, we see a close up view of Mount Nisir, with its twin peaks and a small portion of the main part of the mountain connecting the peaks being the only portion of the mountain above the water level.

MOUNT NISIR[761]

As we scan completely around the twin mountain peaks and small portion of the main part of the mountain protruding from the water, we notice approximately one-tenth, or 900 feet of the 9,000 feet high mountain above the water, with most of the main part of the mountain still beneath the water level, with the slowly moving strong current of the receding waters washing around the sides of the mountain; all as the waters recede at a rate of 1 foot per hour. We can see the highest point of the water, a water mark, only 6 inches above the water level, and the washing effect of the current on the north, east, and west sides of the mountain, where there is almost no debris, due to the

---

761F. Page 86, line 140 On Mount Niṣir the ship landed, and page 250 The place on which Utnapishtim's (Ziusudra's) boat came to rest is given in the Gilgamesh Epic as Mount Nisir; and G. lower page 29 through page 32 We cannot close these brief remarks on the long inscription of King Erridu (Eridu) - pizir of Guti without recalling the fact that, according to a copy of an evidently much older geographical list from the library of Ashurbânapal, it was a mountain of the country Guti, Mt. Niṣir, on which the ship of the Babylonian Noah, Ût-napishtim, landed; and U. page 27 The ark had stopped on Mount Nisir (in the Hellenic story of Deucalion/Noah, the ark landed on Mount Parnassus).

washing effect of the current. Most of the debris is accumulating, in large quantities, on the south side of the mountain, which has a pocket of quiet water next to it, a whirlpool-like suction effect of the strong current, which attracts and holds the debris as it sweeps around the mountain. At this hour, the boat is still 40 miles north of the mountain, and as we scan to the north, it is not clearly recognizable among the thinly scattered debris.

*

It is middle afternoon, 4:00 p.m. of the same day, and from the air and water level, we see the hexagon shaped boat in the strong current, along with some thinly scattered debris, with the steering oar out the north window opening, being guided toward Mount Nisir, still with only a small portion of the main part of the mountain, in addition to both twin peaks, above the water level. The boat is in the strong current of the receding waters, with the enormous mountain very close, directly in front of the boat.

8 Hours Later

Inside the boat, at the north window, we observe the six boatmen struggling with the steering oar, while many people anxiously watch, with Ziusudra and the pilot at the south window, at the opposite side of the boat, also with many people anxiously watching. Many people are also crowding around the east and west window, with a few of them leaning their head out the window for a more favorable view, and the others are praying, as they sit cross legged upon the reed bundle floor, and bow their heads. The pilot is directing the boatmen, as Ziusudra watches the rapidly approaching Mount Nisir, with the high water mark now 8 feet above the water level, exposing dark, wet mud around the mountain. The pilot shouts to the six boatmen, "Left, to the left!"

From the air and water level, we see Mount Nisir, which is still approximately one-tenth, or 900 feet visible above the water, and the hexagon shaped boat, along with some thinly scattered debris, as it is beginning to be closely swept past the east side of the massive mountain.

Meanwhile, back inside the boat, as Ziusudra looks out the south window at the enormous mountain, with the pilot next to him, and as the people anxiously look on, we observe the pilot excitedly directing the six boatmen, who continue to struggle with the steering oar, as he shouts to them, "Right, to the right!"

From the air and water level, we see the hexagon shaped boat very close to the enormous mountain, which dwarfs the size of the boat, just as the boat is closely swept past the east side of the mountain, and caught in the quiet water on the south side of the mountain, in the whirlpool-like suction effect of the strong current, along with the large quantities of debris already caught and accumulated there, while the main flow of the strong current continues to sweep past; all as the waters slowly recede. After a few very long moments pass, suddenly the boat is caught and grounded[762] on the south side of the mountain, and we see the debris filled quiet water slowly and steadily receding away from the hexagon shaped boat, still at the rate of 1 foot per hour. After 15 minutes, we notice a 3 inch water mark that has appeared on the boat, 10 feet from the top, and it is increasing in height each minute, exposing a contrast in colors between the dull black pitch coated weathered hull, and the newly exposed hull; being beneath the water level and thus protected, it is slightly darker in color, although it is also dull black.

*

From the air, we see the brightly shining Sun, and as we scan across the cloudless blue sky, we experience a peaceful and content feeling of relief.

48 Hours Later

We look down and scan, and bathed in the bright sunlight, we see the exterior of the 200 feet diameter by 50 feet high, dull black pitch coated weathered hexagon shaped reed boat, and an area 8 feet below the bottom of the boat, now completely above the debris filled slowly and steadily receding

---

[762]F. Page 86, line 141 Mount Niṣir held the ship fast and did not let (it) move; and V. Genesis 8:4 And the ark rested in the seventh month, on the seventeenth day of the month, upon the mountains (one of the mountains in this mountain range) of Ăr´-ă-răt (within the territory of Armenia (Urartu), similar to this; as I further explain under 'Location Of The Boat' on page 351 in my epilogue).

quiet water level, with the water mark clearly distinguishable 40 feet from the bottom, and 10 feet from the top of the boat. As we scan, we notice the boat is grounded upon a relatively flat area, a comparatively small plateau, and the main flow of the strong current continues to sweep past the mountain. The boat is slightly leaning, with the bottom of the hùll side resting upon the ground, and there are large pools of water, and dark, wet mud all around the boat, as the area is very, very wet. There is also a large quantity of debris in the same area that the boat is grounded upon, with some protruding from the dark, wet mud, and some protruding from the pools of water, but most of the debris is lying slightly beneath the dark, wet mud; all as the waters continue to recede, and the Sun brightly shines upon the scene, reflecting from the pools of water.

We see time passing through a series of alternating cameo views, with the advancing water clock superimposed on them, of normal clouds passing through the sky, and sunrise to sunset sequences, and sprinkled throughout the passing time, we see the slowly receding waters,[763] and the hexagon shaped boat of Ziusudra, grounded upon the mountain plateau, amidst the calm winds, with the days on the water clock, and the spool marking the month, both advancing from the 19th day of the 7th month, 2 days after the boat grounded, through the middle of the 8th month,[764] and the pools of water, some of which reflect the bright sunlight, slowly evaporating under the bright Sun, leaving dark, wet mud, with the water clock advancing through the middle of the 9th month, and the dark, wet mud slowly drying and beginning to crack, and slowly turning to gray, except for the very deep pools of water, which are taking very much longer to evaporate and dry. Then, we observe Ziusudra, wearing his gold God Enki symbol necklace, and the adornments that he had previously worn, gazing out a window at the transformation, as his purple robe increasingly becomes more soiled, with the water clock advancing to the 1st day of the 10th month, when from a window, Ziusudra notices other

---

[763]V. Genesis 8:3 And the waters returned (to their place) from off the earth continually (similar to this).

[764]F. Page 86, line 142 One day, a second day Mount Niṣir held the ship fast and did not let (it) move, 143 A third day, a fourth day Mount Niṣir held the ship fast and did not let (it) move, 144 A fifth day, a sixth day Mount Niṣir held the ship fast and did not let (it) move, and line 145 When the seventh day arrived (I believe not enough land would be exposed in just seven days of receding waters to warrant the release of a bird. Additionally, the Biblical account gives a more detailed chronology at this point, therefore, I prefer to use the Biblical chronology: see 'The Chronology Of The Flood' on page 349 in my epilogue).

mountain tops exposed by the slowly receding waters.[765]

Then, as the waters continue to slowly recede, and expose more land, with the water clock advancing 40 days, to the 12th day of the 11th month, from a window,[766] Ziusudra sends out a dove,[767] and as Ziusudra waits, later on the same day, the dove returns to the window.[768] Then, as the waters continue to slowly recede, and expose more land, with the water clock advancing 7 days, to the 19th day of the 11th month, from a window, Ziusudra sends out a swallow,[769] and as Ziusudra waits, later on the same day, the swallow returns to the window.[770] Then, as the waters continue to slowly recede, and expose more land, with the water clock advancing 7 days, to the 26th day of the 11th month, from a window, Ziusudra sends out a raven,[771] and as Ziusudra waits, later on the same day, the raven does not return to the boat.[772]

Then, with the water clock advancing 1 day, to the 27th day of the 11th month; very few light Earth tremors begin to slightly shake the boat, and the calm winds begin to increase, and become very light, as Nibiru begins its secondary release of the Earth from its gravitational and electromagnetic force

---

[765]V. Genesis 8:3 and after the end of the hundred and fifty days (5 more months) the waters were abated, and 8:5 And the waters decreased continually until the tenth month: in the tenth month, on the first day of the month, were the tops of the (other) mountains seen (similar to these).

[766]V. Genesis 8:6 And it came to pass at the end of forty days, that Nō´-ăh (Ziusudra) opened the window of the ark which he had made (similar to this. It is important to note the past tense of the words, 'which he had made'; the window which was cut into the boat forty days after the flood began, which could have been periodically plugged with the removed section, for how could he see the tops of mountains, in the preceding footnote, if no windows existed at that time, forty days earlier; or this could be out of order, and if it is, it should be included in footnote 749, where the window is made: see page 294 in my comments).

[767]F. Page 86, line 146 I sent forth a dove and let (her) go; and V. Genesis 8:7 And he sent forth a raven (similar to this).

[768]F. Page 86, line 147 The dove went away and came back to me, and line 148 There was no resting-place, and so she returned.

[769]F. Page 86, line 149 (Then) I sent forth a swallow and let (her) go; and V. Genesis 8:8 Also he sent forth a dove from him (similar to this).

[770]F. Page 86, line 150 The swallow went away and came back to me, and line 151 There was no resting-place, and so she returned.

[771]F. Page 86, line 152 (Then) I sent forth a raven and let (her) go; and V. Genesis 8:10 And he stayed yet other seven days; and again he sent forth the dove out of the ark (similar to this. Although Noah releases a fourth bird, Ziusudra does not).

[772]F. Page 86, line 153 The raven went away, and when she saw that the waters had abated, and page 87, line 154 She ate, she flew about, she cawed, (and) did not return; and V. Genesis 8:12 And he stayed yet other seven days; and sent forth the dove; which returned not again unto him any more (similar to this).

fields. Some people, all of whose colorful robes are also increasingly becoming more soiled, begin to become somewhat tense and fearful, as they sit upon the reed bundle floor and cross their legs, and bow their heads and pray, and others take the lashes hanging on the inner horizontal reed beams, and again securely lash themselves to the vertical reed beams: all grasp their symbol necklace pendant. Ziusudra also has a somewhat tense and fearful look upon his face, as he gazes out a window.

Then, as the waters continue to slowly recede, exposing more land, with the water clock advancing approximately 14 days, through the end of the 11th month, and then the spool advancing to the 12th month, and the days on the water clock resume advancing through the beginning of the 12th month, and to the 9th day of the 12th month; distant volcanic activity with thin clouds of smoke, ash, and steam begin, as the very few light Earth tremors increase, and become moderate Earth tremors with fissures, and moderately shake the boat, and the temperatures of both the land and air begin to slightly increase, and the very light winds increase, and become mild, as Nibiru begins its primary release of the Earth from its gravitational and electromagnetic force fields. Now, many people begin to become more tense and fearful, as they join those who already sit upon the reed bundle floor and cross their legs, and bow their heads and pray, and many others now take the lashes hanging on the inner horizontal reed beams, and join those who have already securely lashed themselves to the vertical reed beams: all grasp their symbol necklace pendant. Even Ziusudra is tense and frightened by the activity, as he gazes out a window.

Then, as the waters continue to slowly recede, exposing more land, with the water clock advancing 7 days, through the middle of the 12th month, and to the 16th day of the 12th month; the distant volcanic activity with thin clouds of smoke, ash, and steam cease, and the moderate Earth tremors with fissures diminish, and again become very few light Earth tremors, which again only slightly shake the boat, and the temperatures of both the land and air decrease and approach their normal levels, and the mildly blowing winds diminish and again become very light, as Nibiru concludes its primary release of the Earth from its gravitational and electromagnetic force fields. The people have a somewhat relieved look, as they all grasp their symbol necklace pendant, with some people untying the lashes and releasing themselves from the vertical reed beams, and they are somewhat less tense and fearful, as they again hang the lashes on the inner horizontal reed beams, and join the many people who already sit upon the reed bundle floor and cross their legs, and bow their heads and pray. Even Ziusudra has a somewhat relieved look upon his face, as he gazes out a window.

Then, as the waters continue to slowly recede, exposing more land, and with the very few light Earth tremors and very light winds continuing, with the water clock advancing approximately 14 days, through the end of the 12th month, and then the spool advancing to the 1st month, and the days on the water clock resume advancing to the 1st day of the 1st month; the very few light Earth tremors cease slightly shaking the boat, and the very light winds cease and again become calm, as Nibiru concludes its secondary release of the Earth from its gravitational and electromagnetic force fields, and again all is quiet. Now, all the people are again completely relaxed, as they all grasp their symbol necklace pendant, with the remaining people untying the lashes and releasing themselves from the vertical reed beams, as they again hang the lashes on the inner horizontal reed beams, and then join nearly all the people who already sit upon the reed bundle floor and cross their legs, and bow their heads and pray.

Ziusudra also has a completely relaxed look upon his face, as he directs four boatmen to open the sealed door, which they proceed to do and return it to its former place next to the opening, after which the main deck now becomes brightly illuminated by the sunlight, and Ziusudra gazes out the door opening at the quiet Earth, but notices there are still wet spots near the boat.[773] Then, as the waters continue to slowly recede, exposing more land, with the days on the water clock, and the spool marking the month, both advancing from the 1st day of the 1st month, and progressing through the 2nd month, but before the days reach the 26th day of the 2nd month, the superimposed advancing water clock fades away.

*

With a bright Sun and clear blue sky above, amidst the calm winds, we

[773]V. Genesis 8:13 And it came to pass in the six hundredth and first year (of the life of Noah: see footnote 550), in the first month, the first day of the month, the waters were dried up (drying up) from off the earth; and Nō´-ăh (Ziusudra) removed the (reed door) covering of the ark, and looked, and, behold, the face (surface layers) of the ground was dry (but the deeper layers were still wet, too wet and soft to support mankind, similar to this).

see the dull black pitch coated weathered hexagon shaped reed boat, grounded upon the relatively flat, comparatively small plateau, on the south side of Mount Nisir, with the large quantity of debris slightly beneath the thick coat of dry and cracked gray mud, which covers the entire landscape. All of the plateau area is now dry. There is no animal, bird, or human activity outside the boat, and no signs that there has been, as the boat is still sealed, except for the four, 14 inch by 29 inch window openings, and the nearly 6 feet high by 4-1/2 feet wide door opening, all on the sixth deck, the uppermost, or main deck, near the top and center of the sides of the boat, which leaves only one, 100 feet wide hull section without an opening, the side opposite the door opening. It is a very bleak and quiet scene.

12 Months and 10 Days After The Flood Began

Inside the boat, we see a close up view of the water clock, still securely lashed to the vertical reed beam near the now open doorway, and it shows the 26th day, and the spool is on the 2nd month. We scan the main deck, which is brightly illuminated by the sunlight, now eagerly entering the door opening, in addition to the four window openings, and we see the people, still slightly unkempt looking, and still with slightly dirty faces, groomed and appearing almost as well as they had previously. They are still wearing their now very soiled colorful robes, and symbol necklaces, and all are without their head and ear wear, pendant cylinder seals, and bracelets. Life is still going on as normally as possible, and the people have a good disposition; cheerful and smiling. We notice water now being very carefully rationed, and being very attentively measured and poured into clay cups from a clay vase.

The animals are still being cared for, drinking carefully rationed amounts of water from the large flat clay bowls, and freely eating the large bundles of grass, all neatly placed about them, and goats are being milked, all being performed by the teenage children and young adults.

The birds are also still being cared for, drinking carefully rationed amounts of water from the large flat clay bowls, and freely eating seeds from the large flat reed baskets, all neatly placed about them, also all being performed by the teenage children and young adults.

Then, we observe Ziusudra on the main deck, near the open doorway, with his family, the pilot, boatmen, and other people who are gathering in a group in front of him. He is wearing the same clothing and adornments that he had previously worn, including his gold God Enki symbol necklace, except his purple robe is now very soiled, his wrinkled face is still slightly dirty, and his graying black tight-curly hair and beard are still slightly unkempt. More people are quickly walking up the reed bundle connecting ramps to the main deck, from the five lower decks. Finally, with all 350 people on the boat now gathered in a large group in front of him, Ziusudra fervently reflects, "When we first landed on Mount Nisir, seven months ago, on the seventeenth day of the seventh month, amidst the calm winds, our boat was held fast, the mountain did not let the boat move as the waters receded. Then, after two and one-half months on the mountain, on the first day of the tenth month, as the waters receded, other mountain tops became visible."

"Forty days longer we waited, to the twelfth day of the eleventh month, while the waters were receding, and I sent forth a dove, and let her go. The dove went away, and later came back to me. There was no resting place, so she returned to me. Seven days later, on the nineteenth day of the eleventh month, I sent forth a swallow, and let her go. The swallow, like the dove, went away, and later came back to me. There was no resting place, and so she returned. After again waiting for seven days, on the twenty-sixth day of the eleventh month, I sent forth a raven, and let her go. The raven went away, and when she saw that the waters had abated, she ate, she flew about, she cawed: she did not return."

"Then, beginning the next day, on the twenty-seventh day of the eleventh month, and lasting for about fourteen days, slight tremors shook the boat, and the calm winds began to stir, and we prayed as we became fearful, and we waited. Then, beginning after that period, on the ninth day of the twelfth month, and lasting for seven days, the tremors increased and became more severe, and we noticed fissures, and distant volcanoes became active with thin clouds of smoke, ash, and steam, and the air became warmer, and the winds increased further, and we prayed as we became more fearful, all as the waters were receding. At the end of those seven days, ten months after the flood began, and five months after we landed on Mount Nisir, on the sixteenth day of the twelfth month, the distant volcanoes, with their thin clouds of smoke, ash, and steam, ceased, and the tremors diminished and again became slight, and the fissures ceased, and the warm air relented and returned to normal, and the winds diminished, and we prayed as we became less fearful."

"Then, after about fourteen days, on the first day of the first month, almost two months ago, the slight tremors also ceased, and the winds diminished further and again became calm, and once more the Earth was quiet, and we prayed as we again became relaxed, and we opened the sealed door."

"After all this time, waiting with patience and praying on Mount Nisir, it is now dry and safe enough to go out of the boat." We see a close up view of the relieved and cheerful faces of the people, and then Ziusudra commands, "As we did when we cut the openings for the windows, tomorrow, on the twenty-seventh day of the second month,[774] twelve months and eleven days after the flood began, we shall cut an opening as a door, near the bottom of the boat." The people broadly smile, as their eyes fill with excitement, and then Ziusudra elucidates, "I shall send forth everything to the four winds,[775] and offer a sacrifice to the Gods, and especially God Enki.[776] We shall kill an ox, and slaughter a sheep.[777] We shall pour out a libation on the peak of the

---

[774]V. Genesis 8:14 And in the second month, on the seven and twentieth day of the month, was the earth dried (now the deeper layers, as well as the surface layers, were dry and firm enough to support mankind, similar to this).

[775]M. Page 99, line 30 To the [four] winds he sent forth everything; and F. page 87, line 155 (Then) I sent forth (everything) to the four winds (to all directions; but this I feel strongly could be a reference to the four Winds of Nibiru: see footnote 758) and offered a sacrifice; and V. Genesis 8:17 Bring forth with thee every living thing that is with thee (similar to this).

[776]M. Page 145, line 209 The king Ziusudra, and line 210 Prostrated himself before the Sun-god (the Sun God is God Utu, but this could imply either Ziusudra prayed to God Enki within the sunshine, or prayed directly to God Utu, thanking him for bringing forth the Sun and drying out the Earth enough so mankind could once again walk upon it).

[777]M. Page 99, line 31 He put an oxen and a sheep on the fire; and K. upper page 98 The king kills an ox, slaughters a sheep; and M. page 145, line 211 The king slaughtered a large number of bulls and sheep (should be, an ox and a sheep, for it would be senseless to kill more than needed, after saving the animals from the flood); and F. page 87, line 155 (etc.), and offered a sacrifice (implies a ritual upon slaughter); and V. Genesis 8:20 And Nō´-ăh (Ziusudra) builded an altar unto the LORD (God Enki); and took of every clean beast, and of every clean fowl, and offered burnt offerings (implies a ritual after roasting) on the altar (similar to this. Whether the method of killing the animals and/or roasting the animals constituted the Sumerian/Babylonian 'offered a sacrifice'; or an altar was utilized in order to constitute the Biblical 'offered burnt offerings', is interesting. The question is, do you need an altar, and what was it made of, in order to perform a ritual and make an offering? The answer is no, for it depends on the ritual itself. A sacrificial ritual implies that it would occur prior to or upon slaughter of the animal, before roasting, and an altar would probably not be used to kill the animal upon, however, the blood or a part of the sacrificial animal could have been placed upon an altar, and rituals performed. A burnt offerings ritual implies that it would occur after roasting, and a part of the roasted animal would be placed upon an altar, and rituals performed. Therefore, because the Biblical version is most certainly a variant, both accounts are probably reflecting the same event, but through a different mirror: rituals were performed. However, in either case, evidence of a roasting

mountain.[778] Seven, and yet seven kettles we shall set up,[779] and under them we shall heap up sweet cane, cedar, and myrtle.[780] We shall perform our rituals, and make offerings, and give thanks to the Gods, and especially God Enki, now that our threat has passed. Then we shall feast and rejoice."

*

Inside the boat, as we listen to what sounds like the faint din of music and merrymaking, we scan throughout the main deck, brightly illuminated by the sunlight, and strangely there are no people to be found, and then we zoom to the water clock, and we see a close up view of the water clock, which is still securely lashed to the vertical reed beam near the open doorway, and it indicates the 27th day, and the spool marking the month, is on the 2nd month.

---

pit, an altar, and a hexagon shaped boat, may exist and still be found on a twin-peaked mountain, thus confirming the correct mountain, method of offering, and shape of boat).

[778]F. Page 87, line 156 I poured out a libation (wine) on the peak (top) of the mountain (implies a ritual with wine); and M. page 99, line 32 Providing food and libation on the peak of the mountain (the Biblical account, at a slightly later time in Genesis 9:20, of Noah/Ziusudra planting a vineyard immediately after the flood, as one of his initial acts, if not the first, probably in the highlands, for the lowlands were probably still flooded/wet and not yet habitable, can be better appreciated after we understand the fact that the normal/natural supply of drinking water of mankind, probably stream/river water, had been contaminated by the flood waters, and the supplies he carried with him were limited. Therefore, it was necessary for Noah/Ziusudra to purify the water through fermentation by preparing a low alcohol content beverage, such as in this instance wine, or beer, as had also been performed prior to the flood, in order to drink disease free water: as a means of purification of water. It should be noted that Noah/Ziusudra and his people must have sojourned in the highlands, part of the so-called fertile crescent, for some length of time, for it takes approximately 7 years to fully develop a grape/vineyard, which would confirm the minimum length of stay at that site: see footnote 716).

[779]F. Page 87, line 157 Seven and (yet) seven kettles I set up (very special (7), and yet even more special (7) = very, very highly esteemed/important effort/occasion (14: see footnote 216). We do not know what was prepared in the fourteen kettles, for no specific description of the ingredients is given. However, we can assume it was a stew or soup-like entree, or a supplement to the sheep and ox, for its preparation required the use of kettles. But what was in the kettles, if only one sheep and one ox were sacrificed? There are several possibilities, among which are giblets, from the slaughtered sheep and ox, or fowl (as mentioned in Genesis 8:20), fish, vegetables, or even other much smaller and less significant animals).

[780]F. Page 87, line 158 Under them I heaped up (sweet) cane, cedar, and myrtle (aromatic thus prized special scented wood = very special occasion. Although the accounts report it in the form of the first person, being the high Priest and King, Ziusudra probably assigned the preceding tasks to his subordinates, thus administering them, however, he probably personally performed the rituals associated with them. Through this interpretation, I use the word we, instead of I, after the first sentence in the preceding four sentences).

May 17th
The Next Day

Outside the boat, from the plateau, as we look up and scan across the sky, again there is a bright Sun and clear blue sky above, and amidst the now somewhat increased din, we look down and scan away from the boat and mountain. Off in the near distance, we notice there are still many deep pools of water, in the highlands, some of which reflect the bright sunlight, but the lowlands and valleys are still totally flooded, and wherever the waters have receded from, the landscape appears as a very dreary and colorless gray blanket.

We scan back toward the boat, with the din steadily increasing, and now becoming moderately loud, and we view the banquet area, on the dry and cracked gray mud caked ground, next to the dull black pitch coated weathered hexagon shaped reed boat, upon the comparatively small plateau. It is a joyous occasion for all 350 people,[781] and they are drinking red or white wine, which is being freely poured from pale red and cream decorated clay vases into similarly decorated clay cups. Both adults and young people are merrymaking, happy the ordeal is over, with many already having formed small lines and circle dancing to the music from a few exuberant musicians: all within the bright sunlight. Their appearance is the same as it had been the day before, wearing their very soiled colorful robes, and symbol necklaces, with many now wearing their clean head and ear wear, and bracelets, which had been put away immediately after the farewell banquet at Shuruppak, before the flood began, but all the people still have slightly dirty faces.

As they laugh and run, and make a game of it, the teenage children, along with their pet cats and dogs, are busy locating, unearthing, and collecting the splintered and broken wood, or intact reed bundle debris, and knocking the dry and cracked gray mud off of it. Then, they carry the debris to the adults, who are using it to fuel a large circular fire, within a large circular shallow roasting pit, over which one sheep and one ox are being turned on wood skewers and roasted. There are fourteen, 10 inch diameter, pale red and cream decorated baked clay kettles filled with stew, with two opposite holes near the lip of

---

[781]V. Genesis 8:18 And Nō´-ăh (Ziusudra) went forth, and his sons, and his wife, and his sons' wives with him (similar to this).

each, which are being prepared around the circumference of the large roasting fire. The fire is emitting light smoke, which is being gently lifted, rising high into the calm air. There is also much aroma in the air, some from the various special scented wood being carefully added to the fire and burned, but mostly from the food itself, and all the people are pleasurefully smelling the scent. As we scan over the banquet area, we see 15 wood tables, 30 wood benches, and 20 woven reed mats, which are gathered together; all brought out from the boat. Upon the tables and mats are placed various sizes and kinds of clay dishes, bowls, platters, cups, and vases; all with the familiar pale red and cream decorations.

There is a newly cut, inward tapering nearly 6 feet high by 4-1/2 feet wide door opening, at the bottom and near the center of one of the six, 100 feet wide hull sections, of the 50 feet high hexagon shaped reed boat, with its clearly distinguishable water mark 10 feet from the top. Also, leading to the plateau from the newly cut nearly square door opening, there is a nearly 12 feet long by 4-1/2 feet wide pitch coated reed bundle boat ramp, composed of both removed 6 inch thick door sections, and braced beneath by the 3 inch diameter wood punting-poles, which have been cut in half, to 10 foot lengths, and staggered to extend to the nearly 12 feet length of the ramp. All of the animals and birds are being led, or carried out of the boat, and set free.[782] In addition, most of their food; grass bundles, large flat reed baskets of seeds, and large flat clay drinking bowls, are being brought out of the boat, and all are being neatly placed about. All of the animals gather near the boat, while the birds sit on the roof; all acquainting themselves with their new found freedom, with none of the animals or birds very anxious to leave their proven secure home.

As the people continue their drinking and merrymaking, we continue to scan the banquet area, and from a short distance, again we see the large circular fire, emitting scented light smoke, with the sheep and ox being turned and roasted, and then we zoom to the sheep, and we see a close up view of the roasting sheep, with the flames licking it, which is just beginning to turn from pink to light brown, and has not yet been fully roasted. Then, we look down into the flames of the fire, and we see time passing as the flames diminish,

---

[782]V. Genesis 8:19 Every beast, every creeping thing, and every fowl (etc., that was in the boat), went forth out of the ark (similar to this).

and the fire is reduced to brightly glowing embers, but is still emitting scented light smoke.

*

We look up from the brightly glowing embers to the sheep, which is now a dark brown color, and is fully roasted.

2 Hours Later

After being fully roasted, the sheep and ox are being removed from over the brightly glowing embers of the fire, and carved, and together with the stew, are being served. The meat is offered on large clay platters, from which portions are placed onto individual clay dishes. The stew is carried in its kettle, all fourteen of them individually, by passing a wood stick through the two opposite holes near the lip of each, and poured with a clay ladle into clay bowls. Then, the people begin to eat their first true meal in over 12 months, and they savor every bite of it, as they eat and drink, and laugh and smile. As we scan the banquet area, from a short distance, we see the now moderately glowing embers remaining from the large circular fire, and we follow the scented light smoke, which is rising high into the calm air.

From the air, we see a group of five craft, glistening in the bright sunlight, with their brilliant glowing firefly-like light emitting from the underside of each craft, which are slowly passing over the bleak and harsh looking colorless gray mountain lands, to the northeast of the former Shuruppak area, from which the waters have recently receded. They are in open communication, as they search for signs of survivors, and survey the desolate landscape and debris, which appear as mounds under the blanket of gray mud, now that the waters have receded from them; but there are no visible signs of mankind or of their achievements to be found: as though they had never existed. The lowlands and valleys surrounding the mountains are still totally flooded with very deep water, and some of the highlands still have many deep pools of water in them, some of which reflect the bright sunlight.

We observe God Enki in his craft, with several Gods and Goddesses, 6 Anunnaki God officers, all now in their usual attire, and 4 Igigi Gods, who have continually worn their usual attire. God Enki is looking down, searching the highlands with empathic eyes, again trying to identify the mud blanketed debris with parts of the hexagon shaped reed boat of Ziusudra, as he alternates between the portholes, and the small viewing monitor, for a close up view of the scene below.

Then we observe Goddess Ninhursag in her craft, with several Gods and Goddesses, 4 Anunnaki God officers, also all now in their usual attire, and 4 Igigi Gods, who continue to wear their usual attire. Goddess Ninhursag is looking down with compassionate eyes, also searching the dried and cracked mud covered highlands for survivors. She now wears a necklace of fly shaped, equilateral triangles of lapis lazuli, deep blue stones, in addition to her 4 inch diameter identifying gold symbol pendant, which hangs around her neck on a gold chain, like all the four main Gods wear. She also alternates between the portholes, and the small viewing monitor, for a close up view of the scene below.

Again we observe God Enki in his craft, searching, as he looks at a close up view of a dreary gray mountain side on the viewing monitor. He is now standing next to a computer scanner, which he proceeds to activate by touching the heat sensitive numbers on the computer console, and a fixed light pattern appears on the display. Then he encouragingly declares to Goddess Ninhursag, through the open communication system, "Goddess Ninhursag, now that the waters have receded from the mountains and highlands, let us continue to search through the destruction, for I still believe that mankind may have survived upon these high mountainous lands. I have activated my computer scanner, and increased the sensory range, but there has been no response," and we see a close up view of the computer scanner, with its fixed display light pattern, as God Enki continues, but now in a melancholy tone, "for if there were any survivors, they would have certainly given us a signal by now," and God Enki is filled with sadness by his own words.

Again we observe Goddess Ninhursag in her craft, as she looks down through a porthole, and despairingly responds to God Enki through the open communication system, "God Enki, the more we search, the more I believe there may not be any survivors, for the destruction was far greater than I thought it to be. Look at the debris, it is strewn everywhere. Everything has turned to clay."

From the air, we scan the mud blanketed, debris strewn side of a bleak mountain, all covered with dry, gray mud, which contrasts to the lower areas of wet, dark mud, still surrounding the large pools of deep water. It is a very desolate scene.

Again we observe God Enki in his craft, as he notices the light pattern on the display of the computer scanner, which is beginning to show movement. Suddenly the lights are alive with activity, as they dance on the display, and we see a close up view of the actively flashing display lights. God Enki watches the flashing display lights for a moment, and then enthusiastically exclaims, "Goddess Ninhursag, my computer scanner is detecting smoke, to the northwest, and at a great distance. Let us see what it is," and his eyes fill with joy, as he begins to smile.

From the air, we see the group of five craft very quickly accelerate, as they glisten in the bright sunlight, with their brilliant glowing firefly-like light emitting from the underside of each craft, and they zoom away, toward the northwest, and disappear out of sight over the horizon.

We view the banquet area, and as we scan, we notice everyone is now enjoying the food, for it is the first roast they have had since the farewell banquet at Shuruppak, and the first stew they have had since before the farewell banquet, both over 12 long months ago. The people are halfway through the meal, consisting of roast sheep and ox, and stew, and we observe their joy filled slightly dirty smiling faces, as they converse and laugh with each other, and freely drink the red or white wine.

Then we observe the smiling and joyful, slightly dirty face of Ziusudra, who is clothed in his very soiled special full length purple robe with gold edging, which covers both shoulders, and wearing his gold God Enki symbol necklace, and his gold winged sphere symbol of Nibiru on his robe, above the left breast, and his gold bracelet. As he lifts a cup of wine to his lips, and tilts his head back to drink, his eyes look up to, and simultaneously scan the sky to the south. Suddenly his eyes stop, as he stares at the reflecting sunlight in the far distance: the five craft glistening in the bright sunlight. But Ziusudra is perplexed and hesitant, for the reflections could be from a pool of water upon the side of a mountain, or the craft of the Gods.

From the plateau, we see the five craft glistening in the bright sunlight, as they move closer, following the trail of the food and wood scented light smoke, as they intermittently move and hover over the nearby mountains, searching for the source. In addition to the reflections of the Sun from the craft, which begin to slightly diminish, we now begin to see the brilliant glowing firefly-like light emitting from the underside of each craft, increasing in intensity as the craft approach, now in the near distance.

As soon as Ziusudra sees the reflections move, he knows it is from the craft of the Gods, and the smiling and joyful expression upon his face quickly changes, and fades into a concerned and fearfully anxious look of dismay, as he slowly lowers the cup of wine, and persistently stares at the approaching five craft. As the people at his table continue to converse and laugh, Ziusudra quietly and questioningly declares to himself, "If the Gods are angry, I fear punishment and death for escaping the flood. But if they are not angry, they shall be pleased by the food offerings."

Slowly and individually, after partially overhearing his words, the people at the table with Ziusudra look at him, and notice his fixed stare, and they turn and follow his eyes, and also see the five approaching craft reflecting in the bright sunlight, and they quickly become silent, and also anxiously stare at the five craft; all with slightly dirty concerned faces. As the reflections of the Sun from the craft further diminish and finally disappear, they are now totally replaced by the brilliant glowing firefly-like light emitting from the underside of each craft.

Then, one person at the table with Ziusudra rises, and points to the five craft, and loudly and fearfully shouts, "The Gods are coming to destroy us."

Suddenly everyone at the banquet rises, and looks at the five craft, as they now move very close. All their smiles of joy and laughter instantly fade into a silent, reserved and concerned disposition. Some people force half-smiles, and simultaneously tremble, as they too, now share the anxious fear and uncertainty of Ziusudra, as they also perceive the dangerous circumstance they are in; which could be surviving the destructive flood, only to be destroyed for surviving.

From the air and from the plateau, we see the five craft following the food and wood scented light smoke, as they locate the hexagon shaped reed boat and the people next to it, and quickly move toward them, and hover directly above the people. Then the brilliant glowing firefly-like light emitting from the underside of each craft begins to diminish, as all the silent people stare directly up at the five craft.

We observe God Enki in his craft, relieved and beaming with joy, but showing concern, as he looks at a close up view on the small viewing monitor of the people staring up. Then, he sniffs[783] the now strongly food scented air, which is being emitted from his computer scanner, and appreciatively declares, "The food has such a sweet savor."[784]

Then we observe Goddess Ninhursag in her craft, very satisfied, as she watches the people below her on the small viewing monitor, and smiles as the people look up at her craft. She also sniffs[783] the now strongly food scented air, also emitting from her computer scanner, and approvingly declares, "I can smell the savor."[785]

From the plateau, we observe the people, all anxiously looking up with slightly dirty faces at the five craft directly above them, as the glowing light of each craft continues to diminish, and one person, only knowing from the scenario that his fate is in doubt, and thinking of the worst possible conclusion, loudly and sarcastically remarks, "The Gods gather like flies over the offering."[786a]

From the air and from the plateau, we see the five craft slowly drift away from the banquet area, toward another area on the same plateau, a very short distance from the boat and people, and we watch all the craft hover, descend and prepare to land, as the glowing light of each craft further diminishes, and the 4 pods like cattle hoofs extend 6 feet down from each craft, and the craft begin to land.

Again we observe Ziusudra, as he now cautiously walks toward the five descending craft, with the people closely following behind him, all uncertain as to the temperament of the Gods; friendly or unfriendly, benevolent or malevolent. Then the craft land, and the glowing light of each craft further diminishes, and then completely expires.

Then, from the craft area, we observe all the people, led by Ziusudra, as they stop at a safe distance from the craft, and await their destiny.

After a moment, a side of each craft extends down, forming a 4 feet wide ramp and doorway, and walking through the opaque air flow door appears a concerned God Enki, wearing the same red robe and adornments that he had

---

[783]M. Page 99, line 34 [The gods sniffed] the smell.

[784]F. Page 87, line 160 The gods smelled the sweet savor; and V. Genesis 8:21 And the LORD (God Enki) smelled a sweet savour (similar to this).

[785]F. Page 87, line 159 The gods smelled the savor.

[786a]M. Page 99, line 35 They (the Gods) gathered [like flies] over the offering; and F. page 87, line 161 The gods gathered like flies over the sacrificer (flies metaphorically refers to a large number, and in this circumstance, numerous Gods: see footnote 725).

previously worn, and he is without a helmet or body regulator. He stands at the top of the ramp for a moment, looking for Ziusudra. He quickly locates Ziusudra, and then God Enki runs down the ramp and toward the waiting Ziusudra. Ziusudra and all the people are relieved as soon as they see God Enki, for they now realize they are not in jeopardy, and they bow down, and prostrate themselves, and then they stand,[786b] but remain silent. God Enki stops a few feet from Ziusudra, and they gaze upon each other for a moment. Then, they exchange greetings, and warmly, tightly, and emotionally embrace, as they both shed tears of joy, which streak down their faces, and flow into their beards, and each with a sincere look upon his face, glows with appreciation.

After their long embrace, God Enki steps back and observes the condition of Ziusudra, and then he looks at the boat, and then he again looks at Ziusudra, as Goddess Ninhursag and the other Gods, all without helmets or body regulators, who are all walking and looking at the people, and looking beyond the people, at the hexagon shaped reed boat, with some Gods pointing at the boat, arrive and quietly stand behind God Enki. Then God Enki compliments Ziusudra in a warm and appreciative tone, "You are safe." Again Ziusudra bows down, and prostrates himself on the ground, as do all the people, but God Enki quickly lifts him up, as they gratefully stare at each other, and slowly the sincere look upon both their faces melts away, and they begin to exchange smiles of relief. Ziusudra ardently responds to God Enki, "Thank you God Enki. We have performed our rituals, and made offerings to you," and all the people rise from their prostrating position, and begin to greet Goddess Ninhursag, the Gods and Goddesses, the Anunnaki God officers, and the Igigi Gods, a total of 60 Gods, all crisp and clean, who have just arrived on the plateau. God Anu and God Enlil are conspicuously not present.

Suddenly everyone is smiling and conversing, and the occasion becomes a happy and joyful reunion. Ziusudra invitingly proposes to God Enki and Goddess Ninhursag, "This offering of food is for all the Gods to enjoy," and the people cheerfully shout, "Come join us." All the Gods and people, the people with slightly dirty faces and very soiled robes, and the Gods with crisp and clean attire, immediately go to the tables and mats, and the Gods seat themselves among the people, at the tables.

[786b]M. Page 123, upper line 3 (see footnote 335).

From out of the boat, the teenage children gleefully bring clean dishes and cups for the Gods, and additional woven reed mats to accommodate the displaced adult people from the tables.

God Enki and Goddess Ninhursag sit on each side of Ziusudra and his wife, with God Enki next to Ziusudra, and the people again serve the Gods, as they had done before the flood, and the Gods eat, and drink red or white wine, which is poured into their cups by the people, and together they begin to feast. Now the Gods, in addition to the people, savor every bite, as they eat and drink, laugh and smile; for it has also been over 12 months since the Gods have enjoyed the food and service of mankind. All the people are relieved that the Gods are not angry they survived the flood.

As Goddess Ninhursag eats[787] and drinks, she pensively asserts, "I thought all mankind had been lost. How could God Anu, the president of the assembly, let this occur?"[788] She now grasps the necklace of fly shaped lapis lazuli around her neck, and holds the fly shaped deep blue stones up, and she declares in a loud tone, speaking to all present, but especially to the Gods, "All of you Gods here present, look at these flies of lapis lazuli around my neck. God Anu had made for me, according to my wish,[789] to remind me of the people I saw floating upon the water. As surely as I shall not forget to wear this necklace,[790]

---

[787]M. Page 99, line 36 [After] they had eaten the offering; and F. page 87, line 162 As soon as the great goddess arrived.

[788]M. Page 99, line 39 Where has Anu the president gone? (Possibly implies; how could God Anu, the president of the assembly, acquiesce and allow this to occur).

[789]M. Page 99, line 46 Then she approached the big flies (approached could be in error, and the correct word could be held up or lifted up, which is more appropriate; unless the word approached, in this context, implies being held up or lifted up), and page 163, note v. 46-vi.6 Actual lapis beads in fly-shape are known, and their triangular shape made them particularly suitable for necklaces, if indeed their functional shape did not suggest the fly in the first place (undoubtedly, the fly shaped necklace is a symbolic reminder of the horror of the disaster scene with the drowned floating bloated human bodies: see footnote 725 and 726; but the fly shaped necklace is now also a reminder of perished mankind, or in other words, mankind. However, it is noteworthy that the shape of the flies in the necklace is triangular; as I further explain under 'Water' on page 344 in my epilogue. Since the boat is composed of six equilateral triangles, it could be possible that God Enki, the God of the triangle/triangles, secretly informed Goddess Ninhursag about the boat, and the plan to save mankind, therefore the necklace could also be an esoteric reminder of the boat, and the plan to save mankind, or again, in other words, mankind), and page 99, line 47 Which Anu had made (for her) and (she) was carrying (wearing); and F. page 87, line 163 She lifted up the great jewels which Anu had made (for her) according to her wish.

[790]M. Page 101, line 2 Let [these] flies be the lapis (reminder) around my neck; and F. page 87, line 164 O ye gods here present, as surely as I shall not forget (to wear) the lapis lazuli on my neck.

I shall remember these days, and shall not forget the flood or mankind ever,"[791] and then she resumes feasting.

We scan the banquet area for a very brief period of time, with the southern sky in the background, and we observe the Gods and mankind, with mankind again serving all the Gods, and now sitting side by side, together again; looking and pointing at the boat, conversing, laughing, and enjoying the food and freely drinking the wine. Then, unexpectedly, we notice reflections in the high southern sky. One more craft is approaching, glistening in the bright sunlight. All the Gods and people quickly take notice, and they rise, and again the laughter and smiles instantly fade into looks of anxiety, and silence again prevails at the banquet.

From the air and from the plateau, we see the craft, as it quickly moves toward the banquet area, and hovers directly above the Gods and people. Then the brilliant glowing firefly-like light emitting from the underside of the craft begins to diminish, as now all the silent Gods, and again all the silent people, stare directly up at the craft, as the glowing light continues to diminish. Then the craft slowly drifts away from the banquet area, toward the five craft already on the plateau, a very short distance from the Gods and people, and we watch the craft hover, descend and prepare to land, as the glowing light further diminishes, and the 4 pods like cattle hoofs extend 6 feet down, and the craft begins to land. The group of Gods and people remains silent, as they watch the craft descending, and then land, and the glowing light further diminishes, and then completely expires.

One person apprehensively whispers to another person, "God Enlil, it must be God Enlil," and the other person consolingly responds, "It must be God Anu," as they speculate on who is in the craft, and wait for either of them to appear.

After a moment, a side of the craft extends down, forming a 4 feet wide ramp and doorway, and we observe God Enlil and God Anu[792] walking through

---

[791]M. Page 101, line 4 That I may remember it (the flood and mankind) [every] day [and for ever]; and F. page 87, line 165 I shall remember these days and shall not forget (the flood or mankind) ever!

[792]M. Page 145, line 251 Be conjured by heaven and underworld, let appear, 252 An (Anu, and) Enlil, be conjured by heaven and underworld, appeared, and line 253 They made come up (decreed they be created) the animals which emerge from the earth (animals = people = mankind: see footnote 248).

the opaque air flow door, and down the ramp, followed by their crew, 4 Anunnaki God officers and 2 Igigi Gods, all in their usual attire, and all without helmets or body regulators. All the Gods within the group immediately bow their heads to God Enlil and God Anu, but the people do not bow, they remain standing, frozen in fear. They walk in a slow and deliberate manner, stern faced and silent, toward the standing group of silent Gods and people, and look at the group, and beyond the group, at the hexagon shaped reed boat, as some people discouragingly whisper, "God Enlil," and other people encouragingly whisper, "God Anu."

As God Enlil and God Anu, and their crew, slowly approach from the craft, Goddess Ninhursag very loudly and reproachfully declares,[793] "Has God Enlil come to the incense?[794] Let all the Gods come near to the offering,[795] but God Enlil should not come near![796] For he did not consider the people individually,[797] but consigned all the people to destruction.[798] You decided on total destruction![799] Now their clean faces have become dark,[800] and mankind's grief is my grief! Now determine my destiny! Let him get me out of this distress and relieve me, as he relieved himself of mankind,"[801] as she now openly hurls a contemptuous accusation at God Enlil.

Together, all the silent group of Gods and people are startled, with the people again becoming very fearful, as they are again uncertain of their fate, now that the ruler and supreme commander of the Earth, God Enlil, approaches, for the people know they are still the center of a major dispute among the Gods. The people are watching the confrontation with anxious expressions upon their slightly dirty faces, with some of the Gods empathizing with them,

---

[793]M. Page 99, line 37 Nintu arose to complain against all of them.

[794]M. Page 99, line 41 Has Enlil come to the incense?

[795]F. Page 87, line 166 Let the gods come near to the offering.

[796]F. Page 87, line 167 (But) Enlil shall not come near to the offering.

[797]M. Page 99, line 42 They, who did not consider but brought about a flood (they implies God Enlil and his sons, or the assembly members generally supporting God Enlil); and F. page 87, line 168 Because without reflection he brought on the deluge.

[798]M. Page 99, line 43 And consigned the peoples to destruction; and F. page 87, line 169 And consigned my people (my creation = mankind) to destruction!

[799]M. Page 99, line 44 You decided on total destruction.

[800]M. Page 99, line 45 Now their clean (cheerful) faces have become dark (sad).

[801]M. Page 99, line 48 (She said), His grief is mine! Now determine my destiny!, and line 50 Let him (God Enlil) get me out of this distress and relieve me (as he relieved himself of mankind).

as God Enlil and God Anu, and their crew, silently and slowly walk toward the table of Goddess Ninhursag, God Enki, and Ziusudra and his wife. God Anu is indifferent, while God Enlil is filled with anger.[802] God Enlil equally glares and scowls at the Gods, especially the Igigi Gods, and people as he passes, while they stand in silence. The Gods display emotional expressions of sadness upon their faces, with no physical reaction, but the people cringe and tremble with fear, as God Enlil abhorrently glares at them, as he slowly walks past. One person God Enlil has just passed and stared at, begins shaking and spilling the wine from the cup in his trembling hand.

Finally, God Enlil and God Anu arrive at the table, and especially as he glares at Goddess Ninhursag, God Enlil angrily exclaims, "All we great Anunnaki, decided together on an oath![803] How could have the mortals escaped? No human was to live through the destruction![804] Where did life escape?"[805] Again there is total silence, as God Enlil slowly scans the group standing at the table with Ziusudra, momentarily stopping on each of the people, and on each of the Gods, and we observe their expression filled faces, all of which are sad and pessimistic, except the face of God Enki, which is full of confidence. Then God Enlil returns to, and his eyes stop on God Enki, and he gives God Enki a despising, suspecting, and angry look. Then God Enlil questioningly resumes, "How did mankind survive in the destruction?"[806] God Enlil now turns, and looks at the hexagon shaped reed boat, and God Anu also turns, and looks at the boat; all while there continues to be silence among the group. They now slowly scan the dull black pitch coated weathered hexagon shaped reed boat, and after a few very long moments pass, as they continue to look at the boat, God Anu breaks the silence, and inquires of God

---

[802]F. Page 87, line 171 And (after God Enlil) saw the ship, Enlil was wroth; and M. page 101, line 6 And (God Enlil) was filled with anger at the Igigi; and F. page 87, line 172 He was filled with anger against the gods, the Igigi (perhaps God Enlil thought the Igigi Gods had in some way contributed to the boat: built the boat, help build it, or had given the design of the boat to mankind; or was God Enlil angry because of their revolt, when they initially asked for relief from their heavy labor, and the problems their relief, mankind, created among the Gods).

[803]M. Page 101, line 7 All we great Anunnaki, and line 8 Decided together on an oath.

[804]F. Page 87, line 173 Has any of (how could have) the mortals escaped? No man (human) was to live through the destruction!

[805]M. Page 101, line 9 Where did life escape?

[806]M. Page 101, line 10 How did man (mankind) survive in the destruction?

Enlil,[807] "Who but God Enki could do this?[808] Who can do things without God Enki?[809] For God Enki, alone, understands every matter!"[810] God Anu pauses, and both God Anu and God Enlil turn, and again look at God Enki. Then God Anu concedes, as he turns and looks at God Enlil, "Certainly, I, president of the assembly, did not reveal the command!"[811] God Anu then turns, and again looks at God Enki.

God Enki admonishes God Enlil,[812] in a stern and emotionally toned voice, "God Enlil, how could you without reflection allow the flood to destroy all mankind?[813] Impose your penalty on the criminal![814] On the sinner lay his sin! On the transgressor lay his transgression![815] Be moderating! Pull tight, that mankind may not get too loose; let loose, that mankind shall not be destroyed![816] And whoever disregards your command, the assembly should punish only him![817] For not all mankind was disobedient." God Enki momentarily pauses, and he scans the group, and then he confesses, as he again looks at God Enlil, "I am responsible for saving life for mankind![818] I did it, indeed, in front of you![819] To Ziusudra, I showed a dream, and so he learned the secret of the Gods;[820] the imminent, destroying flood.[821]

---

[807]M. Page 101, line 11 Anu opened his mouth, and line 12 And addressed the warrior Enlil; and F. page 87, line 174 Ninurta opened his mouth and said, speaking to warrior Enl[il] (I believe Ninurta, a son of God Enlil, is a scribal error, and the name should be Anu).

[808]M. Page 101, line 13 Who but Enki could do this?

[809]F. Page 87, line 175 Who can do (complicated) things without Ea (God Enki)?

[810]F. Page 87, line 176 For Ea alone understands every matter.

[811]M. Page 101, line 15 [Verily] I did not reveal the command.

[812]M. Page 101, line 16 [Enki] opened his mouth, and line 17 [And addressed] the great gods; and F. page 87, line 177 Ea opened his mouth and said, speaking to warrior Enlil, and line 178 O warrior, thou wisest among the gods!

[813]F. Page 88, line 179 How, O how couldst thou without reflection bring on (this) deluge?

[814]M. Page 101, line 25 Impose your penalty [on the criminal].

[815]F. Page 88, line 180 On the sinner lay his sin; on the transgressor lay his transgression! (This is probably a variant of footnote 814).

[816]F. Page 88, line 181 Let loose, that he shall not be cut off; pull tight, that he may not ge[t (too) loose].

[817]M. Page 101, line 26 [And] whoever disregards your command, and line 27 Let the members of the assembly punish only him!

[818]M. Page 101, line 19 [I am responsible] for saving life [for mankind].

[819]M. Page 101, line 18 I did it [indeed] in front of you! (In an overt manner, so says God Enki in his admission of culpability; however, this seems to me to be a very clever and covert means to conduct his overt act).

[820]F. Page 88, line 186 It was not I who revealed the secret of the great gods, and line 187 To Atrahasis (Ziusudra) I (God Enki) showed a dream, and so he learned the secret of the gods; and M. page 101, line 20 The secret of the gods [he] learned, and line 22 I showed to Ziusudra in a dream.

[821]M. Page 101, line 21 The imminent destruction of the flood.

He protected himself," and God Enki briefly points to the boat, as he continues, "and the seed of mankind," and we see the hexagon shaped reed boat, Ziusudra, and the people, as God Enki elucidates, "and now, take counsel concerning him in your heart.[822] Certainly, mankind was worth saving. For if the Gods were to lose all mankind, who would provide us with sustenance[823] on this planet, Earth? Who shall toil and work for the Gods,[824] to build the cities, to mine the gold and other precious metals? For without mankind, the Gods would not have servants to prepare our bread, to work the fields, to watch the sheep. The Gods need mankind on Earth!" As God Enki concludes his speech, the group remains in total silence.

Then God Enlil, still with an angry look upon his face, begins to take a second look at mankind, as he slowly scans the group, but only stops and closely looks at the faces of the people, and as he scans, he observes the slightly dirty faces of the people, all very serious, silent, and fearful.

Then we observe the face of God Enlil, as he scans the group, but only scrutinizing the people within it. Slowly, the anger upon his face begins to melt away, and slowly he nods his head up and down, in agreement with God Enki. Shortly, there appears an accepting smile[825] upon the face of God Enlil, and again his eyes stop on God Enki, and God Enlil commands him,[826] in an apologetically toned voice, "In the future, God Enki, I, as administrator and ruler of the Earth, decree that you, our chief scientist, and Goddess Ninhursag, our chief doctor, the birth-Goddess, confer in the assembly,[827] as to the status and function of mankind, and their future, here on the planet Earth."

God Enki and Goddess Ninhursag bow their heads in agreement, and

---

[822]F. Page 88, line 188 And now take counsel concerning him; and M. page 101, line 23 And now take counsel in your heart (accept advice).

[823]M. Page 67, line 339 For food for the (sustenance of the) peoples, (is also food) for the sustenance of [the gods] (nourishment of mankind, will benefit the Gods, or in other words, what is good for mankind, is also good for the Gods: see footnote 286).

[824]M. Page 65, line 337, and line 338 (see footnote 269).

[825]V. Genesis 8:21 and the LORD (God Enlil) said in his heart (pledged to himself), I will not again curse the ground (cause a drought and famine) any more for man's sake; for the imagination of man's heart (conduct) is evil (perverse) from his youth (creation); neither will I again smite (cause pestilence to destroy) any more every thing living, as I have done (similar to this).

[826]M. Page 101, line 41 [Enlil] opened his mouth, and line 42 And addressed Enki the prince.

[827]M. Page 103, line 43 [Come], summon Nintu, the birth-goddess, and line 44 [You] and she, confer in the assembly (plan/discuss/work within the assembly, not in secret).

then we scan the group, as all the Gods, and mankind, look on in amazement and disbelief, digesting the consoling words spoken by God Enlil.

Then, within the group, an Igigi God whispers with astonishment to another Igigi God, "God Enlil is allowing mankind to survive," and again we scan the group of Gods and mankind, and there is a brief pause, as the entire quietly standing group looks at each other in disbelief.

Then God Enki turns to Goddess Ninhursag, and commands,[828] "Goddess Ninhursag, birth-Goddess, creatress of destinies,[829] we shall create seven new categories for the people.[830] Let there be, among the categories, women who bear, and women who do not bear,"[831] and we observe Goddess Ninhursag

---

[828]M. Page 103, line 45 [Enki] opened his mouth, and line 46 And [addressed] Nintu, the birth-goddess.

[829]M. Page 103, line 47 [You], birth-goddess, creatress of destinies.

[830]M. Page 103, line 48 Let us create seven new categories] for the peoples, 49 . . .] . . (first category), 50 . . .] let there be (second category), 1 In addition let there be a third category among the peoples, 3 Let there be among the peoples the Pāšittu-demon (fourth category), 4 To snatch the baby from the lap of her who bore it, 6 Establish Ugbabtu-women, Entu-women, and Igiṣītu-women (fifth, sixth, and seventh category), 8 And let them be taboo and so stop childbirth, 10 [. .]. . . . . , and line 11 [Birth] and life ; and H. upper page 77 (a new method of computing years begins after the flood), and note 37 The sign with which the name of the city Kish is always written represents the head of an ungulate (hoofed) quadruped ("animal city", or perhaps "animal-king"), and page 78, note 46 Since all the following names are animal names, we should perhaps expect an animal name here, and middle page 152 Thus it should be noted that the kings preceding Etana all have Akkadian names and are further unified by the fact that many of their names are those of animals (early rulers were given animal designations, so called "animal-kings", which abruptly changed shortly after the flood, and "human-kings" begin, rulers without the animal designation); and V. Genesis 9:1 through 9:7 (possibly these seven are somewhat related to the seven new categories, similar to these. In Josephus IV, Jewish Antiquities, Books I-IV, page 60, and page 61, the Hebrew account gives a son of Νώχου/Noah,'ΙάΦθα/Japheth, who had seven sons (= Genesis 10: 2 = seven new categories?), and attributes one of these sons named 'Ιαυάνου/Javan, as giving birth to'Ιωνία/Ionia and all"Ελληνες/Hellēnes/Greeks; whereas, in the Hellenic story of Deucalion/Noah, his son Hellen is whom the Hellenes/Greeks were named after: see footnote 73. I believe there were additional creations or adaptations/adjustments to mankind establishing new or different categories or classifications/races before the flood, as well as changes after the flood, for the term "new" is used, and its implication is that if this were the first establishment of categories, the term would not have been necessary; as I further explain under 'Evolution' on page 356 in my epilogue: see footnote 194, 247, 719, and 745. See V. Genesis 10 for continuation of Hebrew account of lineage; and see the book The Wars Of Gods And Men by Zecharia Sitchin, page 295, for his explanation of the origin of the word Hebrew, whose root word Ibri, means originating from Nippur, in effect meaning, of Sumerian origin, pertaining to the birthplace and lineage of Ā´-brăm/Abraham).

bow her head in agreement.

Then God Enlil invites Ziusudra, "Come Ziusudra," as he extends his hands to Ziusudra and his wife, and then he adds, "You and your wife. Come with me, into your boat." God Enlil now takes them by the hand,[832] and in silence, we observe the three, hand in hand, with God Enlil in the middle, and with God Anu, God Enki, Goddess Ninhursag, and a few of the Gods and Goddesses, Anunnaki God officers, and Igigi Gods, all closely and silently following, as they slowly walk toward the nearly 12 feet long by 4-1/2 feet wide pitch coated reed bundle boat ramp, and then go up the ramp, toward the nearly square doorway of the hexagon shaped reed boat.[833]

From inside the boat, we witness them enter through the doorway, and proceed 25 feet into the boat, through two compartment-like trapezoid shaped sections, as they walk upon the pitch coated, tapered downward toward the center, bottom hull of the boat, walking between the bottom hull exposed reed bundle bracing beams. The boat is moderately illuminated by the sunlight, entering through the doorway, as the clay oil lamps, still securely lashed to the vertical beams, all have been extinguished. Then God Enlil lets go of their hands, and God Anu stands next to God Enlil, and together in silence, and very slowly, with a few of the Gods standing behind them, including God Enki and Goddess Ninhursag, they scan the interior of the boat that saved the seed of mankind, and life upon the Earth.

As they scan the interior, they notice the pitch coated shiny black inner side hull and bottom hull; contrasting to the uncoated reed bundle ceiling, and the numerous vertical, horizontal, and diagonal reed bundle bracing beams, including those of the ceiling and hull floor, radiating out from the center. Being the lowest deck, this is one of the two decks where the birds were kept. They look across the 200 feet diameter bottom hull of the hexagon shaped reed boat, which tapers downward toward the center, and notice the 5 feet wide reed bundle ramp next to the hub, and the few remaining large flat reed baskets of seeds, and large flat clay drinking bowls, both neatly placed about. Numerous bird droppings attest they had occupied this area.

Then God Enlil breaks the silence, as he respectfully accepts mankind,

---

[831]M. Page 103, line 2 (Let there be) among the peoples women who bear and women who do not bear (third category).

[832]F. Page 88, line 190 He took my hand and caused me to go aboard (the boat), and line 191 He caused my wife to go aboard (etc.).

[833]F. Page 88, line 189 Then Enlil went up into the ship.

Ziusudra and his wife, and he commands them, "Kneel down beside me," and as God Anu steps aside, Ziusudra and his wife kneel down on each side of God Enlil, and prostrate themselves.[834] First they bow their heads, and then they raise their heads, while they remain kneeling. God Enlil, standing between them, then places each of his hands on their foreheads, and touches them, as he bestows a covenant,[835] "You protected the seed of mankind at the time of destruction.[836] Hitherto, Ziusudra has been but a man,[837] but now, Ziusudra and his wife shall be like unto us Gods.[838] You shall have life, like a God.[839] You shall be elevated to eternal life,

---

[834]M. Page 145, line 254 The king Ziusudra, and line 255 Prostrated himself before An (Anu, and) Enlil; and F. page 88, line 191 He caused my wife to go aboard (and) to kneel down at my side.

[835]M. Page 133, lower line 1 [Giving] the gods life to them, lower line 2 Together your wife and you, and lower line 3 For your help and reverence; and F. page 88, line 192 Standing between us, he touched our foreheads and blessed us; and V. Genesis 9:8 through 9:17 (possibly these are somewhat related, similar to these. Of special interest is 9:11, which reads) And I will establish my covenant with you; neither shall all flesh be cut off any more by the waters of a flood; neither shall there any more be a flood to destroy the earth (which means there will not be any more flood waters to cut off or drown/eliminate all flesh, both people and creatures, or to destroy the habitable lands of the Earth, and 9:15, which reads) and the waters shall no more become a flood to destroy all flesh. (Could it be that Nibiru has changed its orbit, or is the Earth now properly synchronized, at a safe orbital distance, with each orbit of Nibiru. Also of special interest are 9:13, 9:14, and 9:16, which refer to a) bow (a rainbow, to be a) token (or symbol of the) covenant (between God and man/mankind), for perpetual generations/everlasting. (Does the misting up of the waters of the Earth during each orbit of Nibiru actually cause a rainbow to appear in the sunlit sky, and therefore, is the rainbow a manifestation upon the Earth of the passing of Nibiru? I believe it is. Also, it is interesting that the planet Marduk/Nibiru in J. page 83, line 82 is called the 'Bow-star', and on this same page, and on page 63, the word 'bow' is used. Could these be a reference to what our scientists call a 'starbow', caused by the Doppler shift at 99% of the speed of light, as in T. page 219, fig. 112, especially lower illustration? Or could the three references; 'bow, Bow-star, and starbow', be describing the same effect, the shifted or distorted "halo" of the force fields which accompany an orbiting celestial body; and could the Biblical bow, a rainbow, be a reminder to mankind of the appearance of the distorted force fields of that particular orbiting celestial body, Nibiru, the abode of the Gods? I believe they are linked, and this is the correct interpretation, or very close to it: see footnote 38 and 493).

[836]M. Page 145, line 258 At that time, the king Ziusudra, and line 259 Who protected the seed of mankind at the time of destruction (see footnote 211).

[837]F. Page 88, line 193 Hitherto Utnapishtim (Ziusudra) has been but a man.

[838]F. Page 88, line 194 But now Utnapishtim (Ziusudra) and his wife shall be like unto us gods.

[839]M. Page 145, line 256 (Who) gave him life, like a god, and page 133, lower line 4 Life like the gods [you will] indeed [possess] (immortality. The name of the hero of the flood in Sumerian is Ziusudra, which means 'life of long days', lived to possess immortality/extended life; in Akkadian it is Utnapishtim, which means 'I have found life', found immortality/extended life; in Babylonian it is Atra-hasis, which means 'exceedingly wise', wise to attain immortality/extended life; in Hebrew it is Noah, which means 'respite', delay, prolongation of time/life; and in Hellenic it is (Deucalion) Xisuthros or Sisithros, which means 'life to him', attained immortality/extended life).

like a God.[840] With us, in the distance, you shall dwell." [841] We observe the faces of Ziusudra and his wife, both of which are filled with expressions of submission and humility. Then God Enlil motions to Ziusudra and his wife to rise to their feet, and they both rise.

The Gods in the boat, having just listened to the blessing, in anticipation of their exit, all now make a path for them, as does the group of Gods and people outside the boat, who clear themselves from the pitch coated reed bundle boat ramp.

In silence, Ziusudra and his wife follow God Enlil and God Anu, as they turn and slowly walk 25 feet back toward the doorway, through the smiling silent Gods, and through the doorway, appearing within the bright sunlight, closely followed by God Enki and Goddess Ninhursag, who lead the other Gods, as they also exit.

From the plateau, near the boat, we witness them walk down the pitch

---

[840]M. Page 145, line 257 Elevated him to eternal life (immortality), like a god (as I further explain under 'Longevity' on page 353 in my epilogue; whereas the Biblical Noah, as reported in Genesis 9:28 and 9:29, lived 350 years after the flood, and enjoyed a total life span of 950 years, and then he died; and the grave of Deucalion was said to be visible at the city of Athens in the ancient temple of Zeus/God Enlil).

[841]M. Page 145, line 260 They (Ziusudra and his wife) settled in an overseas country, in the orient, in Dilmun (see the book The Stairway To Heaven by Zecharia Sitchin, for an explanation of why Dilmun/Tilmun, the point of transition to Nibiru, is identified as Baalbek in Lebanon. Dilmun/Tilmun means the place/land of life/missiles, and was a spaceport/launch center of the Gods. Lebanon, which in Hellenic means incense burner, for they are the same word, that from which smoke is emitted, reflects and is associated with rising smoke, and was so appropriately named because in the past, the smoke/vapor trail of a rising craft could be observed there, especially at Baalbek: see footnote 653); and F. page 88, line 195 In the distance, at the mouth of the rivers, Utnapishtim (Ziusudra) shall dwell (a reference to Dilmun); and K. page 98 In the mountain (land) of crossing, the mountain (land) of Dilmun, the place where the sun (God Utu, the Sun God) rises (ascends from the spaceport: see footnote 614), They (An and Enlil) caused to dwell (this is an older transliteration of M. page 145, line 260); and M. page 136 (Berossus, circa 290 B.C., a Hellenic variant of the Sumerian flood story, the Hellenic account according to Polyhistor, circa 100 B.C.) and then (Xisuthros/Ziusudra) disappeared (etc.). Xisuthros (Ziusudra) himself did not appear to them any more, but there was a voice out of the air (probably emanating from the craft of the Gods) instructing them (the people on the mountain) on the need to worship the gods, seeing that he was going to dwell with the gods (going to their abode and/or planet) because of his piety, and that his wife, daughter, and pilot shared in the same honour (according to this version, the daughter and pilot were also taken away), and (Berossus, the Greek account according to Abydenus, circa 100 B.C. - 400 A.D.) the gods took him (Sisithros/Ziusudra) away from mankind (away from the Earth: see footnote 719), and page 137 In addition to Berossus and Gilgameš XI, another attestation of this item occurs on the Babylonian Mappa Mundi, where under the name Ut-napištim (Ziusudra) the flood hero is described as living in a remote corner of the universe.

coated reed bundle boat ramp, through the smiling silent group of Gods and people, and toward the craft of God Enlil, with the group of Gods and people now following them. The crew of God Enlil, the 4 Anunnaki God officers and the 2 Igigi Gods, who are now quickly preceding them, run up the craft ramp, and into the craft through the opaque air flow door. When God Enlil and God Anu, and Ziusudra and his wife are near the ramp of the craft, God Enki and Goddess Ninhursag stop, as does the group following them, and they watch the four of them walk up the craft ramp, turn around at the top of the ramp, and raise their hands in a good-bye gesture, and wave to the group. Then, they turn and enter the craft[842] through the opaque air flow door, and after a moment, the ramp retracts to form a side of the craft, the underside of the craft begins to glow at a very low level, the craft rises off the plateau, and the 4 pods like cattle hoofs retract. Then the craft drifts over and hovers directly above the group of Gods and people for a moment, as the glowing light increases, and then the craft ascends into the sky, as the glowing firefly-like light becomes brilliant.

We observe the faces of the group, as they watch the craft ascend, and now they begin to see the craft glistening in the bright sunlight, as reflections of the Sun replace the brilliant glowing firefly-like light emitting from the underside of the craft, with some people grasping their symbol necklace pendant, and we follow the craft, as it quickly disappears into the distant bright blue sky.[843]

*

---

[842] T. Page 337 through page 373.

[843] F. Page 88, line 196 So they (the Gods) took me (Utnapishtim/Ziusudra) and caused me to dwell in the distance, at the mouth of the rivers (a reference to Dilmun. According to this epic, the boatman/pilot is also at the abode with Utnapishtim/Ziusudra and his wife, but there is no reference made to taking the boatman/pilot there, or when and how he got there), and page 224 through page 269, Chapter IV The Story Of The Flood; and E. page 82 through page 153 (Marduk/Nibiru); and G. page 33 through page 65, Chapter V The Earliest Fragment Of The Deluge Story; and K. page 97, and page 98 (which is an earlier transliteration and is nearly identical to M. page 143, line 201, 202, and line 203, and page 145, line 205 through line 260); and U. page 20 through page 47; and L. Chapter 8, and especially page 281 through upper page 284; and I. lower page 116 through upper page 121; and O. page 11 through upper page 22; and R. page 99 through page 105; and X. page 373 through page 389; and C. page 358 through page 361.

From space, we observe the planet Nibiru, in all its glory, with its two rings of clouds, within which orbit its four moons, appearing as a radiating and glowing cross, bathed in a sunset red glow, and with star-filled space surrounding the planet.

*

According to my calculations,[844] the heavenly planet Nibiru, shall again be passing near the Earth, approximately between the year 2,000 A.D. and the year 2,100 A. D..

THE END

844(See the chronological schedule under 'Next Orbit' on page 319 in my epilogue).

COMMENTS, DIAGRAMS, and ILLUSTRATIONS

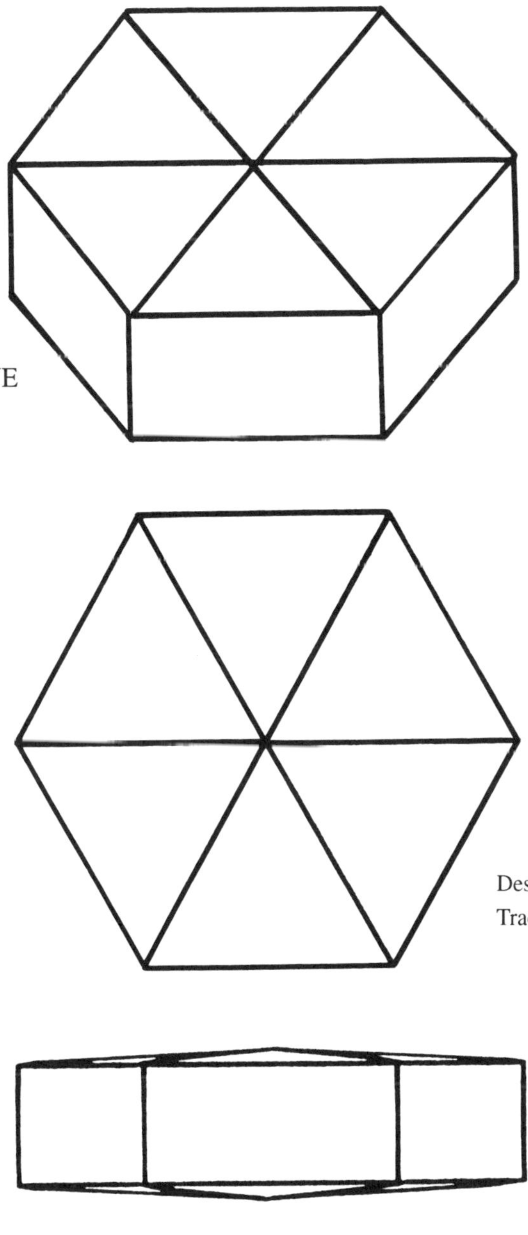

ERSPECTIVE VIEW

TOP VIEW

SIDE VIEW

Design Patent: DES. 291,881
Trademark: Reg. No. 1,278,945

THE BOAT OF ZIUSUDRA

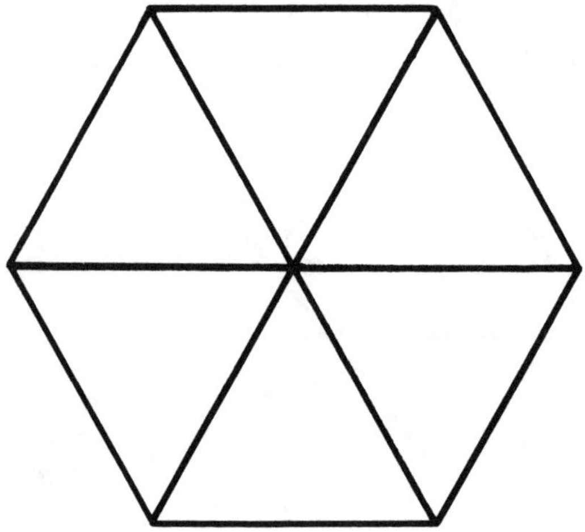

Source: ATRA-ḪASĪS, page 89, TABLET III, reads;

Line 25. The boat which you build
Line 26. . . .] be equal [(. .)]

Refers to the measurements of the boat, as Ziusudra (Atra-ḫasīs), inside the reed hut, listened to the instructions of God Enki. The sentence could read:

Line 26. Its width to be equal to its length.

The measurements shall be elaborated upon in the interpretations of line 30 of The Gilgamesh Epic, which follows after several pages, for they are both identical descriptions of the width and length of the boat.

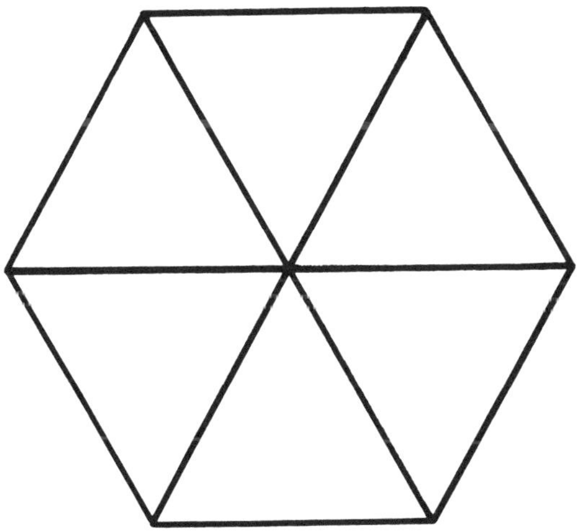

Source: ATRA-ḪASĪS, page 127, TABLET III, fragment CBS 13532 (F), reads;

Line 2. . . .] I will explain
Line 3. . . . a flood] will seize all the peoples together
Line 4. . . .] . before the flood sets out
Line 5. . . .] . . . all that there are. . . . .
Line 6. . . .] build a big boat.
Line 7. Let its structure be [. . . . .] entirely of reeds.
Line 8. . . .] let it be a *maqurqurrum*-boat with the name, The Life Saver.
Line 9. . . .] roof it over with a strong covering.
Line 10. [Into the boat which] you will make
Line 11. [Send. . .] . wild creatures of the steppe, birds of the heavens
Line 12. . . .] heap up

Refers to the construction and material used to construct the boat, as Ziusudra, inside the reed hut, listened to the instructions of God Enki. The sentences could read:

Line 2. Your dream last night I will explain
Line 3. Waters of a flood will seize all the peoples together
Line 4. Prepare yourself before the flood sets out
Line 5. Assemble all that there are close to you
Line 6. Together build a big boat.
Line 7. Let its structure be made entirely of reeds.
Line 8. Truly let it be a maqurqurrum-boat with the name, The Life Saver.
Line 9. Completely roof it over with a strong covering.
Line 10. Into the boat which you will make
Line 11. Send all the wild creatures of the steppe, birds of the heavens
Line 12. Supplies heap up

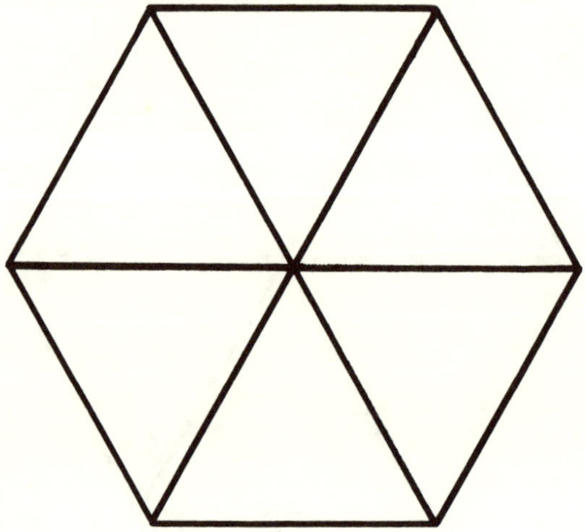

Source: ATRA-ḪASĪS, page 129, TABLET III, fragment DT 42 (W), reads;

Line 1. [. . .] . let it . [. . .
Line 2. [. . .] . like a circle [. . .
Line 3. Let [the pitch] be strong above and below,
Line 4. [. .] . . caulk the [boat].
Line 5. [Observe] the appointed time of which I will inform you,
Line 6. Enter [the boat] and close the boat's door.
Line 7. [Send up into] it your barley, your goods, your property,
Line 8. [Your wife], your kith, your kin, and the skilled workers.
Line 9. [Creatures] of the steppe, all the wild creatures of the steppe that eat grass,
Line 10. [I] will send to you and they will wait at you door.'

Line 11. Atra-ḫasīs opened his mouth to speak
Line 12. And addressed Ea, [his] lord,
Line 13. 'I have never built a boat . [.]
Line 14. Draw the design on the ground
Line 15. That I may see [the design] and [build] the boat.'
Line 16. Ea drew [the design] on the ground.
Line 17. '[. .] my lord, what you commanded [. . .

Refers to the design of the boat. Line 2, [. . .] . like a circle [. . . , is the only surviving clay tablet clear reference to the shape of the boat, to my knowledge; however, thinking along the thoughts of style and habit of scribal continuity of repeating words and blending one line into the next, one of the other possible Babylonian references is line 1, immediately preceding this clear reference, which I render as Six decks let it contain, could have been originally rendered [Six] sides let it contain; and its probable variant, line 60 of The Gilgamesh Epic, which has been scribally rendered as Six decks I built into it, as I further explain on page 288 within this section, after correct interpretation, especially if line 1 above was originally intended to reflect shape instead of number of decks, could read Hexagon decks I built into it, although the following line, line 61, would tend to somewhat weaken this position; with the remaining reference to the shape of the boat, although somewhat ambiguous, the Hellenic description given in terms of "stades," or stadium/circular shaped. All other explicit references; Babylonian, Hebrew, and Hellenic, are only to the dimensions of the boat. Once I read line 2, 'like a circle', not a circle, but similar to a circle, a circular-like shape, which I consider the most profound clue to discovering the true shape of the boat, I knew the design was within my grasp, and from that moment on, I also knew that I would not rest until I solved the enigma.

From pictographs, a rectangular shaped reed bundle boat was in common use, but this was a special boat, a unique boat, it was uncommon in shape: a hexagon, for 'like a circle' is certainly not a rectangle. If it was not uncommon in shape, God Enki (Ea) would have simply instructed Ziusudra (Atra-ḫasīs) to further brace and board an existing rectangular shaped boat, for there would have been no need to build a special hexagon shaped boat.

The hexagon shape was from the advanced technology of God Enki, perhaps a previously proven structurally sound and appropriate design for withstanding normal high waves of water elsewhere, on some other water planet, or perhaps it was similar to a space station or artificial satellite, a design for withstanding the various forces of exposure in space, possibly far greater than the force of the high waves, utilizing a variant of the applicable flat cylinder design of the Duku satellite of his advanced technological civilization, or perhaps it was an expression of water, as I further explain in my footnote 539.

For whatever the reason, God Enki gave Ziusudra a design he could quickly and easily be taught to mentally grasp and implement; an advanced and uncommon design which could be built using the common materials and existing methods of fabrication, with which Ziusudra was familiar: thus a hexagon shaped reed bundle boat. But even then Ziusudra had to ask God Enki to draw a diagram, so he could actually see the unusual design of the boat, as Ziusudra, inside the reed hut, listened to the instructions of God Enki. The sentences could read:

Line 1. 'Six decks let it contain
Line 2. Let it like a circle be
Line 4. Thoroughly caulk the boat.
Line 13. 'I have never built a boat like that before
Line 17. 'Ea, my lord, what you commanded I shall build.'

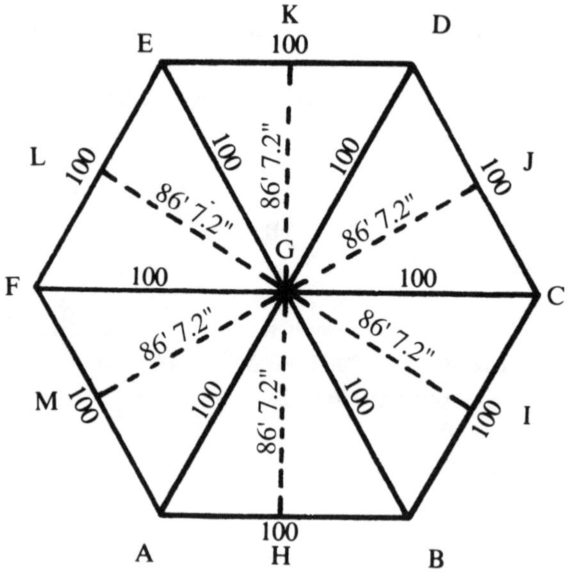

Source: THE GILGAMESH EPIC, page 81, Tablet XI, reads;

Line 30. Its width and its length shall be equal.

Refers to the measurements of the boat:

F to C, which is the width, and A to D, B to E, which both are the length, are equal; 200 feet.

G to A, G to B, G to C, G to D, G to E, G to F, which all are the width, cumulatively equal 600 feet, and A to D, B to E, F to C, which all are the length, cumulatively equal 600 feet, both are equal; 600 feet.

G to A, G to B, G to C, G to D, G to E, G to F, which all are the width, cumulatively equal 600 feet, and A to B, B to C, C to D, D to E, E to F, F to A, which all are the length, cumulatively equal 600 feet, both are equal; 600 feet.

A to D, B to E, F to C, which all are the width, cumulatively equal 600 feet, and A to B, B to C, C to D, D to E, E to F, F to A, which all are the length, cumulatively equal 600 feet, both are equal; 600 feet.

M to J, I to L, which both are the width, and H to K, which is the length, are equal; 173 feet 2.4 inches.

A to D, B to E, F to C, which all are the length, are equal; 200 feet.

H to K, M to J, I to L, which all are the width, are equal; 173 feet 2.4 inches (see my footnote 538).

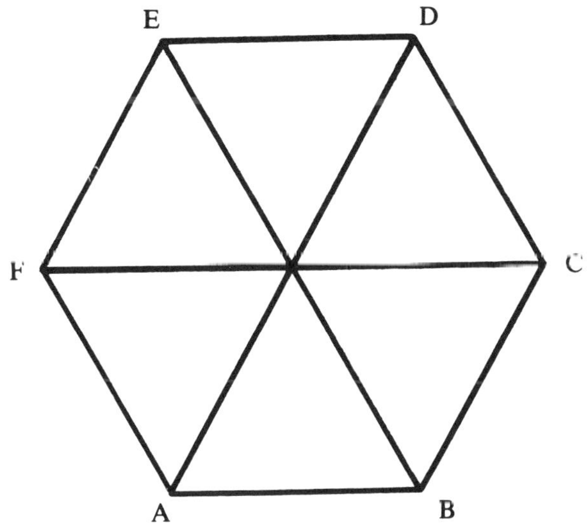

Source: THE GILGAMESH EPIC, page 82, Tablet XI, reads;

Line 56. On the fifth day [I] laid its framework.

Refers to the framework, and the order of construction of the boat:

After four days of bitumen and reed preparation, the framework was laid out upon the ground. A to D, B to E, F to C, A to B, B to C, C to D, D to E, E to F, F to A, which constitute the framework of the boat, were built first, and then the hull bottom and sides were added. The boat was built one level, or deck, at a time; upwards (see my footnote 586).

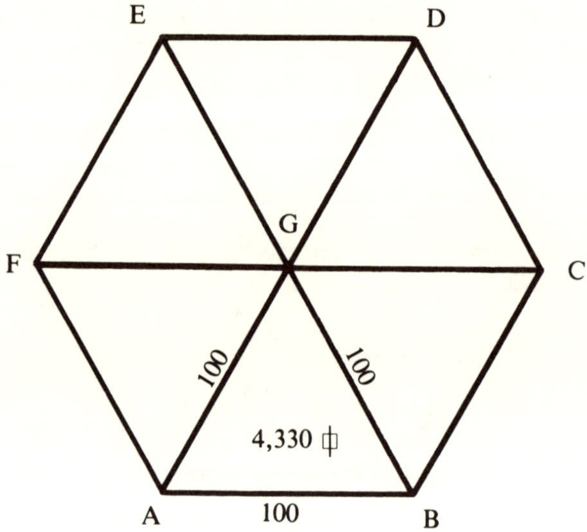

Source: THE GILGAMESH EPIC, page 82, Tablet XI, reads;

Line 57. One ikû was its floor space,

Refers to the framework area of the boat:

The area of equilateral triangle AGB is 4,330 square feet (1/2 x B x H = A, or .5 x 100 x 86.6 = A). Therefore, each deck, composed of 6 equilateral triangles, contains 25,980 square feet, and the entire boat, being 6 decks in height, composed of 36 equilateral triangles, contains 155,880 square feet.

The exact area of one ikû is not known at this time, but according to existing sources, it is approximately 3,600 square meters, which equals 38,750.40 square feet. The difference between 25,980 square feet and 38,750.40 square feet is great. However, until we can determine the exact area of one ikû, this aspect of comparison shall remain open, but I would suspect that it is closer to 2,413.26 square meters, which would give each deck an area of 25,980 square feet (see my footnote 587b).

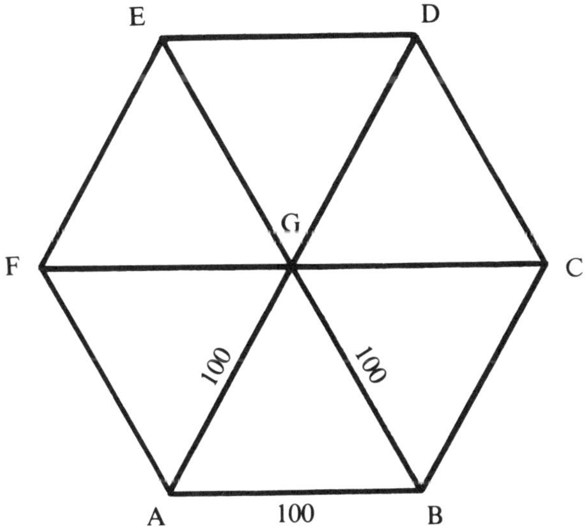

Source: THE GILGAMESH EPIC, page 82, Tablet XI, reads;

Line 57. one hundred and twenty cubits each (triangle) was the height of its walls;

Refers to the framework measurements of the boat:

In equilateral triangle AGB, in addition to all the other equilateral triangles shown, A to G, B to G, considered together, is the height of the walls. The internal diagonal members, walls, sides, or height of each equilateral triangle are equal; 200 feet. The term height is used instead of radius. Therefore, the remaining member, the outside member, or hull, A to B, considered separately, is the base of each equilateral triangle, with a span of 100 feet; not only is this line indicating that the walls/sides of the triangles are equal, it is also implying and thus giving us the perimeter of the boat, 600 feet.

Since there is some disagreement as to the type of cubit, the distance from the tip of the middle finger to the elbow, used to express the ancient measurements, the Sacred Cubit being 25.2 inches, the Royal Cubit being 20.63 inches, and the Hebrew Cubit being 17.58 inches, throughout, where applicable, I have used the accepted cubit; therefore, using the Babylonian cubit of 20 inches, which probably was inherited from the Sumerians, 120 cubits equal 200 feet (see my footnote 587b).

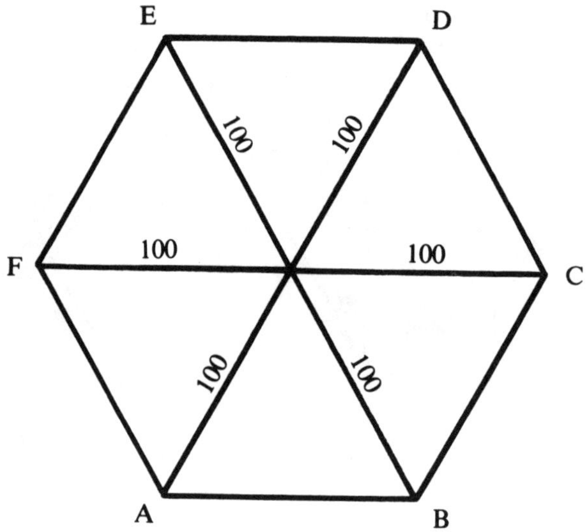

Source: THE GILGAMESH EPIC, page 82, Tablet XI, reads;

Line 58. One hundred and twenty cubits measured each side (to side) of its deck.

Refers to the framework measurements of the boat:

A to D, B to E, F to C, which are the measurements from each side of each deck, to the other side of each deck, directly across each deck, are equal; 200 feet.

120 cubits equal 200 feet (see my footnote 588).

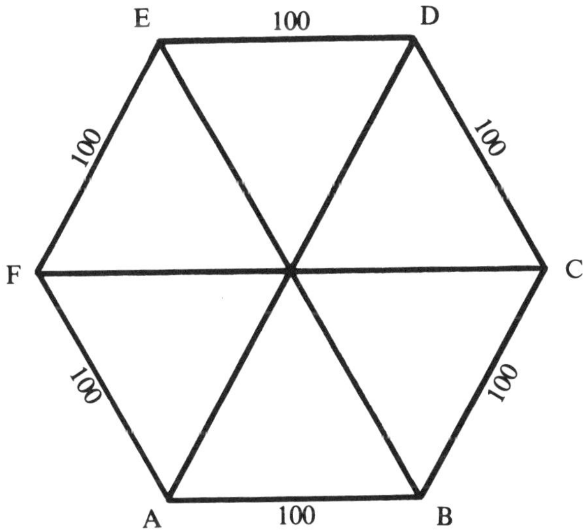

Source: THE GILGAMESH EPIC, page 82, Tablet XI, reads;

Line 59. I 'laid the shape' of the outside (and) fashioned it (built the boat).

Refers to the outside framework, or hull, and the order of construction of the boat:

A to B, B to C, C to D, D to E, E to F, F to A, which all are outside members, perimeter, or hull, with a base, or span of 100 feet each.

I do not believe this is a reference to the composition of the hull, or "skin" of the boat, as remote as the possibility may be.

The reference in this line to 'laid the shape' could be a variant of 'like a circle', as on page 280 within this section (see my footnote 586).

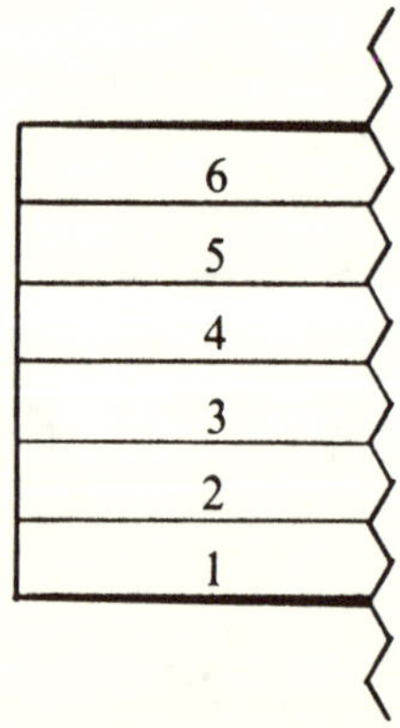

Source: THE GILGAMESH EPIC, page 82, Tablet XI, reads;

Line 60. Six decks I built into it,

Refers to the number of decks built into the boat, and the order of their construction:

Lower decks were first built into the boat, and then the uppermost of six decks, or main deck was built. Therefore, the boat was built from the bottom upwards, in the water. However, in order to waterproof the bottom of the boat, the lowest level, or deck, was built on land, and while the boat was at this stage, light in weight, it was moved into the water.

In the above source book I cite as a reference, the author Heidel includes in parentheses his own words, or as he states on page ix "elements not in the original but desirable or necessary for a better understanding in English." These words merely misled me, as I am certain they have many other serious readers, for he interprets line 60, by inserting the word (lower) between Six and decks, as a reference to the height of the boat in terms of number of decks, which he gives in the next line, line 61, as seven (stories) or decks. This is a critical mistake, for line 60 is actually one of the clearest, if not the clearest and most explicit reference in all the clay tablets to the number of decks, and no word need be inserted, and does not need any desirable or helpful words, it is very clear, for the most important word "Six" is there, boldly staring up at us from the page. Sometimes it is more rewarding to just read the clay tablets, without helpful words by anyone. Line 60 should read Six decks I built into it, and this line is probably a variant of line 1, which I render as Six decks let it contain, as on page 280 within this section; or better yet, it could be that the ancient word six, or perhaps its equivalent/variant, was also the word for hexagon, for the two are very closely related, and if so, line 60 and line 1 would read Hexagon, or its equivalent/variant; or perhaps with the duplicity of meaning ascribed to words employed by the ancients, Six could also mean Hexagon (see my footnote 587a).

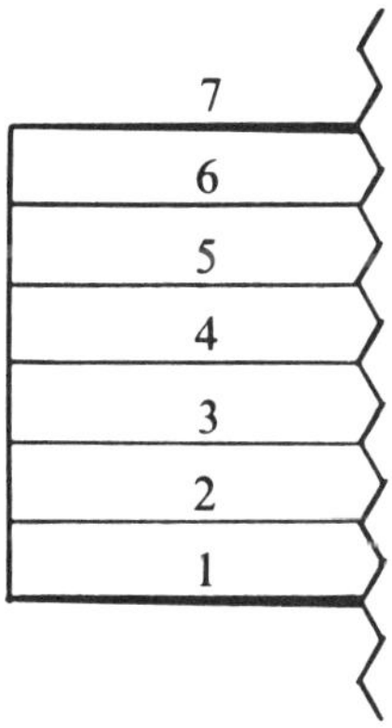

Source: THE GILGAMESH EPIC, page 82, Tablet XI, reads;

Line 61. (Thus) dividing (6, the framework members of the 6 decks) into seven (splitting/increasing 6 members/decks into 7 members/levels).

Refers to the total number of framework members built into the boat:

With the addition of the sixth deck, the uppermost, or main deck, the boat was then divided into seven levels (see my footnote 587a).

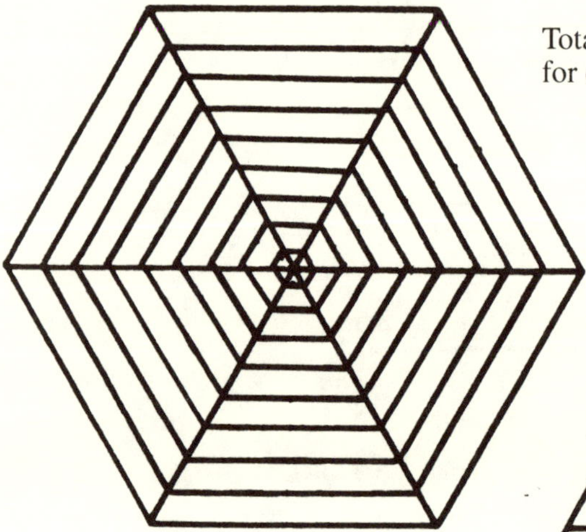

Total ground plan for each deck

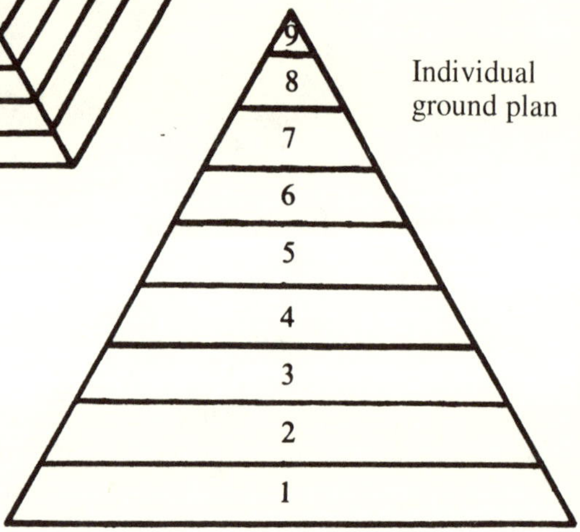

Individual ground plan

Vertical truss bracing beams

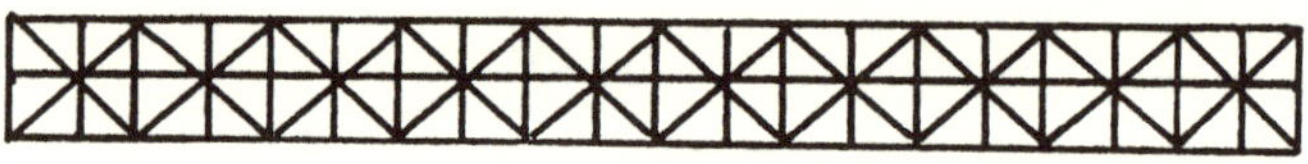

Source: THE GILGAMESH EPIC, page 82, Tablet XI, reads;

Line 62. Its ground plan I divided into nine (sections).

Refers to the truss bracing of each equilateral triangle:

The ground plan, or floor plan, was divided into nine sections. Starting at the hull, and extending vertically toward the center of the boat, connecting each equilateral triangle, from floor to ceiling, and supporting each deck, reed bundle structural vertical triangle truss bracing beams, composed of vertically and horizontally divided triangles, were installed every 10 feet on center, with the center most truss being 5 feet 1 inch from the hub; thus dividing each equilateral triangle into nine trapezoid shapes, or open compartment-like sections, which look somewhat like a spider's web, throughout the six equilateral triangles per deck.

6 cubits equal 120 inches, which equal 10 feet (see my footnote 591).

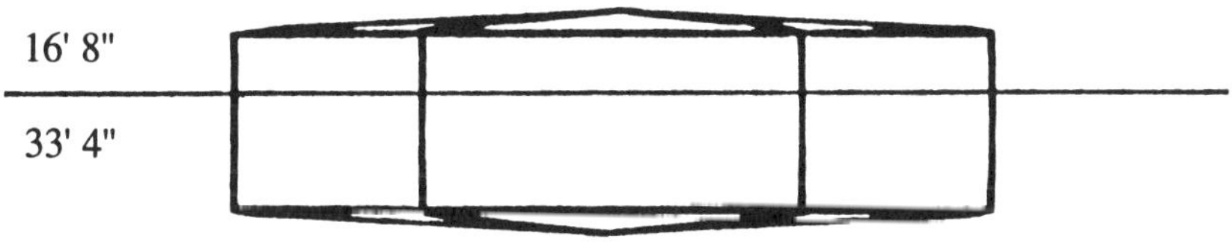

Source: THE GILGAMESH EPIC, page 83, Tablet XI,
reads;

Line 76. [. . . .] . . thc ship was completed.
Line 77. Difficult was [the . . . .].
Line 78. . . . . above and below.
Line 79. [. . . .]. . its two-thirds.

Refers to the completion and the draft of the boat, two-thirds, or 33 feet 4 inches, of which upon completion, empty of contents, was below the water level. Therefore, leaving one-third, or 16 feet 8 inches, above the water level (based on a framework of 50 feet). Thc sentences could read:

Line 76. Finally the ship was completed.
Line 77. Difficult was the task.
Line 78. Sealed above and below.
Line 79. Below water its two-thirds.

This two thirds figure may not have any relevance to the number of decks inside the boat, but I am very suspicious of this line (see my footnote 540).

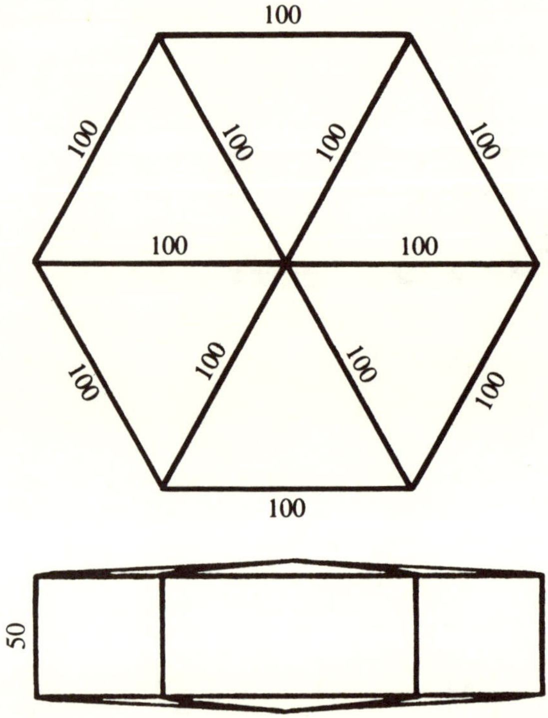

BABYLONIAN BOAT

Babylonian cubit = approximately 20 inches.

Length = 360 cubits x 20 inches =7,200 inches ÷ 12 inches per foot = 600 feet.

Width = 60 cubits x 20 inches = 1,200 inches ÷ 12 inches per foot = 100 feet.

Height = 30 cubits x 20 inches = 600 inches ÷ 12 inches per foot = 50 feet.

All framework dimensions.

Formula: 3 to 1 ratio between the perimeter and the diameter of the boat.

Only the height figure is not given in the clay tablets, however, an explanation of how I arrived at this figure shall be given in the following detailed construction probabilities notes, under letter F.

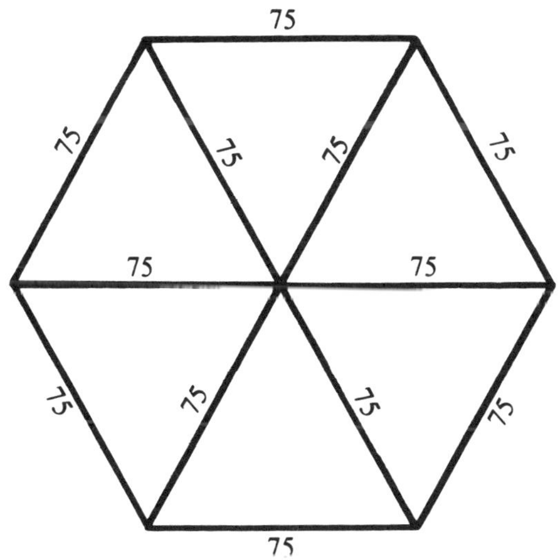

HEBREW BOAT

Hebrew cubit = approximately 18 inches.

Length = 300 cubits x 18 inches = 5,400 inches ÷ 12 inches per foot = 450 feet.

Breadth = 50 cubits x 18 inches = 900 inches ÷ 12 inches per foot = 75 feet.

Height = 30 cubits x 18 inches = 540 inches ÷ 12 inches per foot = 45 feet.

All framework dimensions.

Formula: 3 to 1 ratio between the perimeter and the diameter of the boat.

The word breadth is of major interest, for the word width is not used, and I believe if the original Biblical Hebrew word, which has been transliterated as breadth, is researched and correctly transliterated, its true meaning would more closely identify and indicate a radius. Also, the Biblical term for ark, teba, which in Exodus 2:3 is appropriately applied to the ark of the infant Moses, constructed of bulrushes coated with slime and pitch, or in other words, a pitch coated reed-like vessel, stems from the root word for sunken, which implies a submersible boat, a type of submarine; and the ancient clay tablets called the boat a sulili, which today is the same Hebrew term called soleleth, which means a submarine (see my footnote 549).

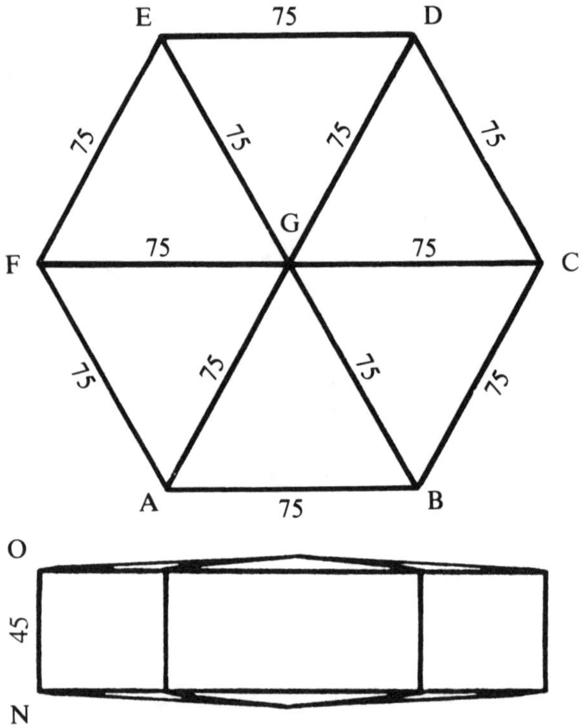

Source: THE HOLY BIBLE, GENESIS,
reads;

6:15 And this is the fashion which thou shalt make it of: The length of the ark shall be three hundred cubits, the breadth of it fifty cubits, and the height of it thirty cubits.

6:16 A window shalt thou make to the ark, and in a cubit shalt thou finish it above; and the door of the ark shalt thou set in the side thereof; with lower, second, and third stories shalt thou make it.

Refers to the measurements, window, door, and decks of the boat:

Although the Biblical Hebrew boat of Noah is smaller in size, which could be due to the different unit of measurement, interpretation, or transliteration from the original clay tablets, the shape of the boat fits the same measurement 3 to 1 ratio formula as the Babylonian/Sumerian boat of Atra-ḫasīs/Ziusudra.

Therefore, using the 18 inch Hebrew cubit, the perimeter of the boat, A to B, B to C, C to D, D to E, E to F, F to A, equals 300 cubits, which equal 450 feet.

The radius of the boat, G to A, G to B, G to C, G to D, G to E, G to F, each equal 50 cubits, which equal 75 feet. The height of the boat, N to O, equals 30 cubits, which equal 45 feet.

Noah is given instructions for a window, its size and location, either to be made and sealed before the flood, which I consider very unlikely, or to be cut open/opened or made at a later date. Genesis 8:6 that Nō´-ăh (after resting on the mountain) opened the window of the ark which he had made (prior to resting on the mountain, in order to release the birds), for what would be the purpose of making a window during construction of the boat, only to seal the window shut, or to shut him in, a short time later, as is pointed out in Genesis 7:16 and the LORD (Nibiru; caused effects that) shut him in, or as I additionally explain in my footnote 664, 749, and 766, Genesis 8:6 could be out of order.

Also, he is instructed to make a door in the side of the boat, to permit entry; or in the manner in which to seal the doorway, to set in, or put a door into the side entry doorway, to seal it shut.

The boat is to have lower, second, and third stories, which does not agree with the 6 decks of the Babylonian boat. However, if only 3 decks are installed, they would divide the 45 feet height into 3 stories, each 15 feet in height, which would be an obvious waste of deck to deck space, in addition to a possible hull bracing problem, necessitating extra supports at the hull, which would not be as strong as side to side, additional deck support at the hull; which if added in a 3 deck ark/boat, then 3 extra decks could have been as easily installed, solving both problems simultaneously. It could be possible the word stories, which is not a nautical term, refers to a unit of measurement, or floor to floor height in dwelling construction, composed of perhaps two, 7-1/2 feet units, or stories. The plural form of the word stories, could refer to 2 lower, 2 second, and 2 third decks; or collectively, since the number of lower decks is not given, we could assume 4 lower, or below the water level decks, and the second and third, or above the water level decks, which would then total 6 decks, each approximately 7-1/2 feet in height. Although the number of decks is not explicitly given, they are implied, but their location in terms of stories is clearly defined (see my footnote 540 and 590).

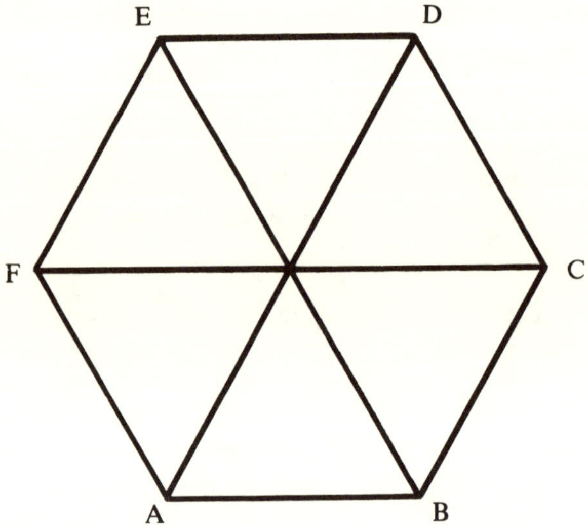

HELLENIC BOAT

On page 135 of the book Atra-ḫasīs, The Babylonian Story Of The Flood by Lambert and Millard, Berossus, a priest of Babylon in circa 290 B.C., who wrote the book called Babyloniaka, is quoted by Alexander Polyhistor, a Greek of the first century B.C., the book is lost to us, and then Polyhistor is quoted by Eusebius, in his book of the fourth century A.D., Chronicles, which in part states that Xisuthros (Ziusudra) is instructed by Kronos (God Enki) to build a boat with the following dimensions:

"five stades long and two stades wide"

The accepted figure for one Greek stade is 600 feet, therefore, the boat is 3,000 feet long and 1,200 feet wide.

Since the Hellenic dimensions of the boat were probably acquired from the Babylonians during the time of their conquest by Alexander the Great in 331 B.C., the Hellenic boat was give to us in terms of the Babylonian cubit and dimensions: 3,000 by 1,200, both divisible by 600, which indicates the Babylonian connection, and makes the Hellenic boat a variant of the Babylonian boat. How can this description be valid? How can these large dimensions be true, and still agree with the Babylonian boat? It can be valid only if each of the 6 decks perimeter and diameter dimensions are collectively expressed. In order to match the Hellenic account, the most recent report, to the older Babylonian account, and taking into consideration the fact that the Hellenes were known as excellent mathematicians, and that hexagon is a Greek word meaning 6 corners, or angles, many interesting combinations of dimensions are possible, of which my favorites I will render.

The most obvious, is the term "stades," or stadium/circular, as I previously explained on page 280 within this section, as a reference to the shape of the boat: the shape is contained within the dimensional terms.

If we assume long refers to the perimeter of each inner level (stade), A to B, B to C, C to D, D to E, E to F, which cumulatively equal 600 feet, and exclude the top and bottom level, as shown on page 289 within this section, and multiply by 5 inner levels (5 stades), 2, 3, 4, 5, 6, we get 3,000 feet; and then if we assume wide refers to the perimeter of each outer level (stade), A to B, B to C, C to D, D to E, E to F, which cumulatively equal 600 feet, and consider only the top and bottom level, and multiply by 2 outer remaining levels (2 stades), 1, 7, we get 1,200 feet (cumulatively 7 stades: this very well could be a variant of line 61 on page 289 within this section); or if we assume wide refers to the diameter of the boat, A to D, B to E, F to C, which cumulatively equal 600 feet, and consider only the top and bottom, and multiply by 2 outer diameters (2 stades), 1, 7, we get 1,200 feet.

If we assume long refers to the perimeter measured up the 50 foot side, across the 200 foot top, down the opposite 50 foot side, and across the opposite 200 foot bottom, cumulatively we have 500 feet, measured from A, B, C, D, E, F, 6 corners, and multiply by 6, we get 3,000 feet; and then if we assume wide refers to the diameter of the boat, A to D, B to E, F to C, which cumulatively equal 600 feet, and consider only the top and bottom, and multiply by 2, we get 1,200 feet. In this example, the height of the boat is included within the length figure.

Therefore, regardless of the dimensions used, the Babylonians, Hebrews, and Hellenes are all describing the same hexagon shaped boat (see my footnote 588).

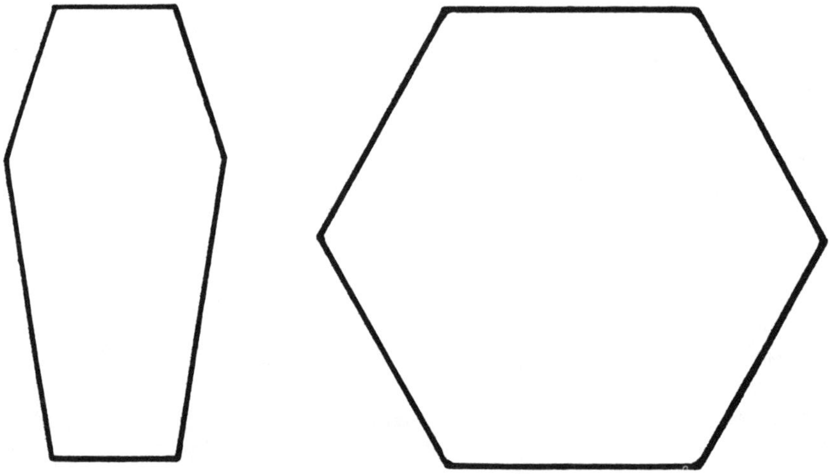

HEXAGON DESIGN

Throughout the centuries since the great flood, definitions of the boat have been rendered in Sumerian, Babylonian, Egyptian, Hebrew, and Hellenic. Out of all of them, several common obscure references to the protective function of the shape of the boat have been drawn, among which are chest, box, and coffin, as on page 55 of The Babylonian Expedition Of The University Of Pennsylvania by Hilprecht, and as on page 233 of The Gilgamesh Epic And Old Testament Parallels by Heidel (see my footnote 539 and 549).

These do not necessarily refer to a rectangular shape. If I may, as shown above, the oldest coffins could very well have been composed of six sides, a polygon, as many are today; as well as a chest or storage box for valuables, similar to the Greek wedding crown storage box of today, which is also a polygon in shape.

The fact that the boat was composed of six sides, a hexagon shape, a design from an advanced technology, at a time when the boats of mankind were rectangular in shape, of what could be called the hollowed out tree trunk, the dugout, or canoe design, of which mankind basically continues to use, even in the sophisticated boat designs of today, supports the belief that the Gods of the Earth, and therefore of mankind, are extraterrestrials, astronauts, sailors of the stars, space travelers from an advanced civilization: GODS OF THE UNIVERSE™

Furthermore, the significance and implications of the hexagon shaped design are far-reaching, for they also add credence to the belief that an advanced technological race of beings, supreme beings, visited the Earth, because the design of the boat is from an advanced technology not indigenous to Earth, their advanced technology, and not the primitive technology of early mankind; therefore, the supreme being who gave the design to the human being was what today we call an astronaut.

Even today, the outstanding exception to the rectangular shaped boat is the flat circular shaped boat of the Near East, called gufa, which is used on the Euphrates River, the same river the hexagon shaped boat was built in. These gufa boats could be the last vestiges of the similar legendary hexagon shaped boat of the great flood (see my footnote 5 and 531).

In Lebanon, aside from the fact that the capital of ancient Phoenicia, Sidon, namesake of God Enki/Poseidon, still exists today, it is a local tradition among the area inhabitants that the ancient ruins of Baalbek are associated with the great flood. Is there a link between the ruins and the flood? I believe there is, and the link is still partially visible! Acknowledging that more recent temples have been built upon the original site, one of the extant central features of an ancient temple is a hexagon. I believe that an important function of this temple was to venerate the God/astronaut/supreme being and the hexagon shape of the boat, which he gave to mankind in order to escape the flood. The hexagon feature is visible proof of a flood connection (see my footnote 841).

According to The Encyclopedia of Judaism and Encyclopedia Judaica, magen David, or literally shield of David, also called the Star of David, "formed by two equilateral triangles which have the same center and are placed in opposite directions," form a hexagram border and thereby outline the central feature, a hexagon. Circa 1800 A.D., the figure was adopted as a symbol of Judaism "in contradistinction to the Christian use of the cross," which predates Christianity, as I illustrate on page 310 within my comments. However, it also predates Judaism, and was used "as early as the Bronze Age (circa 3,000 B.C.)," and "appears in early Byzantine and many medieval European churches," and "was often used in Europe and the Middle East for decoration, possibly with magical connotations, especially in Muslim and Christian countries, by Jews and non-Jews alike, without having any special and recognizable significance (its true meaning long forgotten)." The term magen David "occurs as a designation of God," and shield of David designates "seal of the God of Israel," and "was used for certain amulets, among which one (water) for putting out fires," and as a "protective shield (boat/ark)," and "symbol denoting the harmony between the antagonistic elements of water (Earth) and fire (Nibiru)." I believe the figure symbolizes God Enki/Poseidon, the God of the triangle/triangles, the God who gave the hexagon design to mankind, to protect mankind from the flood waters of the great deluge; and the shield or Star of David would be more appropriately called the shield (boat/ark) of mankind (Ziusudra/Noah) or Star (hexagon) of God (God Enki/Poseidon). Now that the shape of the boat/ark has been discovered, and we understand what this symbol represents, we can more clearly understand its widespread use.

Additionally, it should be noted the hexagon shape is a very basic and natural shape, and is appropriately linked to water, for it is actually the shape of water. When we observe a drop of water, as a drop of rain on a window, close up while squinting, it is a hexagon shape: the shape of water is the shape of the boat. Again, peculiar as it may seem, a boat in the shape of a hexagon was the means utilized to save mankind from the hexagon shape of water. Moreover, frozen water, water which has crystallized in the form of a snowflake, also is the shape of the boat, for all snowflakes are hexagon in shape; and of more interest, snowflakes will form hexagon cylinders identical to the shape of the boat of Ziusudra, at temperatures colder than 12 degrees below zero fahrenheit.

Incidentally, some genes are hexagon in shape. Could the design of the boat be based on the shape of a gene, which floats within the fluid or "water" of our bodies?

Also, a beehive consists of many hexagon shaped wax cells, which are extremely strong structures, capable of supporting about thirty times their own weight. It is peculiar but fitting the way bees use a sealed hexagon shaped depository for the protection of their future generations, analogous to the way mankind was protected in the sealed hexagon shaped boat.

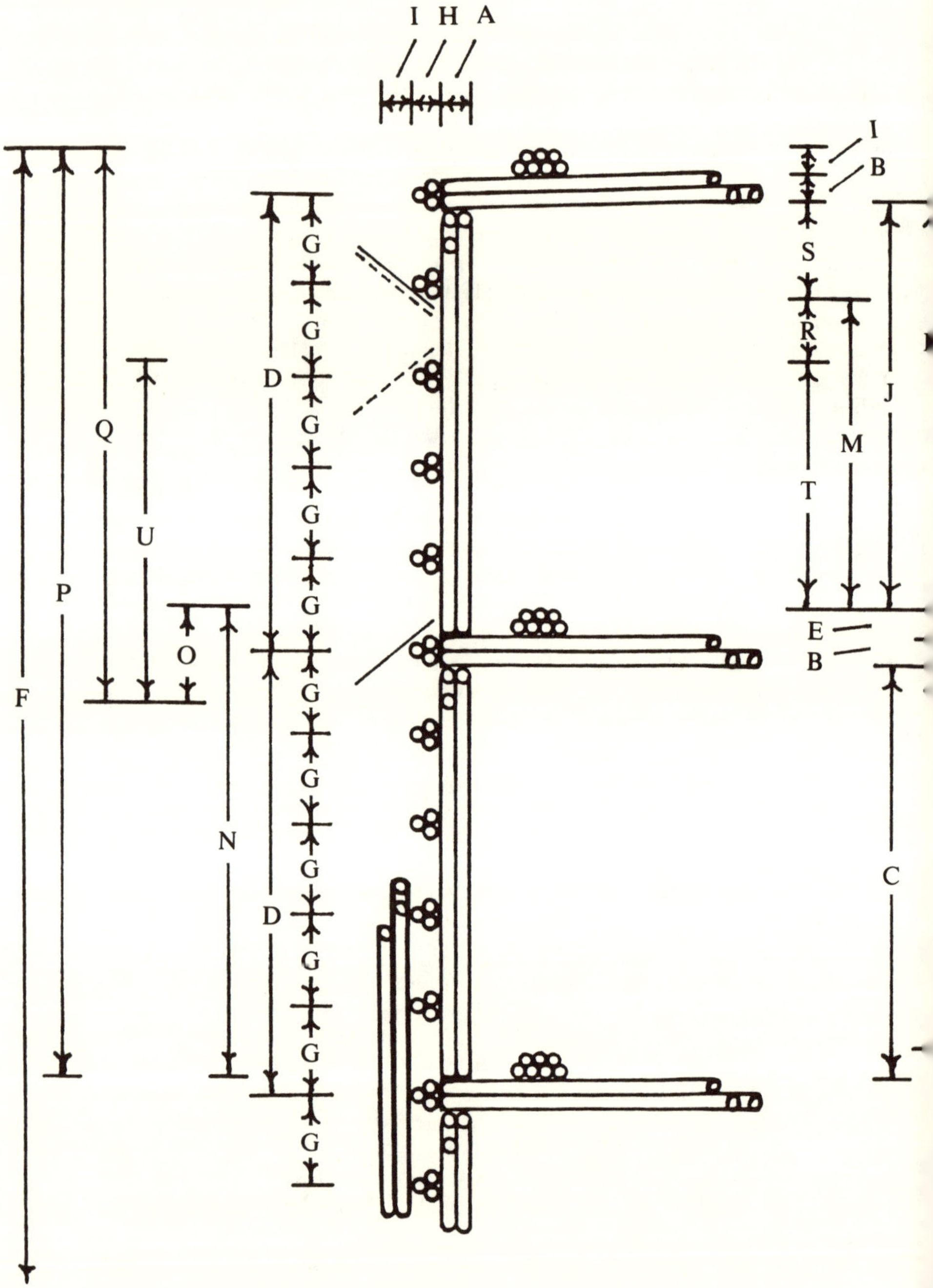

Cross section view of reed bundle detailed construction probabilities; see next page for detailed construction probabilities notes.

Detailed Construction Probabilities Notes

A. The framework beams, or vertical triangle truss bracing beams with dividing beams, are 6 inches in diameter, and are composed of three 3 inch diameter reed bundles, with staggered ends to fit together, and provide the primary framework for the hull.

B. The horizontal floor and ceiling bracing beams are 6 inches in diameter, and are composed of three 3 inch diameter reed bundles, 20 inches on center, with staggered ends to fit together. They extend outward from each equilateral triangle, and connect each of the nine vertical divided triangle truss bracing beams, and provide the primary framework for the floors, ceilings, and hull.

C. Top of the 20 inches on center, 6 inch floor beams to the bottom of the 20 inches on center, 6 inch ceiling beams; 7 feet 10 inches. This is the height of the structural vertical truss bracing beams, which are composed of vertically and horizontally divided triangles; nine vertical truss bracing beams per each equilateral triangle, spaced 6 cubits, or 10 feet apart on center, except the nearest to the hub, which is 6 feet 7 inches on center from the hub, between which are located the open compartment-like trapezoid shaped sections. They are 7 feet 10 inches in height throughout the boat, on each deck, except the top and bottom deck, where the height at the hull is 7 feet 10 inches, and increases at every 10 feet on center location, until the height of the truss beams nearest the hub reach 9 feet 6 inches. Secondary diagonal bracing beams further brace the hull at every 5 foot location.

D. The framework height of each deck at the hull is 5 cubits by 20 inches, or 100 inches, which equal 8 feet 4 inches on center; and the minimum height of the top and bottom deck, beam to beam at the hull, is 7 feet 10 inches, and the maximum height, beam to beam at the hub, is 9 feet 6 inches. All of the other decks are uniform in height, beam to beam, being 7 feet 10 inches throughout. Any additional deck height would have been a waste of space, and after all, the boat was not provided for luxury and comfort, it was designed for utility, and a specific function: to protect human beings from the huge flood waves.

E. The floor is 6 inches thick, and is composed of three 3 inch reed bundles, staggered, as are their ends, and alternately reversing them to fit together.

F. The framework length and width of the boat, which are 600 feet and 100 feet respectively, are given in equal 20 inch cubits, as are the balance of the framework dimensions. Therefore, the framework height of the boat, in order for the boat to be symmetrical, which I am certain it was, must equal 5 cubits by 6 decks, or 30 cubits, which equal 50 feet. Because we do not know exactly how thick the hull was, and normally we add a hull onto a framework, thereby enhancing the framework dimensions; or perhaps in this instance, the bracing beams and hull may have been included within the dimensions given for the framework, thereby making the framework dimensions also the completed dimensions: therefore, since I gave framework dimensions within the text, I shall give enhanced dimensions within these notes, which only affect the exterior dimensions of the boat. When the boat is completed, due to the 6 inch bracing beams and the 6 inch hull, 12 inches on each of the six sides are added to the framework dimensions; and due to the one-half beyond center of the 6 inch bracing beams, or 3 inches, and the 6 inch hull, 9 inches on the top and bottom are added to the framework dimensions. The enhanced finished exterior diameter is 202 feet, and the enhanced finished exterior height at the side hull is 51 feet 6 inches, and at the hub it is 54 feet 10 inches. Also, when empty, being 34 feet 4 inches below the water level at the side hull, and 36 feet at the hub, the boat must be able to float in the Euphrates River.

G. The horizontal 6 inch hull bracing beams are 20 inches on center.

H. The horizontal hull bracing beams, 20 inches on center, are 6 inches in diameter, and are composed of three 3 inch reed bundles. They provide the secondary framework for the hull.

I. The hull, or skin, is 6 inches thick, and is composed of 3 inch diameter by 5 feet long reed bundles, staggered, as are their ends, and alternately reversing them to fit together, over which is applied a thick coating of pitch.

J. Top of 6 inch floor to bottom of the 20 inches on center, 6 inch ceiling beams at hull, on all of the decks; 7 feet 4 inches.

K. Top of 6 inch floor to bottom of the 20 inches on center, 6 inch ceiling beams at hub, top and bottom deck only, due to the 20 inch taper, or larger height of the top and bottom hull at the hub; 9 feet.

L. Top of 6 inch floor to bottom of the 20 inches on center, 6 inch ceiling beams at hub, on all of the decks except top and bottom; 7 feet 4 inches.

M. Doorway dimensions; 5 feet 8 inches high by 4 feet 6 inches wide, inward tapering perimeter, represented by solid tapered lines through the hull.

N. Bottom of doorway to water level, before boat is loaded; 8 feet 10 inches.

O. Bottom of doorway to water level, after boat is fully loaded; 1 foot 8 inches.

P. Top of roof to water level, before boat is loaded; 17 feet 2 inches.

Q. Top of roof to water level, after boat is fully loaded; 10 feet.

R. Window dimensions; 14 inches high by 29 inches wide, inward tapering perimeter, represented by broken tapered lines through the hull.

S. Top of window to bottom of ceiling or roof bracing beam; 1 foot 8 inches.

T. Bottom of window to top of 6 inch floor; 4 feet 6 inches.

U. Bottom of window to water level; 6 feet 2 inches.

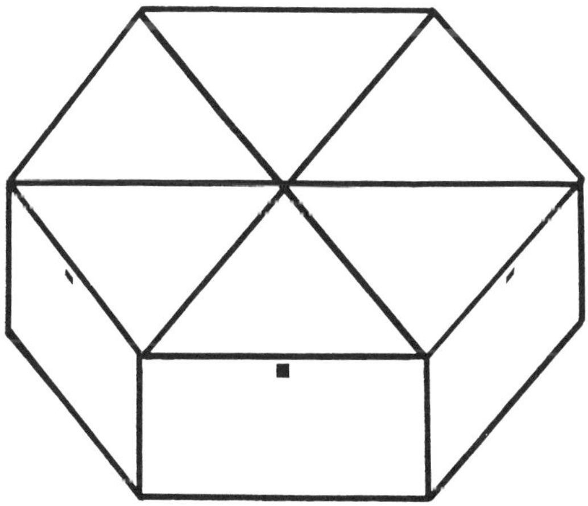

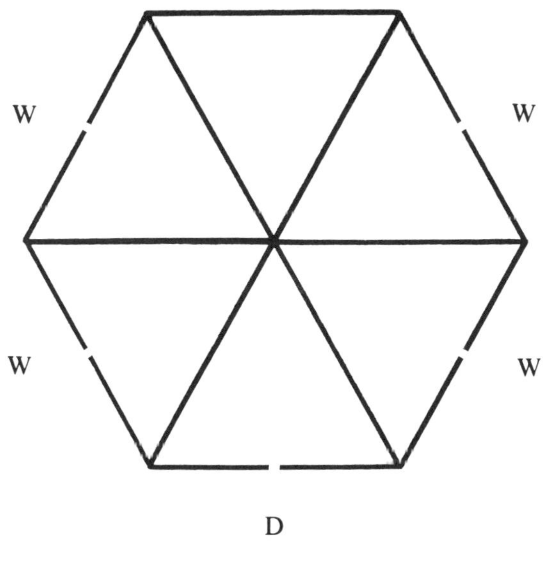

Probable window and door location.

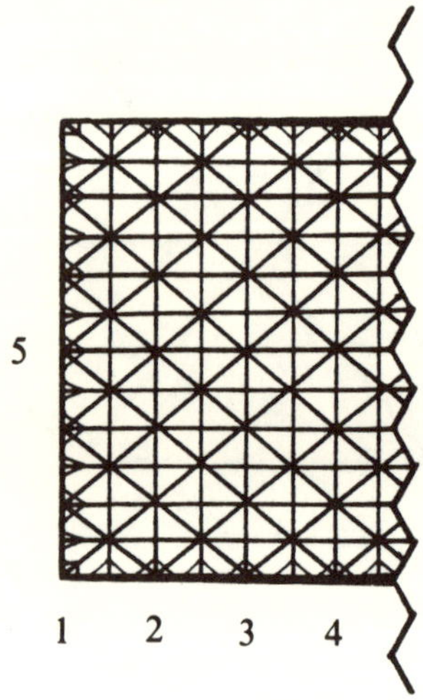

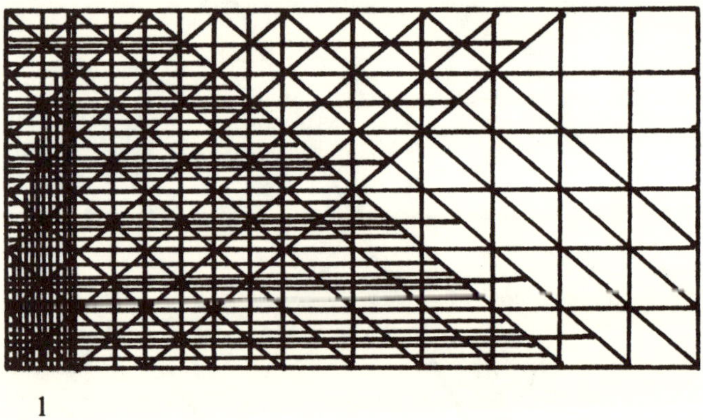

Probable top, bottom, and side hull, or outer (1) truss bracing. The vertical truss bracing beams, or framework beams, all nine of them, of which four are shown, were composed of vertically and horizontally divided triangles, braced around the perimeter by the diagonal truss bracing beams (5).

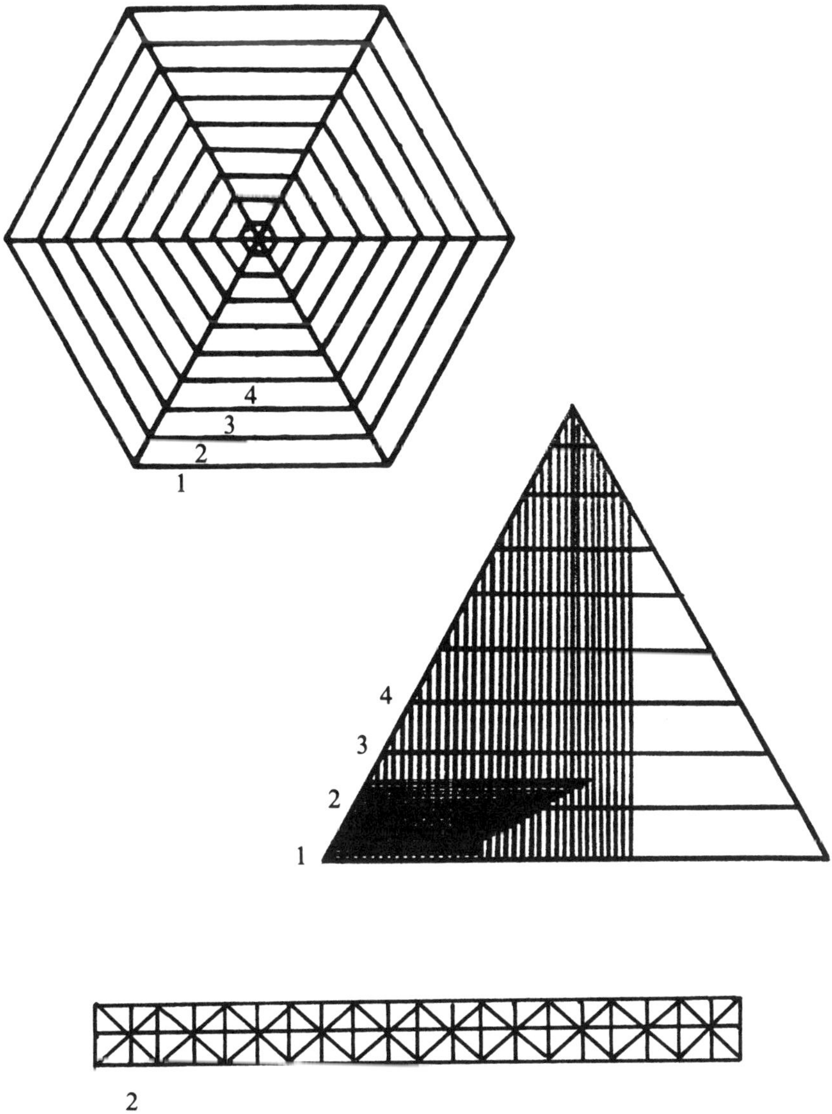

Probable floor, and inner (2) truss bracing. The vertical truss bracing beams, or framework beams, were composed of vertically and horizontally divided triangles.

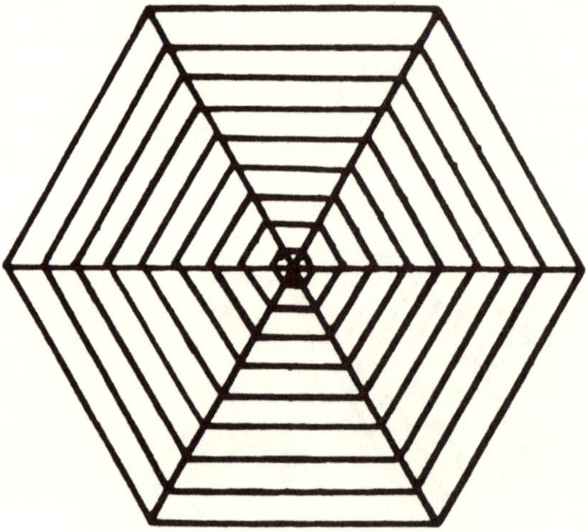

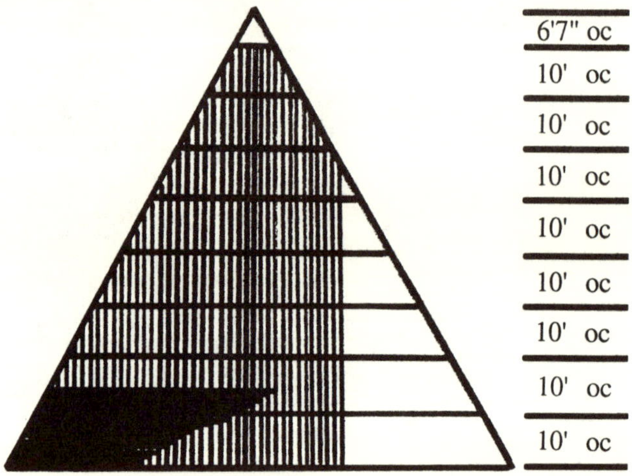

Probable ramp location was through a floor opening between framework members, next to the hub of one of the six equilateral triangles composing each deck. On each of the decks, all except the lowest deck floor, or hull, and staggered around the hub, a 5 feet wide reed bundle ramp allowed movement and permitted central ventilation within the six decks of the boat.

The structural vertical triangle truss bracing beams, starting at the hull, were installed at every 10 feet on center (oc) location, thus dividing the 86 feet 7 inch distance from the base to the apex of each equilateral triangle into nine open compartment-like sections. After the division, 6 feet 7 inches remain from the center of the ninth truss beam, the innermost truss beam, to the center of the hub. If we subtract one-half of the 6 inch diameter truss beam, 3 inches, and one-half of the 30 inch diameter hub, 15 inches, the space remaining between the innermost truss beam and the hub is 5 feet 1 inch; just wide enough for a 5 feet, or 3 cubits wide reed bundle ramp.

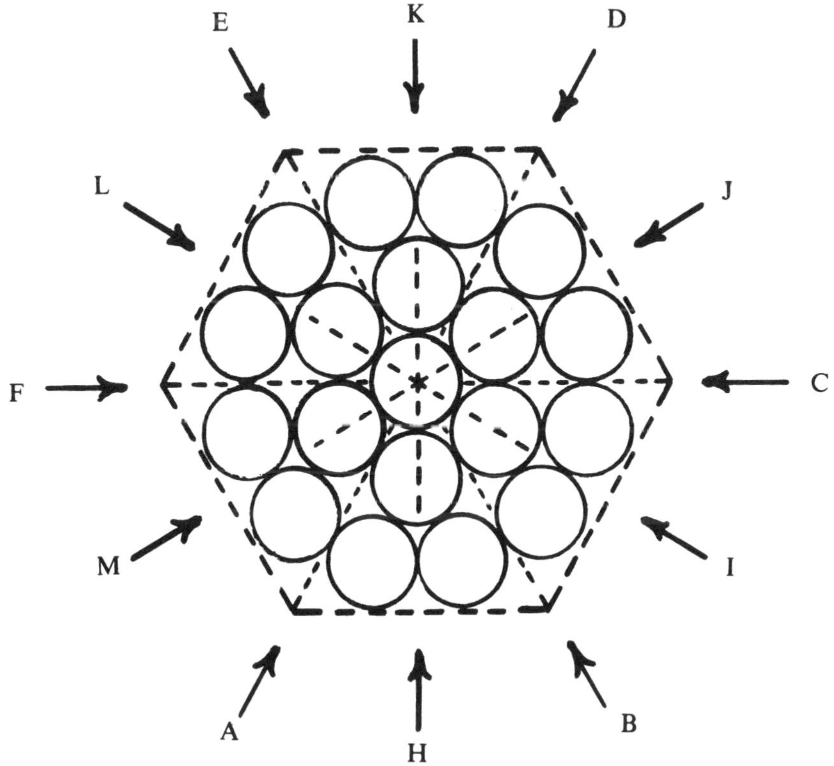

Probable hub detail. The hub probably was composed of nineteen, staggered 6 inch diameter reed bundles, which are solid reed bundles, and are not made up of three smaller, alternately staggered 3 inch diameter reed bundles, as are the framework beams. This arrangement gives the hub a diameter of 30 inches. At A, B, C, D, E, F, the 6 inch diameter framework beams of the equilateral triangles are secured, and at H, I, J, K, L, M, the 20 inches on center, 6 inch diameter beams dividing the equilateral triangles are secured, except where the ramp is located. Thus, all of the forces being exerted upon the boat, including pressure and stress, are equally distributed.

It should now be apparent that the entire boat, including the main component parts: 6 inch reed bundle framework, truss design of the bracing beams, and the hub, were all designed upon an equilateral triangle formula.

ANU

NIBIRU

ENLIL

EARTH

ENKI

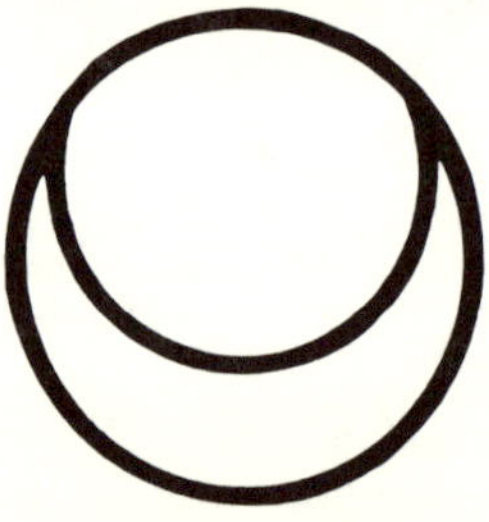

MOON

NINHURSAG

BIRTH

Likely symbol pendants of the four main Gods, shown separately, and probably worn by them on a necklace. Throughout ancient times, the four main Gods were known by Sumerian, Babylonian, Hellenic, and more recently by Roman/Latin names, and were often represented or depicted by ideograms or symbols.

AN or Anu, 'Ανός, Οὐρανός or Ouranos, Uranus; whose cross-like symbol depicts the radiating planet Nibiru, the abode of the Gods, with its two rings of clouds, within which orbit its four moons, upon which God Anu is the president of the assembly, the governing body.

Alalg̃ar or [á]-làl-gar, EN.LIL or Enlil, 'Αλαπαρος or ''Ιλλινος, Zeus (Eagle), Jupiter; whose symbol depicts the Earth within a seven pointed star, the seventh planet as you enter our solar system and proceed toward the Sun, upon which God Enlil is the administrator, supreme commander, and ruler.

A-lulim(ak) or [á]-lulim, E.A, EN.KI or Enki, 'Αλωρος or 'Αός, 'Κρόνος or Kronos/Poseidon (Trident), Neptune; whose symbol depicts the crescent shape of the Moon, through its influence upon the Earth, especially the waters of the Earth, symbolizes the power and wisdom of God Enki, the chief scientist.

NIN.H̲UR.SAG or Ninhursag, Nintu, Ninmaḫ, Mämi, Athena (Owl), Minerva; whose symbol depicts birth through the shaver and umbilical cord cutter, the instrument which was used in the capacity of the birth coordinator by Goddess Ninhursag, the chief doctor.

NIBIRU

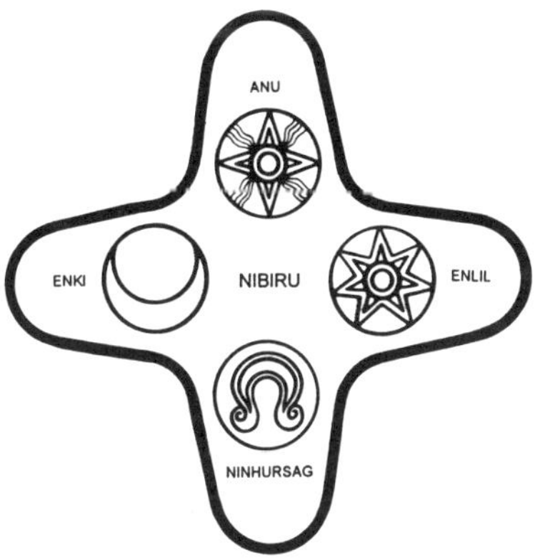

Likely symbol pendants of the four main Gods, shown as a composite, within the outline of the cross-like appearance of Nibiru; and the winged sphere symbol of Nibiru, within a rainbow-like outline, similar to the one the Gods wear above the left breast, all taken from ancient sources.

ANCIENT
NIBIRU

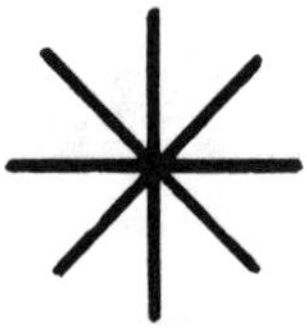

SUMERIAN
AN

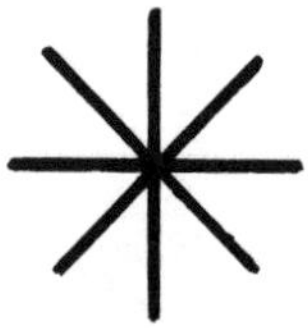

AKKADIAN
STAR

BABYLONIAN
HEAVENS

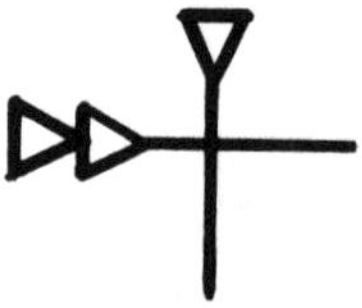

ASSYRIAN
GOD

EGYPTIAN
LIFE

Pictured above are symbols, with the name of the people who used them, and the interpretation of the meaning of the symbol below their name.

Could all of these symbols have a common meaning? Do they depict the planet Nibiru, as it crosses through the solar system near the Earth, with its force fields affecting, or pulling upon the Earth? I believe this is the correct interpretation of their common meaning. With this perspective, it should be quite clear and obvious that the progressive use and intent of these symbols throughout thousands of years, all represent the same event: the crossing of Nibiru near the Earth.

As with the other symbols, the Egyptian symbol for everlasting life or immortality, ankh, as shown above, is no exception. It could very well have an older meaning and a far greater importance. The loop or oval could represent the Earth, and the cross could represent the planet Nibiru, which crosses through the solar system near the Earth, and figuratively touches, or affects the Earth with its force fields, pulling/stretching/distorting the Earth and the fluid components of the Earth toward it.

King Ashurbânipal of Assyria (668-626 B.C.), as illustrated courtesy The Trustees of the British Museum, London, who gathered many clay tablets of Sumerian and Babylonian writings in his famous library at Nineveh. He is quoted in Atra-ḫasīs, The Babylonian Story Of The Flood by Lambert and Millard, page 26, (d) 'I study stone inscriptions from before the flood, which are obtuse, obscure and confused', and as is described in The Babylonian Expedition Of The University Of Pennsylvania by Hilprecht, page 40, number 1 "This version (of the Deluge Story) is an Assyrian copy of a Babylonian original"; and for his efforts, deserves credit for rendering much of the ancient information of our creation and early history to us. Note his pendant earring, and his winged sphere symbol above his left breast, and his tall, fez-like hat: all symbols of his authority and probably vestiges of an earlier period, when the Gods were upon the Earth and among mankind.

EPILOGUE

WHY

Ever since I was an altar boy in the Eastern/Greek Orthodox Church, I have wondered about the mysteries of our creation, our creator, and our purpose in life. While holding a candle during the services, I would gaze at the numerous religious paintings called Icons, which embellished the altar. There were Angels standing upon billowing clouds, with auras around their heads, and with wings upon their backs; and Saints, all with auras surrounding their heads. With my young and inquisitive mind, I would always ask myself "Why," which incidentally is still one of my most favorite words, but may seem rhetorical to some people, and I thought, "Someday I may find the answers."

I believe I now know what the Icons depict. By somewhat lifting the veil of mystery myself, and opening my eyes and mind, I believe the Angels embody the Anunnaki Gods, the astronauts of the Gods, as the recently excavated and interpreted clay tablets seem to indicate, and the auras depict the force field around our bodies, like that newly found through Kirlian photography; both of which we only recently rediscovered.

But where did the ancient people who first painted these pictures obtain their information? The question has several obvious answers. The information was handed down to us directly from the Gods, through our most distant ancestors, who dutifully recorded it; or some of our ancient ancestors were in contact with the Gods, and genetically inherited a special talent from the Gods for seeing the aura, and actually saw it around the Anunnaki Gods/Angels, as well as saw their flying apparatus, which they compared to the wings of birds; or during the last, or a recent orbit, passing, or window period of planet Nibiru, the information was revealed, and thus imparted to a highly enough advanced civilization of mankind capable of comprehending it. In all three examples, the information was bestowed upon mankind by the Gods themselves, so the answers really come from the same source. Thus, the Gods keep mankind primed, and probably lead us by imparting a slightly higher degree of information than we can understand, for they know, that within the next 3,600 Earth years, until the next window period, we will certainly advance intellectually and be able to understand the lead information.

Also, because the ancients believed in many Gods, we must assume the experience of revelation and contact, during a window period, was fresh in their memories when they pressed the information into their soft clay tablets.

Incidentally, I use the currently accepted religious term "God", which I recognize and employ as a synonym for alien, although our most distant ancestors, especially the Sumerians, not only properly identified them as aliens, but additionally identified the aliens as astronauts, DIN.GIR.AN.KI, "Divine Beings/Gods of/from Heaven and/to Earth", or "GODS OF THE UNIVERSE™" and the pictograph sign for Gods, dingir/DIN.GIR, has been literally interpreted as meaning "Righteous/Pure/Bright Ones of the Blazing Rockets/Rocketships", and the Babylonians called them i-lu or ilu/I.LU, "Lofty Ones" (shepherders), all of which did not necessarily connote a religious characterization in the same sense that the term "God" does today. I believe they are all metaphorically veiled references to astronauts/supreme beings, for the ancient term I.LU, which means "Lofty Ones", which I equate to supreme beings, could very well have been used to imply a being with a high technological state of development, and/or designate a being from a lofty or physically higher place: another planet.

Additionally, I frequently find myself asking myself "Why am I the one who discovered the shape of the boat, and untied the Gordian Knot of the flood?" It all started with "Why," and since there is a lot we don't know, perhaps it will also end with "Why?" The question is, "Why are there no Gods on Earth today, where are they; and if there are, where are they?"

INFLUENCE

Since our earliest days, mankind has been taught by the Gods, whom we then imitated, most of the time intentionally, and perhaps sometimes unintentionally, without being aware of it. Today, even though we are not in a window period, or close to one that has recently passed, for those of us who do not believe in arbitrary actions, such as arbitrary evolution, we could still be influenced by, and be imitating the Gods; through our subconscious mind, especially during certain periods most people call dreaming, both day and night, what is professionally referred to as hypnogogic imagery, for I assert that when you couple hypnogogic imagery (dreaming/visualizing) with research, what I experienced while writing this book, you get a phenomenon which I call perceptive interpretive research; or through deliberate programming of our genes; or through some unimaginable means of remotely triggering our brain. For today, with our "advanced" knowledge, we do not even know why we sleep, let alone why we dream.

Thus, the Gods could still be revealing and imparting information through us, when we are ready to accept and understand it. I believe the writing of this book and its revelations have been rendered through such a manner, perhaps to better prepare mankind for the approaching event: the return of the Gods. For I have an affinity, an obsession, there is something within me, a kind of driving force, a curiosity, which influenced me to eliminate the mundane daily distractions of life, what I like to call being up on the mountain, which by the way is not a bad place to be, and gather and channel all of my energies into the task of sorting out this story, and reinforming mankind of its existence and importance, and as a result, I now consider the Earth as my classroom, and all mankind as my pupil. It may not be mere coincidence: it may be fate. Some people call it inspiration or impulse, I call it selection. I believe I was selected to reveal this discovery. How do we really know? For if we, mankind, genetically created a surrogate or extension of ourselves on some distant planet, a mutant type of sub-human, living robots that would replicate themselves, would not we want to retain some degree of control over their actions through programmed or controlled evolution, not always on a day by day basis, but in a general direction; and monitor and prompt them surreptitiously; and then regularly, during the window period, make precise disciplinary corrections or changes in any deviations from the programming? Could this be what the Gods have been doing to mankind? And if so, the Gods may even today be monitoring mankind in this same manner, perhaps through our heartbeat, as is metaphorically alluded to in Atra-ḫasīs, The Babylonian Story Of The Flood by Lambert and Millard, on page 59, line 214 So that we (the Gods) may hear (monitor) the drum (the heartbeat of mankind) for the rest of time (until mankind or the Earth expires), and line 227 For the rest [of time they heard the drum] (see my footnote 175 and 193a).

Subconsciously, I have always felt close to, or been sensitively tuned in to the Gods. From almost constantly thinking about them in the sense of being extraterrestrial astronauts, to even many years ago drawing their sunset red abode on a wall, what I then troublesomely and reluctantly thought was the ancient Egyptian solar disc, which I even then doubted was the Sun. Incidentally, while on this theme, it may also be of interest to know that I unsuspectingly remodeled my office, where I researched and wrote this book, to look somewhat like the interior of the boat of Ziusudra, with its vertically divided triangular truss bracing beams, nearly 6 years before I discovered the triangular component design of the boat.

Furthermore, if the Gods still have control of mankind, the proof shall be upon actually witnessing their reappearance, from the return of their planet Nibiru to our visible eye within the sky, to observing the Gods as they walk upon the Earth. Only then will we know they still have undisputed control of mankind. But what are they grooming us for? Could it be to colonize other planets, or maybe just a curious kind of genetic experiment, to ascertain what the parameters of our programmed evolution are on the planet Earth, and how the forces of the Earth affect our development?

The clay tablets state that mankind was created to work for the Gods, and provide them with sustenance. Is this the reason, and what exactly does sustenance mean, certainly not just food? Maybe we are not quite ready for the answer. But one way or another, the purpose and function of our existence is strictly for the benefit of the Gods, and not ours. But to those people who are alive when the window period reoccurs, it shall be a submissive event for mankind, as it has always been in the past: a glorious day for mankind, and the Gods.

MYTHOLOGY

Some individuals may think this book and my comments about it are all based upon mythology, fictitious stories serving to explain some abstract or inaccessible phenomenon. But let me say this: sometime in the not too distant future of this planet Earth, perhaps 2 or 3 thousand years from now, say about the year 5,000 A.D., hopefully mankind, our future generations, will stand upon some flat and desolate areas of our planet, and according to very old fragments of ancient things called books, declare that large cities once existed upon those sites; and they had depositories called libraries that contained volumes of books with knowledge, which told about supreme beings/Gods/astronauts, visiting the Earth shortly after the year 2,000 A.D..

But others will surely challenge the declaration, and argue they were just part of the mythical past, mythological cities and stories, cities and events that never existed in reality. For in the future, all of our cities, and most of our achievements and experiences, shall be cast into the mythological abyss, where time is the diluting agent. To us today, the ancient Sumerians, Egyptians, Hellenes, and Romans are all in this abyss. But wait, we have recently found and proven the existence of some of their mythological cities; however, their literature and stories remain to be proven. We must wait for time to prove their ancient stories.

MOST RECENT ORBIT

If my calculations of the next orbit, between 2,000 A.D. and 2,100 A.D., are correct, then the most recent orbit occurred 3,600 years prior to that, approximately between 1,600 B.C. and 1,500 B.C.. Is there a correlation between what happened on Earth during this period that would indicate an orbit of the supreme planet, Nibiru, and a visit by the supreme beings, the Gods? I believe there is, for they occurred simultaneously. Two events come to mind.

First, the bull worshipping Minoan civilization on the Mediterranean island of Crete was suddenly destroyed, it is believed by flood waves or fire, or both, caused by unusual volcanic activity, attributable to the volcanic eruption on Santorini/Thera, near Crete (see my footnote 199).

Secondly, and probably far more important, was the period during the end of the Middle Kingdom of Egypt, which coincided with the Hebrew Exodus (after all, who was Moses talking to, and from whom did he receive the clay tablets with the Ten Commandments), when the king named Taui Thom ruled in Egypt. This king was later called Timaeus by the Hellenes (see my footnote 719), and Tutimaeus by the Egyptian historian Manetho, as described by Velikovsky on page 96, note 3, middle page 97, and page 113, note 18 of Worlds In Collision. It was during this period that the disturbances or so-called "plagues" took place, with the effects upon the Earth so thoroughly explained by Velikovsky. But the primary cause of these effects, the catalyst, the triggering agent, I believe was the planet Nibiru.

These disturbances upon the Earth deeply affected the polytheistic religious beliefs, in which the supreme celestial God, or planet, called Ra, was identified as Amon (hidden or invisible) or Amon-Ra (hidden or invisible celestial body), during the non-humanly visible period of its orbit, could have very well been an Egyptian variant of the Sumerian An, whether we reverse the letters in the name Ra or not, as was sometimes done at other times translating other names from language to language, or the Babylonian Anu; all of which during various time periods represented the supreme celestial God, or planet, as well as the leader or highest of the supreme beings upon that planet. These events later, during the reign of Amenhotep IV, approximately between 1,367 B.C. and 1,350 B.C., shortly after the last orbit, precipitated the Amarna Revolution in Egypt, when he instituted the belief and worship of the life giving power of a certain celestial body, what many today interpret as the Sun, which I believe was the planet Nibiru. He called this celestial body Aton-Ra (apparent or visible celestial body), although he probably did not actually see the rare stellar object, which was probably visible 130 to 230 years before his reign, and what we may have been misinterpreting as the rays of the Sun, or Aton-Ra, could actually be drops of water; after misting up and then falling back down to Earth as rain from the water vapor laden sky as the planet Nibiru passes, as is depicted on page 184, Ill. 146, and page 197, Ill. 149 in The Art Of The Ancient Near East by Lloyd. Additionally, it is clear that Ra was associated with a celestial body other than the Sun, for it has been described as a "bloody star", the color of Nibiru, as described by Velikovsky on page 109, note 13 of Worlds In Collision (see my footnote 16). Certainly the Egyptians would have only called the Sun by its proper name, if it was Ra, and not included in its identification an abstract reference as merely a colorful star, "bloody star", for it is highly unlikely they considered the Sun as a star.

The belief in Ra was the belief in one God, the supreme celestial body itself, as well as what I believe was a concurrent belief in the leader or the highest in authority of the supreme beings upon the supreme celestial body, which today we interpret as monotheism. As a result, many changes occurred in artistic, literary, religious, and social practices, and the Egyptians began calling their king Pharaoh, and he changed his name to Akhenaton, which means the son of the apparent or visible God, progeny of the visiting supreme celestial body itself, or son of God, progeny of a supreme being from the supreme celestial body, in other words, product of the planet or its beings, or both, which we all are; a type of sub-God, or demigods, or more specifically stated, a type of sub-Igigi God, when created (see my footnote 165 and 184). And now, besides the attainment of "knowledge", or generative capabilities, perhaps we have advanced beyond that stage, but either way, we are still sons and daughters of the highest Gods, in a very special way.

Could this period, between 1,600 B.C. and 1,500 B.C., have been the most recent window period, a time during which the Gods bestowed gifts of information upon mankind; and could the "flowering" or high degree of cultural achievement of all the past great civilizations be attributable to the imparting of information by the Gods, during this or previous window periods? Again, only time can prove my calculation and beliefs.

NEXT ORBIT

There is an ancient text which is given on page 363 of The 12th Planet by Sitchin, which in effect reads that during the zodiacal "Age of the Constellation Lion," between 10,860 B.C. and 8,700 B.C., there was an orbit of the celestial body, if I may call it the Lion orbit. I believe the Lion orbit was the flood orbit. Keeping in mind its normal southern to northern orbital trajectory and the total period of passing being about 16 months, at this time of observation, the text describes the planet Nibiru as "Supreme, Supreme, Anointed," the "Lord whose shining crown (halo of satellites) with terror is laden," or a glowing celestial body full of terror. The text also states that the "Supreme planet," hence my affinity to the designation of supreme being, the largest and most powerful or influential planet, "a seat he has set up," or established an orbit "Facing the confined orbit of the red planet [Mars]." This clearly states that during this orbit, at this time of observation, the "Supreme planet" appears to be in an established orbit near the orbital path of Mars, and Mars appears to be in its normal, or confined orbit, and the "Supreme planet" is not yet excessively disturbing to the Earth, but only appears threatening. Later, at this time of observation, as the planet orbits closer, the text states that the planet "Daily within the Lion he is afire," appeared daily within this period as a fireball (see my footnote 16), and states that the planet now sheds "His light his bright kinships on the lands pronounces," sheds his light and influences upon the Earth.

Why does this not yet sound like a description of the destructive effects of the flood orbit? If this reflection of the planet was made immediately after the flood orbit, I believe the scribe would not have favored the planet so kindly. This could only be a description of the earliest part of the flood orbit, first an observation from a great distance, and then secondly a closer observation, both during a somewhat peaceful and tranquil appearing part of the Lion orbit. Therefore, I believe these observations were made of the planet and inscribed on the clay tablet immediately preceding the occurrence of the disruptive effects upon the Earth during the Lion orbit. Only then, could the Lion orbit have been the flood orbit, and if this is correct, then the flood occurred between 8,800 B.C. and 8,700 B.C..

But if the Lion orbit was not the flood orbit, and this description was only an observation of the first return orbit after the flood, when the planet again appeared peaceful and tranquil throughout its orbit, then the flood occurred between 12,400 B.C. and 12,300 B.C., and all the dates I have given in this book would have to be adjusted accordingly, and carried backward 3,600 years, for the orbit between 5,200 B.C. and 5,100 B.C. is probably too recent for the flood to have occurred; but I do not totally rule it out, although the evidence is lacking, except for the gnawing fact that this orbit is closer to the 2,888 B.C. year in which Gilgamesh ruled, sought, and eventually, either directly or indirectly, spoke with Utnapishtim/Ziusudra, as I elaborate upon later under 'Prophecies'. Whatever the case, the date of the flood may have to be adjusted, but the story basically remains the same.

However, this text does point out quite clearly that there was an orbit during the period of the Lion. It was during this same period that mankind was re-establishing themselves upon the mountainous lands of the so-called "fertile crescent", a home necessitated by the yet flooded valleys, for this was another period as well, the diluvial period of mankind, the end of the antediluvial period of mankind, and immediately prior to the beginning of the postdiluvial period of mankind: our current period.

Also, the Lion orbit must have occurred during the very latest part of the Lion period, for we have no oral or written accounts of an orbit between 100 B.C. and the current year, the most recent correlative period, a historical period, and a period which has been well documented.

Furthermore, it has been reported that, under normal conditions, according to a kind of "rule of waves", every third wave is a larger wave; and it has been reported that 10,800 was a sacred number of ancient astrologers (astronomers). Is there a correlation? I believe there is. Just like every 3rd wave is a larger and thus more powerful wave, so is every 3rd orbit or "wave" of Nibiru a more perturbing and thus more destructive orbit. One orbit of Nibiru equals 3,600 Earth years, and 3 orbits equal 10,800 years, therefore, this sacred number equals 3 orbit cycles of Nibiru. Does a conjunction occur among all or some of the planets and Nibiru, or is perigee somehow different between Earth and Nibiru, perhaps a different phase of either planets' orbit every 3rd orbit, and cause more than normal perturbations of Earth? If the flood orbit between 8,800 B.C. and 8,700 B.C. was one of these sacred number orbits, and caused more than normal effects, and the orbits between 5,200 B.C. and 5,100 B.C., and 1,600 B.C. and 1,500 B.C. were normal orbits, then the next orbit between 2,000 A.D. and 2,100 A.D. will be a sacred number or 3rd orbit, for 8,800 B.C. or 8,700 B.C. plus 2,000 A.D. or 2,100 A.D. equals 10,800 years.

Below, I have prepared a chronological schedule, Illustration 1, of what I believe indicates the past and future orbits of the planet Nibiru, which occur every 1 Shar year, which equals every 3,600 Earth years, for I believe the great flood occurred between 8,800 B.C. and 8,700 B.C., and the most recent return orbit of Nibiru occurred between 1,600 B.C. and 1,500 B.C., and the next orbit shall occur between 2,000 A.D. and 2,100 A.D..

Illustration 1

ORBITS OF NIBIRU

Orbit	Flood and Lion Orbit	Orbit	Most Recent Orbit	Next Orbit
14,500 B.C.	10,900 B.C.	7,300 B.C.	3,700 B.C.	100 B.C.
14,000 B.C.	10,400 B.C.	6,800 B.C.	3,200 B.C.	400 A.D.
13,500 B.C.	9,900 B.C.	6,300 B.C.	2,700 B.C.	900 A.D.
13,000 B.C.	9,400 B.C.	5,800 B.C.	2,200 B.C.	1,400 A.D.
12,500 B.C.	8,900 B.C.	5,300 B.C.	1,700 B.C.	1,900 A.D.
12,400 B.C.	8,800 B.C.	5,200 B.C.	1,600 B.C.	2,000 A.D.
12,300 B.C.	8,700 B.C.	5,100 B.C.	1,500 B.C.	2,100 A.D.

CATALYST

Throughout Worlds In Collision by Velikovsky, he eloquently describes the phenomenal effects upon the Earth of a celestial body orbiting near the Earth. I agree with him in respect to the effects upon the Earth, but I disagree with his primary cause, which he claims is a planet that is expelled or erupts from Jupiter, that planet being Venus. He also claims that Mars is a primary cause. The formation of Venus occurred earlier than the formation of Jupiter, according to the clay tablets, as is described by Sitchin in Chapter 7 of The 12th Planet. Therefore, Venus certainly could not have sprung from Jupiter, but Venus may have first "brushed" Jupiter, and acquired comet-like characteristics, and then encountered the Earth, giving rise to the conception that Venus sprang from the head of Jupiter, and appeared as a comet.

I believe the planet Nibiru is the catalyst, the external agent, the primary cause, and especially the small planets Venus, Earth, and Mars, and perhaps even the large planet Jupiter, suffer the effects as Nibiru, at perihelion, orbits through the Asteroid Belt region, between Mars and Jupiter, and more specifically, between Mars and the Asteroid Belt; although the planets may cause secondary disturbances to each other, particularly while being temporarily pulled or stretched out of orbit.

The clay tablet story which I have presented, only describes in some detail the effects of one passing of Nibiru, if I may call it the flood passing. More remote or more recent passings, including the last passing at the time of the Egyptian King Tutimaeus, during the Hebrew Exodus, between 1,600 B.C. and 1,500 B.C., could possibly have been at a different proximity, and with different orbital locations of the nearby planets in relation to the Earth; and the consequences and degree of disturbance to the Earth could have varied greatly within wide parameters, being far less or far greater than those which I describe.

In the future, as perhaps in the past, because of the proximity of the orbits of Venus, Earth, Mars, and Jupiter at the time of passing of the "Supreme" celestial body Nibiru, Venus and Mars could be influenced into close contact with the Earth, or each other, or with Jupiter, as well as with Nibiru itself. Such a close passing of Nibiru could have a devastating effect upon the Earth, and cause a stasis of the Earth, a retardation or slowing down of its rotational spin, or an advance or speeding up of its rotational spin, or a reversal of the poles, as well as the flood passing effects: a displacement of the fluid components of the Earth; the water, air, and molten core, and a temporary tilting of the terrestrial axis, and a temporary pulling or stretching out of orbit, as I describe in my interpretation of the clay tablets. For certainly, if a celestial body came sufficiently close to the Earth to displace the fluid components of the Earth during the interim, to the degree described in the clay tablet flood accounts, those same force fields could also simultaneously act upon and temporarily tilt the terrestrial axis, and temporarily pull the Earth out of orbit. This could account for the present wobble of the axis of the Earth; and if a 90 degree shift in the axis were to occur, we could account for the evidence of a past nearly tropical climate at the current axis. Also, if, without slowing the rotational spin of the Earth, the centrifugal force of the Earth was neutralized by the forces of Nibiru, the equatorial bulge of waters would recede/fall downward toward the axis from their 13 mile high level. This would result in an effect similar to slowing the rotational spin: the displacement of the equatorial waters toward the axis (see my footnote 744).

However, the evidence for such a close passing, closer than during the flood passing, and the extremely severe disruptive effects upon the Earth it would cause, a slowing down or speeding up of its spin, or a reversal of the poles, or other effects, is lacking at the particular passing I have written about, the flood passing, but this does not preclude their occurrence, or a past or future occurrence of such magnitude. We do have evidence of a close passing to Earth of Nibiru, Venus, or Mars, and possibly all three, also perhaps involving Jupiter, during the most recent orbit, between 1,600 B.C. and 1,500 B.C., when time was affected through a reversal of sunrise and sunset, and the length of day and night was altered, as described by Velikovsky in Chapter 5 of Worlds In Collision. But during the flood passing, or orbit, between 8,800 B.C. and 8,700 B.C., Ziusudra maintained a water clock, which gave him an independent means of measuring time, other than sunrise to sunset, and no mention is made of any discrepancy in days after Ziusudra saw the Sun, 40 days after the flood began.

Therefore, it is my opinion that by using the evidence presently available, a slowing down or speeding up of the rotational spin, or a reversal of the poles, did not occur during the flood passing. However, as I have already stated, we should not preclude their occurrence, including other effects we are not presently aware of during the flood passing, even with lack of evidence.

Furthermore, displacing the equatorial waters toward the axis would, if anything, cause the Earth to spin faster, in much the same way as an ice skater would draw her arms in toward the center, or axis, in order to spin faster.

Additionally, in reference to the planets, especially the Earth, being out of orbit, it is interesting to note the Biblical Isaiah in 13:13, who describes the phenomenon as follows; Therefore I (Lord/Nibiru) will shake the heavens, and the earth shall remove out (be pulled out) of her place (normal orbit). And in reference to the reversal of the poles, it is also interesting to again note Isaiah in 24:1, who describes the event as follows; Behold, the LORD (Nibiru) maketh the earth empty (displaces the water and/or displaces some of the molten core to the surface of the Earth), and maketh it (the surface of the Earth) waste, and turneth it (the Earth) upside down (a reversal of the poles).

QUANTUM MECHANICS

Is the universe, and therefore our solar system, an example of quantum mechanics? Some people, including myself, believe it is possible the macrocosm, the very large, and the microcosm, the very small, are alike, as described by Velikovsky on page 387, and page 388 of Worlds In Collision.

On the microcosm level, where our understanding is far greater than on the macrocosm level, but where we have by no means acquired total comprehension; in an atom, electrons orbit the nucleus, held in orbit by electricity, and when a substance radiates or absorbs energy, due to some outside force, all the atoms of the substance are disturbed and begin to vibrate. As they vibrate, electrons "jump" orbits, into smaller orbits as energy is given off, and into larger orbits as energy is absorbed. These units of energy are called quanta, or photons. Briefly, this is quantum mechanics, as I believe is allegorically and metaphorically put forth on the clay tablets, which is transliterated by King in Enuma Elish, The Seven Tablets Of Creation, and brilliantly explained by Sitchin in Chapter 7 of The 12th Planet.

On a macrocosm level, our solar system can be compared to an atom, with the Sun being the nucleus, and the planets the electrons orbiting the nucleus. If the planet Nibiru were the outside force acting upon the solar system, could Nibiru cause the planets to "jump" or be pulled or stretched out of their normal orbits, radiating or absorbing units of energy, like photons?

I believe change is the only constant, therefore this, or a phenomenon very close to this, a kind of "planetary billiards", occurs when Nibiru nears the perihelion of its orbit and the other planets. Also, Nibiru could cause perturbations of the Sun itself, in addition to its force fields; and Nibiru may not be the only celestial/extraterrestrial force which directly or indirectly perturbs/influences our solar system.

Applying the electron/planet correlation, after planet Tiamat ("Maiden of Life"), here compared to an electron, is vibrated/fissured (the clay tablets could be describing the proper procedure for splitting a sub-atomic particle), and a lower portion of its energy/mass is cleaved and given off/reduced to form the Asteroid Belt, the then existing orbit of the remaining upper portion, the newly formed Earth, then orbiting near where the Asteroid Belt is today, in its original 4th orbit position from the Sun, was then correctly decreased/reduced to its present Earth 3rd orbit position (into smaller orbits as energy is given off), precisely what is predicted and occurs to electrons/planets according to quantum mechanics, thus confirming the theory. Also, the new Earth orbit position, now closer to the Sun, is an orbital location whose proximity/distance from the Sun now makes it environmentally an ideal "garden" for the creation and nurturing and sustenance of our form of life; whereas the former original orbital location (near the Asteroid Belt) was not conducive to life, as we know it. That may have been the purpose and reason for the cleaving and orbital relocation. As we gaze upon our solar system, we know it was the result: the "Maiden of Life" provides US with life.

But planets are held in orbit by gravity, not electricity, or so our knowledge today leads us to believe. Could planets be held in orbit by an electrical field, as well as by a gravitational field? I believe they are, and possibly other force fields unknown or undetectable by our present day technology.

Although I use the terms gravitational and electromagnetic force fields to describe some of the physical forces of Nibiru influencing the planets, they may be characteristics of the same force, of an electrical nature. At this time, we do not know, for we do not fully understand these forces, but as Einstein said, "The two forces may be related." I believe the microcosm is the model for the macrocosm, and if this theory is accepted, then the electrical force, the force that holds the microcosm together, not gravity, would also prevail in the macrocosm.

CREATION OF PLANETS, SATELLITES, AND RINGS

Why do some planets have satellites and rings, others just satellites, and others neither? Cosmologists and astronomers today have a variety of answers. I also have some thoughts on this topic.

According to the clay tablets, after the creation and formation of our solar system, of which the planets are all satellites of the Sun, some of the planets themselves acquired major satellites, with no explicit mention being made of the formation of rings: see Illustration 2 through 15 within this section, which can be better understood by simultaneously reading the address of God Anu, at the beginning of this book. However, I believe concurrent with some of the planets acquiring major satellites, all of the four original outer planets, Neptune, Uranus, Saturn, and Jupiter, formed rings, as well as major satellites, and the rings were created and exist because they are smaller debris material that was later pulled away from these originally ringless planets, at the same time the major satellites were created, and not smaller debris material that failed to gravitationally compress toward the planets and coalesce when they were forming, as in the nebulous theory.

I believe the planet Nibiru, which orbits through our solar system near the Asteroid Belt every 3,600 Earth years, was and continues to be the driving and shepherding force, a kind of push/pull force, the external agent, or catalyst.

Initially, as the planet Nibiru, being guided into a new southern to northern orbital trajectory, approached the outermost original planets, Neptune, Uranus, Saturn, and Jupiter, in the same orbital direction, and passed beneath and near their orbital paths, very close to each of the planets, within what I call the scene of close contact, bits and chunks were literally ripped out, forcefully and violently pulled away from them, by the gravitational and other forces of Nibiru. These chunks or large debris material became major satellites, some of which were attracted and interchanged between Nibiru and the outer planets, which would account for their wide variety of shapes, sizes, and characteristics, while seven others were retained and remained as major satellites of Nibiru; and this could account for the fact that presently Uranus has no large major satellites, only medium and small, unlike the other outer planets of Neptune, Saturn, and Jupiter, for the initial four of the seven satellites of Nibiru came from Uranus, and were probably its largest satellites. Simultaneously, rings were formed from the bits or smaller debris material, and the rings and major satellites are both relatively recent material; occurring after the formation of the planets, and not forming with them, and evidence of this is the rings are composed of strata, especially the rings of Saturn, because the material in them, including satellites, was pulled away at intervals, during different orbits of Nibiru: see Illustration 16 within this section. Again, I give the scenario of these events in the speech of God Anu, at the beginning of this book.

Also, it is possible that the major satellites of the outer planets, Neptune, Uranus, Saturn, and Jupiter, and the major satellites of the inner planet, Mars, are composed of, in part or whole, in addition to the aforementioned, larger debris material from the shattering and scattering of the ten satellites formed from within Tiamat, and shattered and scattered by Nibiru, which became diverse orbit asteroids and comets, larger debris material, or meteoroids, smaller debris material; or larger debris material from the lower half of Tiamat, crushed by Nibiru, which later became the Asteroid Belt, various size debris material. The Asteroid Belt is composed of nearly 3,000 large asteroids, some as large as 485 miles in diameter, with a total of approximately 50,000 asteroids (see my footnote 32).

Additionally, the location of the outer planets as Nibiru orbits near them, either during the very early orbits, or even recently, affects or determines the direction of rotation of their satellites, or even the planets themselves: see Illustration 17 and 18 within this section. This is why some of the satellites revolve retrogradely, even belonging to the same planet. This could have occurred upon the initial pulling away of the material to form the major satellites, or even during a later or recent orbit, interchanging or forming minor satellites, for the right or left and below or above location of the outer planets in relation to Nibiru can change with each orbit of Nibiru, as can the direction of rotation of their satellites: see illustration 16 within this section.

Upon entering the solar system, Nibiru actually crosses the orbits of the outer planets, in the same orbital direction, but on a different plane of orbit, similar to the 17 degree angle of Pluto. However, while crossing, Nibiru is initially below the orbital plane of the outer and inner planets, and passes under the general vicinity of the south axis of the Earth, until it reaches its perihelion, and thereafter it is above the orbital plane of the outer and inner planets, as is Pluto, and then passes over the general vicinity of the north axis of the Earth. This different orbital plane of Nibiru could easily account for the variety in the tilt of the axis of all the planets in our solar system. Also, the orbital path of Nibiru is so wide and elliptical, far greater than that of Pluto, that it can only possibly encounter and disturb the outer planets, especially Jupiter, once, if at all, either on its southern approach toward the perihelion of its orbit, or afterwards, as Nibiru begins its northern return toward deep space; for the length of orbit, or length of year of the outer planets is greater than the approximate 1.33 Earth year period of passing of Nibiru, using the length of year of Earth as 1: Jupiter 11.87, Saturn 29.48, Uranus 84.07, Neptune 164.90, and Pluto 248.49. This chronology of occurrence, only once, may also apply to the encounter and disturbance of the inner planet Mars, 1.88, but the other inner planets affected could possibly suffer multiple encounters and disturbances, the Earth twice and Venus three times, for their length of orbit, or length of year is less, Earth 1 and Venus .62, than the period of passing of Nibiru, which is 1.33 Earth years.

The orbital path of Nibiru must be located between Mars and the Asteroid Belt, where the gravitational and other forces of Jupiter and Nibiru would somewhat neutralize themselves, because they are both massive planets, thereby only slightly affecting the stabilized orbits of the asteroids, which would be caught in a kind of neutral force zone: see Illustration 15 within this section. Also, Nibiru is plainly visible with the naked eye, as on page 222, Fig. 113 of The 12th Planet by Sitchin, therefore, its orbital path must be toward the inside of the orbital path of the asteroids, or toward the inside of the Asteroid Belt, relatively close to the Earth, or else Nibiru is much larger or more radiant than I render. For if the orbital path of Nibiru were between the Asteroid Belt and Jupiter, the gravitational and other forces, the effects of both massive bodies, acting in conjunction upon the Asteroid Belt would greatly distort it, pulling and stretching the asteroids even closer to Jupiter, and especially toward Nibiru, than where they presently orbit, with both planets attracting and capturing great numbers of the asteroids as minor satellites, evidence of which is lacking orbiting Jupiter.

Additionally, in Enuma Elish, The Seven Tablets Of Creation by King, the clay tablets indicate the outer planets are to act as a stabilizing factor upon the Asteroid Belt, which my theory agrees with, through: page 77, line 137 He (Marduk/Nibiru) split her (Tiamat) up (divided the planet) like a flat fish into two halves; line 138 One half (crushed remains of the lower half) of her he stablished (established/drove into a new orbit, forming the Asteroid Belt) as a covering for heaven (space); line 139 He (Marduk/Nibiru) fixed a bolt (fixed or locked the asteroids into a confined orbit), he stationed a watchman (a stabilizing factor, or an equalizing force, composed of Jupiter, and in conjunction with Jupiter, perhaps the other outer planets, Saturn, Uranus, and Neptune, whose purpose it was to keep the asteroids, various size debris composed of rocks, solidified molten rocks and frozen water from Tiamat, in place or within their fixed/confined orbit); and line 140 And bode them (the stabilizing factor, or the equalizing force) not to let her (Tiamat's frozen) waters come forth (be issued or removed out from the Asteroid Belt).

The orbital paths of the innermost planets of Venus and Mercury are far inside the orbital path of Nibiru, therefore their orbital paths are not crossed by Nibiru: they are not within the scene of close contact. This is why they do not have major satellites. Being far inside the orbital path of Nibiru, in a somewhat protected sanctuary, they were not instrumental in attracting and changing the initial and early orbits of Nibiru through our solar system. However, at times during their orbits, at specific locations, the actual proximity of each planet is close enough to the orbital path of Nibiru to be somewhat affected by its forces.

The orbital path of Mars, another inner planet, is also not crossed by Nibiru, it is not within the scene of close contact, but we know Mars has two major satellites. These could have been formed in the same way the outer planets major satellites were, through ripping out chunks, or acquired through interchange. Also, they could have been formed during a later orbit of Nibiru, for the planet Mars is in the first inner orbit slot next to Nibiru, within the somewhat protected sanctuary, although its exposure is far greater than Mercury, and much greater than Venus.

Mars, being the nearest inner planet to the orbital path of Nibiru, and Jupiter, being the nearest outer planet to the orbital path of Nibiru, are both especially vulnerable and have the maximum exposure to the forces of Nibiru. However, because of the great difference in mass, Mars suffers far greater consequences than Jupiter; as the massive Jupiter is capable of somewhat neutralizing the forces of Nibiru, Mars is not. This criteria would also apply to Venus, although Venus has more mass than Mars, it is further away from the orbital path of Nibiru. Also, the effects could vary due to the location of each planet within its orbital path as Nibiru passes, or in other words, the precise location of the planet in relation to Nibiru determines the magnitude of the effects.

The orbital path of the Earth, as we know it today, another inner planet, is also not crossed by Nibiru, for it is not within the scene of close contact, and we know the Earth has one major satellite, the Moon. But the Earth originally was the upper half of the cleaved or split planet Tiamat, which orbited near where the Asteroid Belt is today, before it was caused to "jump" or was driven, along with the Moon, to its present orbit by one of the four remaining of seven original satellites of Nibiru; and the Moon was, and still is, the main satellite of Tiamat, now cleaved and called the Earth, the largest and only remaining satellite of eleven original satellites of Tiamat that were formed from out of Tiamat, without the direct aid of an external agent, or Nibiru. Therefore, the original upper half of Tiamat/Earth is older than the Moon, and the newly formed lower portion of the Earth is younger than the Moon. Although the orbital path of the Earth is also inside the orbital path of Nibiru, within the somewhat protected sanctuary, the Earth, being a planet of small mass, and depending upon its location, its proximity to Nibiru when Nibiru passes, could suffer varying effects from the forces of Nibiru, just as I described for Mars and Venus.

Pluto, an outer planet today, the last member of our known solar system to be accounted for, initially had no major satellites, because it originally was itself a satellite of Saturn, but we know Pluto currently has one major satellite. Its present orbital path is crossed by Nibiru, and therefore, it is within the scene of close contact, but after the initial fact that it was a satellite of Saturn, due to its 17 degree different plane of orbit, similar to that of Nibiru, it may be more closely and safely synchronized with Nibiru, thus protecting it.

Also, the variations and separations in the concentric rings of Saturn, may be a result of the formation of the rings on more than one occasion. Initially, when the chunks were ripped out of the planet to form its major satellites, and simultaneously its original satellite Pluto was attracted and loosened by Nibiru; and then three orbits later, when Pluto was "whipped" into its present orbit: both of which could very easily account for the formation of and separations in the rings, as I point out later. These are two occurrences the clay tablets mention, but there may have been other occurrences more recently than at the time of creation of our present solar system, or the initial encounters with Nibiru, as described by the clay tablets, including a pulling apart or stretching of the rings during recent orbits of Nibiru.

The theory I have proposed and described above is similar to the well known tidal theory for the creation of our solar system, but it has an outstanding major difference: it is a reoccurring or cyclical tidal theory.

It could be possible the clay tablets are indicating that our solar system was out of symmetry with itself when first formed, and Nibiru was the synchronizing factor, creating reasonable stability. Also, could the major satellites be stabilizing agents for their assigned planets, much the same way Nibiru could be the stabilizing agent for our solar system today?

With regard to the initial pass, first orbit, second orbit, and third orbit of Nibiru, perhaps a more lucid insight might be gained by referring to my Illustrations in this section, and simultaneously reading the following scenario of unfolding events given in Enuma Elish, The Seven Tablets Of Creation by King.

Initial pass, Illustration 11; page 71, line 93 Then advanced (orbited forward in the same direction, the rapidly rotating) Tiamat (with eleven satellites and accompanying smaller debris. At this point, the Moon was already partially neutralized, occurring earlier on page 67, line 67 As (Marduk/Nibiru) gazed, (Kingu/the Moon) was troubled (became erratic) in his gait (orbit), and line 68 His will (the particular force causing the erratic orbit) was destroyed and his motions (erratic orbit, but not rotation) ceased) and Marduk (Nibiru, with seven satellites and accompanying smaller debris), the counselor of the gods (the other planets); line 94 To the fight (scene of close contact) they came on, to the battle they drew nigh (Nibiru approached Tiamat in the same orbital direction, from beneath and behind, and drew progressively closer towards converging and juxtaposing. This could be the source of the renowned expression of fate, seven/eleven; the seven satellites of Marduk/Nibiru, versus the eleven satellites of Tiamat); line 95 The lord (Marduk/Nibiru) spread out his net (force fields) and caught her (the rotation of Tiamat is abruptly/suddenly retarded/slowed, due to the prevailing/more powerful force fields of Marduk/Nibiru); line 96 And the evil wind (one of the seven satellites of Marduk/Nibiru) that was behind (him) he let loose (was released from orbit around Marduk/Nibiru, and hurled as in a sling) in her face (and forcefully orbited toward the surface of Tiamat); line 97 As Tiamat opened her mouth (began to fissure because of sudden slowed rotation) to its full extent (increasingly wide); line 98 He drove in the evil wind (into the fissure), while as yet she had not shut her lips (thus, one of the satellites of Marduk/Nibiru, the evil wind, is driven into the fissuring Tiamat); line 99 The terrible winds (two more of the six remaining satellites of Marduk/Nibiru, the planets themselves do not collide, only three of the seven satellites of Marduk/Nibiru actually collide with Tiamat) filled her belly (are driven deep into the fissuring Tiamat); line 100 And her courage was taken from her (the force fields of Tiamat were neutralized), and her mouth she opened wide (the fissure opened even more wide); line 101 He seized the spear and burst her belly (discharged a huge bolt of lightning, or electricity, into the fissured depths of Tiamat); line 102 He severed her inward parts, he pierced (her) heart (the core of the planet Tiamat); line 103 He overcame her (force fields) and cut off her life (caused Tiamat to cease being a threatening factor to the stability of the solar system by extinguishing her force fields); page 73, line 104 He cast down (subdued) her body (fissured or cleaved the planet throughout its diameter) and stood upon it (the force fields of Marduk/Nibiru seized and crushed Tiamat); line 105 When he had slain (conquered or gained control of) Tiamat, the leader (of her eleven satellites, all of which, including Tiamat, composed the aggregate factor of instability in the solar system); line 106 Her might was broken (her power was destroyed), her host (the ten smallest of her eleven satellites, all except the largest, the Moon) was scattered (were shattered and driven/whipped into deep space in diverse elongated orbits as larger and smaller debris, the larger as asteroids and comets, the smaller as meteoroids); line 107 And (this is how it was accomplished, after Nibiru met and cleaved Tiamat) the gods (the eleven satellites of Tiamat) her helpers, who marched by her side (orbited around Tiamat); line 108 Trembled, and were afraid, and turned back (reversed or changed the direction of their orbits, which also then became erratic); line 109 They took to flight (began to orbit away from Tiamat, and the force fields of Marduk/Nibiru) to save their lives (to correct their erratic orbits, and again have fixed congruent orbits); line 110 But they were surrounded (by the force fields of Marduk/ Nibiru), so that they could not escape (from the control of Marduk/Nibiru); line 111 He took them captive (seized them in their erratic orbits), he broke their weapons (extinguished their force fields); line 112 In the net (force fields of Marduk/Nibiru) they were caught (a particular force field, possibly electromagnetic or gravitational) and in the snare (they were held in very tight or close erratic orbits to each other) they sat down (were compelled to orbit within the control of the particular force field); line 113 The [. .] . . (four regions) of the world they filled with cries of grief (they were nearly colliding into each other in their erratic orbits around Tiamat, and perhaps some of the small debris did impact); line 114 They received punishment from him, they were held in bondage (temporarily held in erratic near collision orbits); line 115 And on the eleven creatures (satellites) which she (Tiamat) had filled with the power of striking terror; line 116 Upon the troop of devils (mischievous satellites), who marched at her [. . . .] (command); line 117 He (Marduk/Nibiru) brought affliction, their strength [he . . .] (broke); line 118 Them (the ten smallest of the eleven satellites) and their opposition (the force fields of the satellites) he trampled under his feet (the force fields of Marduk/Nibiru shattered, into larger and smaller debris, and scattered the ten satellites into deep space in diverse elongated orbits, with the larger debris becoming asteroids and comets, and the smaller debris becoming meteoroids. This could account for some, if not most, of the craters upon the Moon, which was not shattered; as I further explain later within 'Age Of The Earth'); page 75, line 119 Moreover, Kingu (the Moon), who had been exalted over them (had dominance over, or had the power to control or influence the other ten satellites, probably because the Moon was the largest of the eleven satellites); line 120 He conquered (Marduk/Nibiru neutralized the force fields of the Moon), and with the god Dug-ga (pot of lead, which is the equivalent of a dead object) he counted him (the Moon was considered comparable to a lifeless planet, or god. This is probably where the rotation of the Moon ceases); line 121 He took from him the Tablets of Destiny (power to control other celestial bodies) that were not rightly his (the power should not have been given to the Moon by Tiamat); line 122 He (Marduk/Nibiru) sealed them with a seal and in his own breast he laid them (secured and drained the powers from the Moon, and added them to the existing powers of Marduk/Nibiru); line 123 Now after the hero Marduk (Nibiru) had conquered and cast down (subdued) his enemies (Tiamat and her eleven satellites); line 124 And had made the arrogant foe (Tiamat and her eleven satellites) even like . . . (dead); line 125 And had fully established Anšar's triumph over the enemy (accomplished the objectives of Saturn; to subdue the threatening Tiamat); line 126 And had attained the purpose of Nudimmud (achieved the purpose of the altering of its original orbit by Neptune); and line 127 Over the captive gods (Tiamat and the Moon) he strengthened his durance (tightened the imprisoning hold on their orbits).

First orbit, Illustration 12; page 75, line 128 And unto Tiamat, whom he had conquered, he (Marduk/Nibiru) returned; line 129 And the lord (Marduk/Nibiru) stood upon (seized with its force fields and crushed) Tiamat's hinder parts (rear parts, which is also the lower half of the fissured planet. Only if Marduk/Nibiru were approaching Tiamat from beneath and in the same general orbital direction is this possible, to approach the hinder or rear parts first, thus confirming the same orbital direction: see my footnote 30 and 33); line 130 And with his merciless club (one of the force fields, possibly electromagnetic or gravitational) he smashed her skull (totally cracked and loosened the upper half of the fissured planet); line 131 He cut through the channels of her blood (severed the molten mass of the remaining core, and separated the totally cracked upper half from the crushed remains of the lower half, thus dividing the two); line 132 And he made the North wind (the force fields of one of the four remaining satellites of Marduk/Nibiru) bear (drive and direct while remaining in orbit) it (the totally cracked upper half of Tiamat and the Moon) away into secret places (into a new orbit); line 133 His fathers (the outer planets, Jupiter, Saturn, Uranus, and Neptune, who were instrumental in creating the encounter) beheld, and they rejoiced and were glad (for the threat to them was extinguished, and removed further from their orbital paths); and page 77, line 134 Presents and gifts they brought unto him (the outer planets, Jupiter, Saturn, Uranus, and Neptune, bestowed some of their power to Marduk/Nibiru).

Second orbit, Illustration 13; page 77, line 135 Then the lord (Marduk/Nibiru) rested, gazing upon her dead body (crushed remains of the lower half of lifeless Tiamat); line 136 (this is how it was accomplished) While he (Marduk/Nibiru) divided (into two halves) the flesh of the . . . . (body; fissured Tiamat), and devised a cunning plan; line 137 He split her up (divided the planet) like a flat fish into two halves; line 138 One half (crushed remains of the lower half) of her he stablished (established/drove into a new orbit, forming the Asteroid Belt) as a covering for heaven (space); line 139 He fixed a bolt (fixed or locked the asteroids into a confined orbit), he stationed a watchman (a stabilizing factor, or an equalizing force, composed of Jupiter, and in conjunction with Jupiter, perhaps the other outer planets, Saturn, Uranus, and Neptune, whose purpose it was to keep the asteroids, various size debris composed of rocks, solidified molten rocks and frozen water from Tiamat, in place or within their fixed/confined orbit); and line 140 And bode them (the stabilizing factor, or the equalizing force) not to let her (Tiamat's frozen) waters come forth (be issued or removed out from the Asteroid Belt).

Third orbit, Illustration 14; page 77, line 141 He (Marduk/Nibiru) passed through the heavens (orbited through space), he surveyed the regions (thereof; and was selecting a new orbit for Pluto, which had been loosened from Saturn during the initial pass, occurring earlier on page 39, line 1 Anšar (Saturn) opened his mouth, and; line 2 [Unto Gaga] (Pluto), his [minister] (satellite), spake the word; line 11 "[Go], Gaga, stand (orbit) before (closer to) them (the other planets)"); and line 142 And over (further than the orbits of the outer planets) against the Deep (near/facing deep space) he set (fixed the orbit of Pluto; beyond) the dwelling (orbit) of Nudimmud (Neptune; thus Pluto became a planet. An epithet for Pluto is "one who shows the way", and could mean, the planet which shows/indicates the way/orbital path of Nibiru. Also, Pluto may have acquired its current major satellite either: during the initial pass, by fissuring and breaking up into two unequal portions, somewhat similar to Tiamat, except on a much smaller scale, of course, after it was loosened/partially pulled away, as a single body, from Saturn by Nibiru; or debris material was attracted/collected from Nibiru, Saturn, or especially the simultaneously forming rings of Saturn, as Pluto was loosened/partially pulled away, and especially during the third orbit, as Pluto was swept/completely pulled away, and then cast/whipped into its new orbit by Nibiru, and this double pulling away of Pluto could very easily account for the unique/complex rings of Saturn, both the diameter/largeness, and stratification/contortion, as I previously mentioned; or throughout the encounter, by attracting other previous satellite interchange debris).

Illustration 2

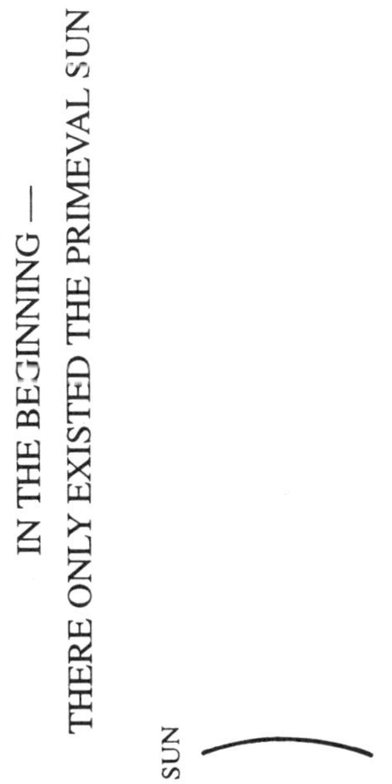

Illustration 3

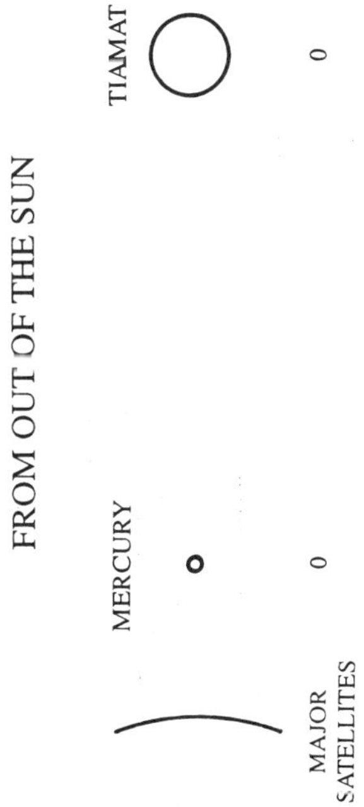

Illustration 4

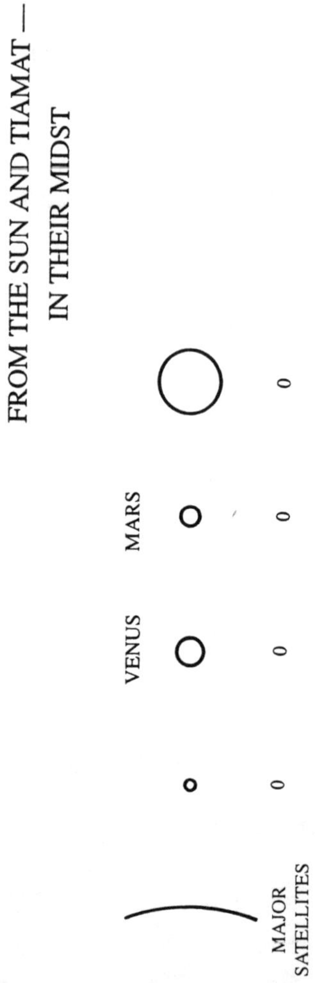

Illustration 5

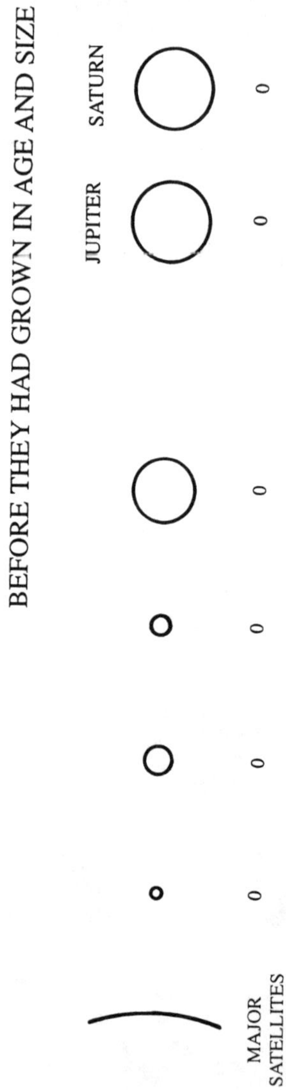

Illustration 6

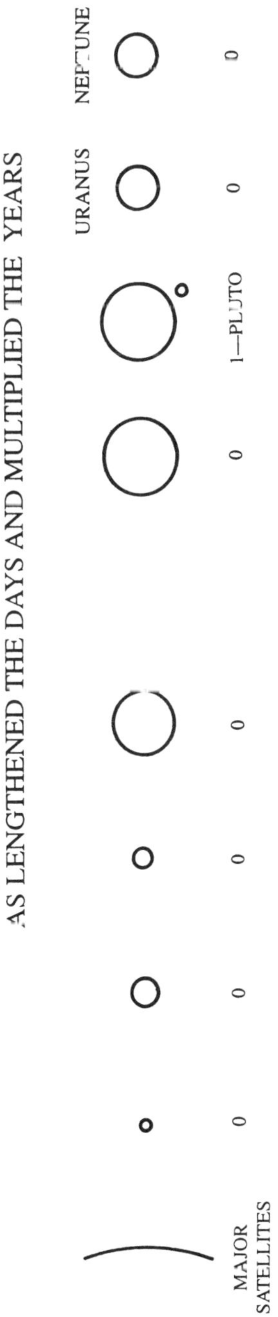

Illustration 7

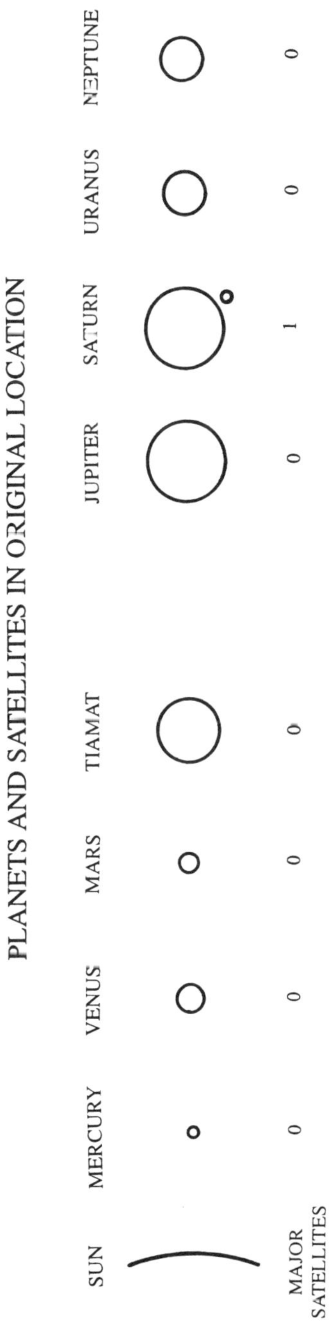

Illustration 8

PLANETS DEVELOP ERRATIC ORBITS —
TIAMAT FORMS SATELLITES AND FURTHER THREATENS PLANETS

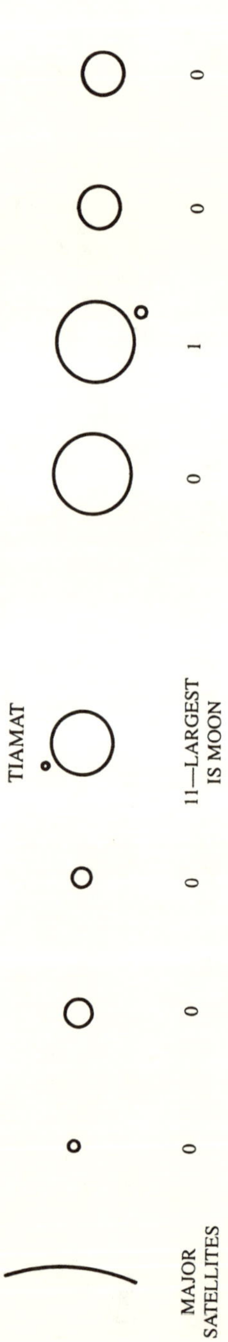

Illustration 9

BEGINNING OF INITIAL PASS — ENTER NIBIRU —
WHILE CROSSING BENEATH ORBITAL PATH OF OUTER PLANETS

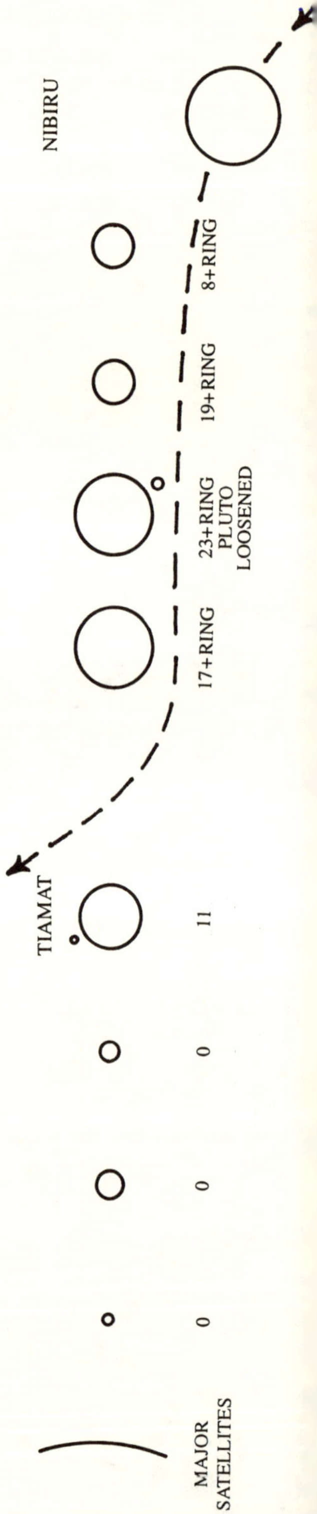

Illustration 10

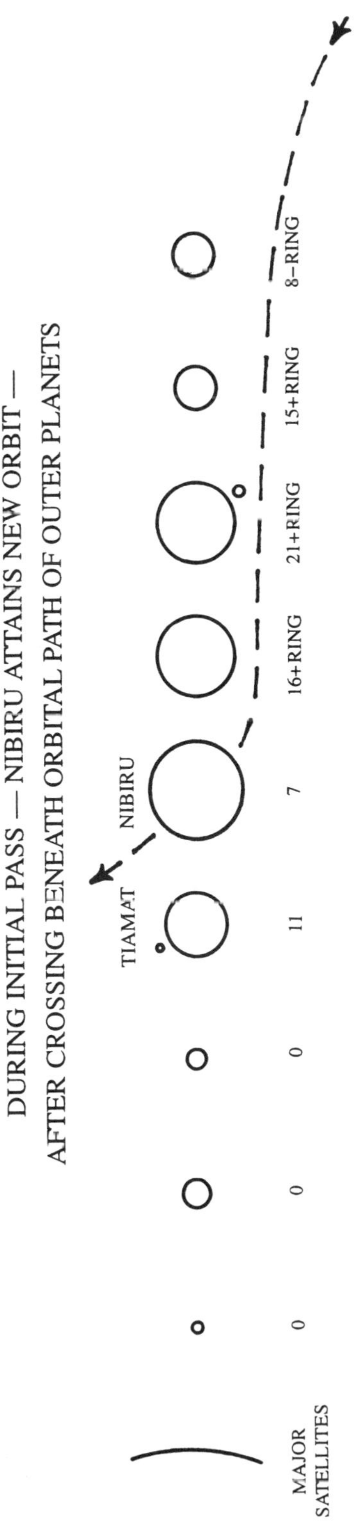

Illustration 11

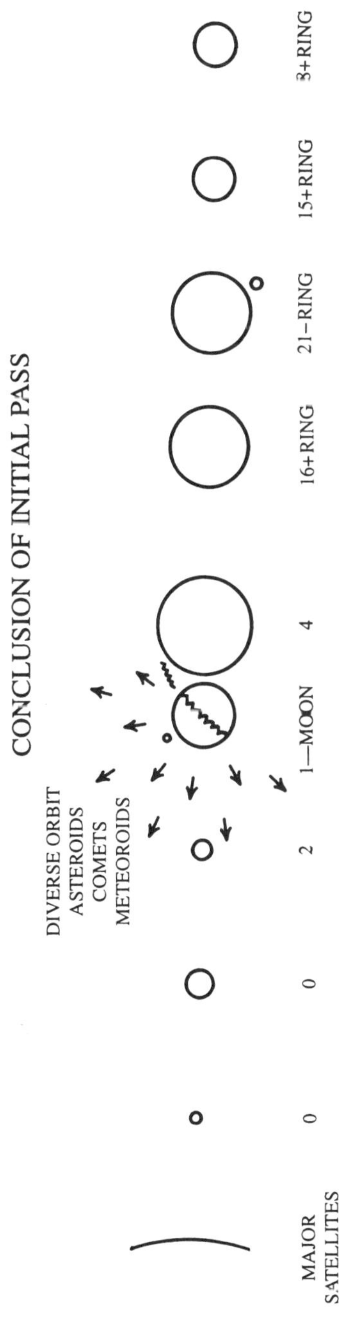

Illustration 12

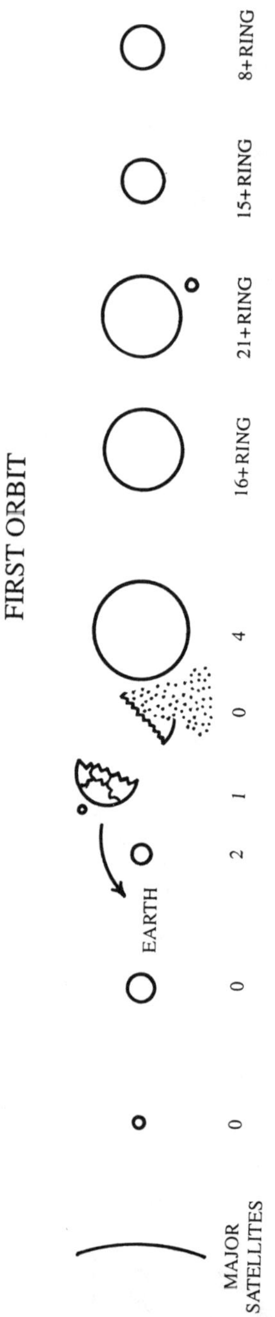

Illustration 13

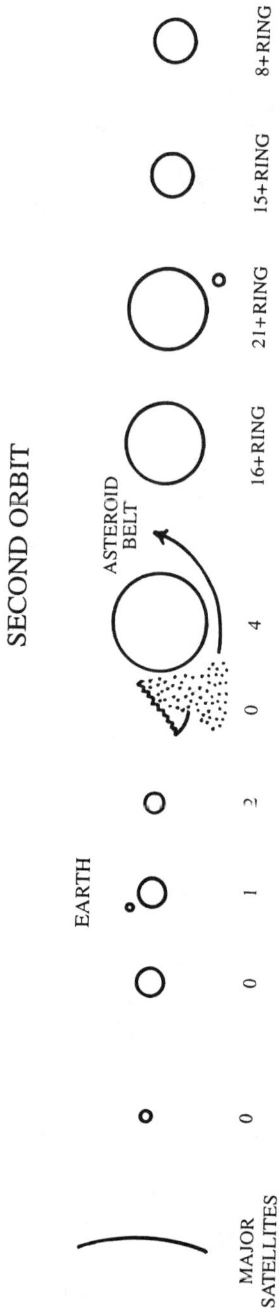

Illustration 14

THIRD ORBIT

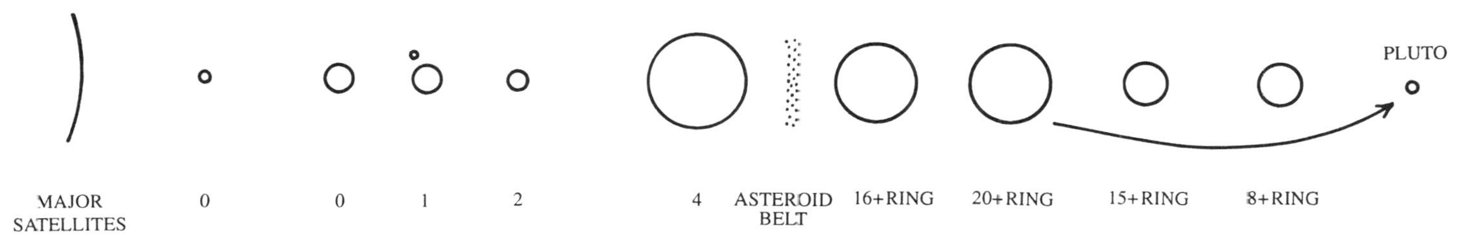

Illustration 15

PLANETS AND SATELLITES IN PRESENT LOCATION

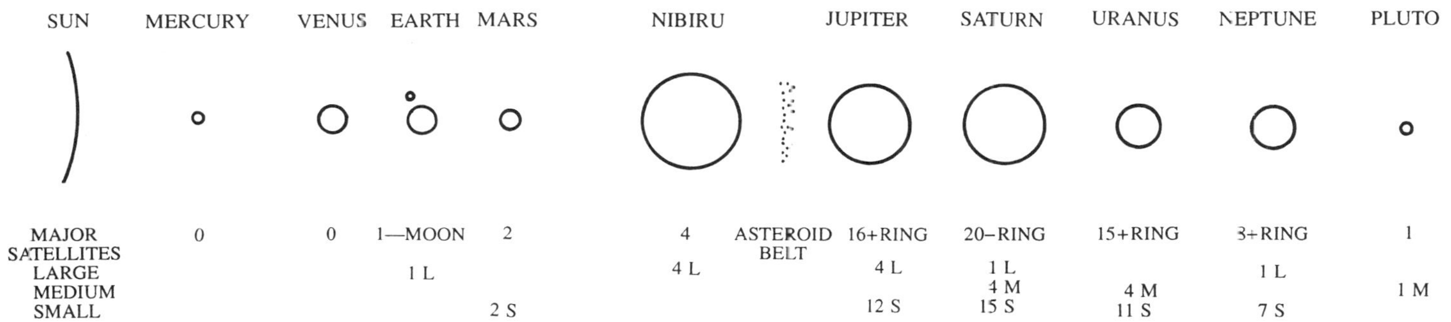

Illustration 16

EXAMPLE OF ROTATIONAL DIRECTION OF SATELLITES DURING INITIAL PASS AS NIBIRU ATTAINS NEW ORBIT

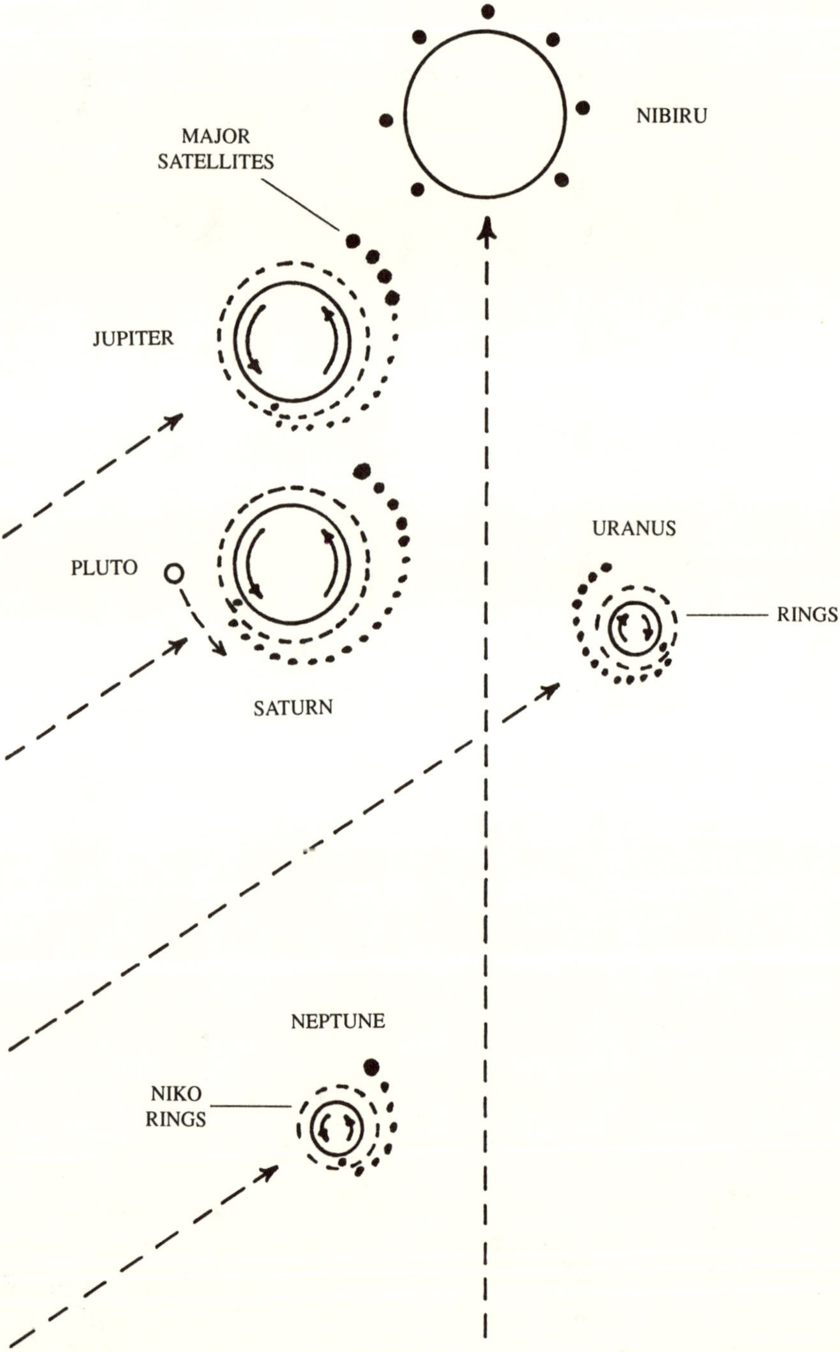

RECENT ORBIT Illustration 18

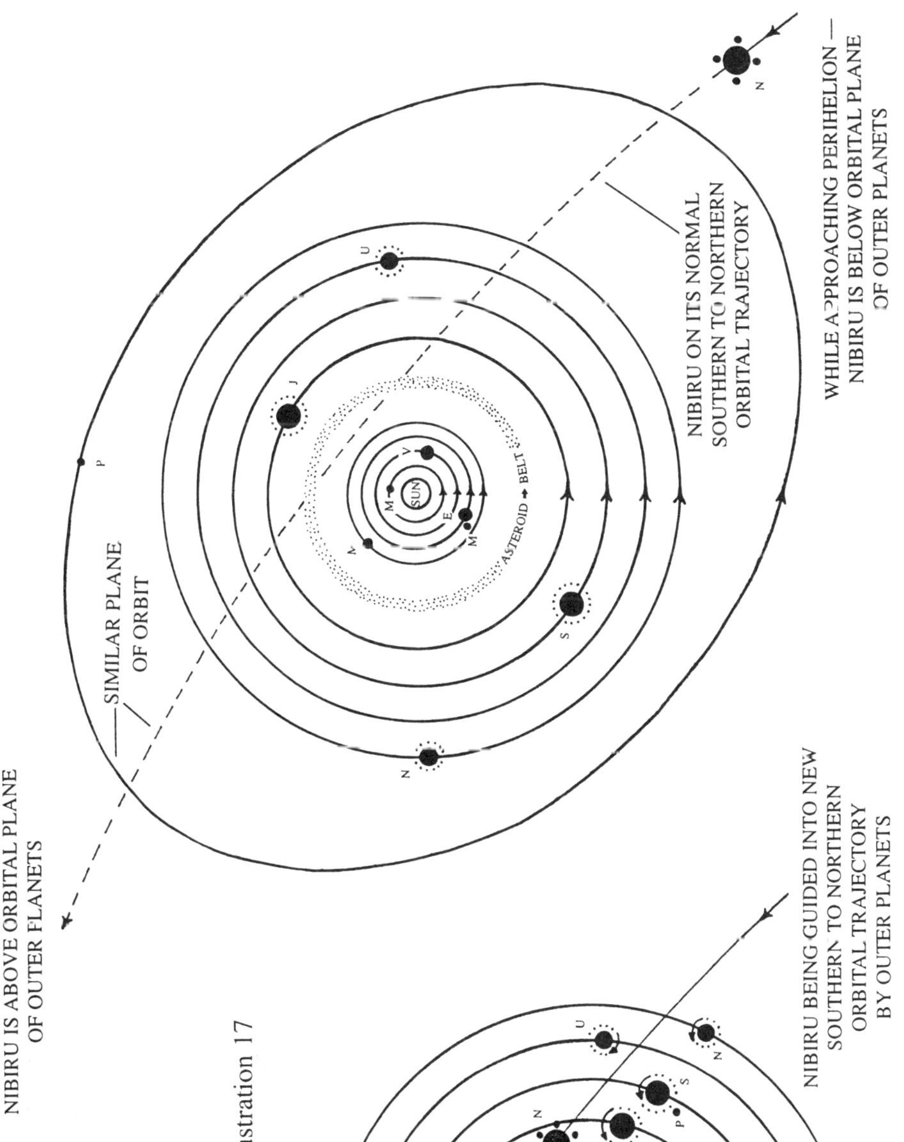

DURING INITIAL PASS Illustration 17

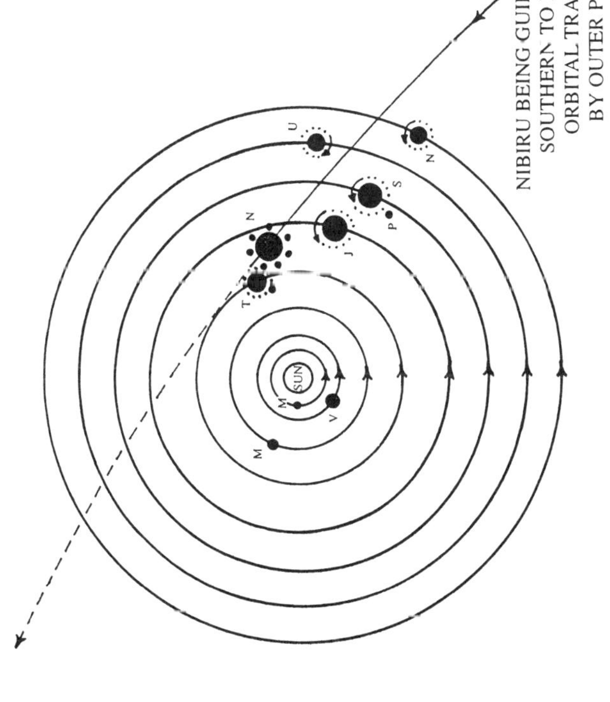

WINDS

Since one of the four satellites, moons, or Winds of Nibiru, called the North Wind, which I refer to in my footnote 18, 31, and 758, drove the upper half of the split planet Taimat, the Earth (as well as the Moon), from the Asteroid Belt region into its present orbit; it may be this same moon of Nibiru, or the other moons, and not Nibiru itself, that triggers the disturbance of the Earth and the other planets. Or, it may be a combination of Nibiru and its four remaining moons, or Winds, as on upper page 124 of Worlds In Collision by Velikovsky, who quotes Timaeus by Plato, who describes the effects of a collision of the Earth "overtaken by a tempest of winds," as well as the proximity of the nearby planets, which could also partially account for the varying degree of disturbance during each orbit, or passing.

WATER

Water plays an important role in the clay tablet narrative I have rendered, but it may play a far more important role, and in many ways we do not understand its significance, for there appears to be a water link, the element of water, throughout the common fabric of the story. I shall mention some of the more interesting facets of what I call the water enigma.

In Sumerian, the name of God Enki is A-lulim(ak), which could mean "sheep of the glittering waters", and the name of God Enlil is Alalĝar, which means "he who in raintime brought rest", as described on page 263 and page 264 of The 12th Planet by Sitchin.

Is all the water from the crushed lower half of the original water planet Tiamat frozen in the Asteroid Belt, as is indicated by the clay tablets in Enuma Elish, The Seven Tablets Of Creation by King, on page 77, line 140 And bode them (the stabilizing factor, or the equalizing force) not to let her (Tiamat's frozen) waters come forth (be issued or removed out from the Asteroid Belt); along with the rocks and solidified molten rocks? And if the water is not frozen in the Asteroid Belt, then where did it go? Where is it? Could the rings of Jupiter, Saturn, Uranus, and Neptune contain some of Tiamat's water, as chunks of ice within the rings, or could some of their satellites, or the planets themselves, especially Neptune, have been "splashed" or "sprinkled" with the water, and have coats of ice, or liquid water, or clouds of water molecules, water vapor, in their atmospheres?

Could it be possible that Nibiru periodically tilts the Earth's axis and triggers the process of production of new ice, and therefore the advance of the ice sheets?

Why is Neptune associated with water, and thought of as a water planet, like the original Tiamat, and the current Earth? Neptune was the first outer planet to attract Nibiru into our solar system and perfect its orbit into its present orbit; because of this role, was some of the water of Nibiru bestowed upon Neptune during their initial encounter?

Why are snowflakes, frozen water, all hexagon shaped; and why are snowflakes at temperatures colder than 12 degrees below zero fahrenheit, and the boat of Ziusudra, which was constructed to survive a flood of water, both hexagon cylinders in shape? Now, snow has a special new profound meaning, for when I experience a snowfall with its hexagon shaped snowflakes, as if they were a reminder, I think of them in terms of the great flood: the hexagon shaped boat upon the hexagon shaped water.

Incidentally, because carbon has 6 neutrons, 6 protons, and 6 electrons, there could be a link between human beings based on carbon 6 molecules; and the sexagesimal system employed on Earth from ancient times based on 60, which I believe is actually based on the unit of 6; and the inherent shape of water, which is represented by the number 6. Therefore, the entire Earth system of our water planet could be based on one key or master governing number, by which 6 is divisible: the number is 3. A hexagon, which embodies the number 6, is composed of 6 triangles, a form with 3 points; and a trident, with its 3 points, one of the symbols of God Enki/Poseidon (as well as the crescent Moon and entwined serpent/snake), is an expression of the elements of a triangle (3 points of his trident are triangular in shape, especially when depicted as 3 double barbed points, and could represent and thus be symbolic of the 3 pyramids at Giza, Egypt). Consequently, all initial forms which incorporate the use of a triangle/triangles (see my footnote 789), symbolically represent and are thus attributable to God Enki (trident, triangle, temple gable design, pyramid, hexagon boat, embodiment of the number 6 and its multiples); while manifestations which are expressions of the number 3, appear to be guided by a synchronous Earth and solar system organization based upon the number 3 (every 3rd wave/orbit is more destructive, mystical number 10,800 equals 3 orbits, Earth is now 3rd planet from Sun): which all contain the common element of 3, and which may not be mutually exclusive. Therefore, the trident scepter of God Enki/Poseidon not only symbolizes authority/control of the waters of the Earth, the traditionally accepted interpretation, but all initial forms, including life forms, upon the Earth, the remaining watery half of the original water planet Tiamat; or, considering and combining all of the aforementioned, if the trident is not one of the exclusive symbols of God Enki/Poseidon (the so-called "candelabra" or loosely called trident, which is actually an embellished trident, boldly displayed on the side of a mountain at the Bay of Paracas, Peru, has been attributed to Ishkur/Adad/Viracocha), then the trident and its attributes could very simply symbolize the Earth, being the 3rd planet from the Sun (thus, the "candelabra" could represent control of electrical forces/lightning surrounding the "trident" Earth, one of the attributes of God Adad, the storm God, in charge of rain, lightning, and thunder).

We must remember that we are living on a water planet, and water is a very volatile substance, as can easily be observed along the shoreline after a severe storm or hurricane, which are produced under normal conditions, without outside influences. If we multiply the intensity or forces of our normal destructive processes by a factor of just 2, the effects would be equally devastating; and then imagine what the waters of the Earth could and would do if the factor was increased more, and then possibly substantially more, if triggered by the forces of Nibiru, or any other celestial body, if a close passing/orbit occurred today.

THE BEGINNING OF THE FLOOD

In my footnote 613b, in reference to Atra-ḫasīs, The Babylonian Story Of The Flood by Lambert and Millard, which states on page 91, line 37 He (God Enki) announced to him (Atra-ḫasīs = exceedingly wise/Utnapishtim = I have found life/Ziusudra = life of long days) the coming of the flood for the seventh night, I believe "seventh night" could be a scribal error, or a misnomer, and could lead to some controversy or be interpreted as a sign of nonconformity between the three accounts given herein, Atra-ḫasīs and Gilgamesh, Gilgamesh and Biblical, and Atra-ḫasīs and Biblical, and it should be either sixth night, or seventh day, for normally we consider day as coming before night. Also, it may be helpful to refer to Illustrations 19 and 20 within this section, which are chronological schedules, as you read the following.

In the clay tablet accounts, it can be assumed the day was counted from sunrise to sunrise during the time of Ziusudra, as it is today in the same region. Thus, the old day ends at sunrise, and a new day begins, or is counted at sunrise. Therefore, if the flood began as in The Gilgamesh Epic And Old Testament Parallels by Heidel, page 84, line 96 As soon as the first shimmer of morning beamed forth; or sunrise, the flood began at the end of the sixth night, or as the seventh day began. Thus, this clay tablet account is speaking of seven units, in terms of days, or from day to day, and the word day is implied.

In the Biblical account, the word day is directly emphasized, as put forth in Genesis 7:4 seven days; 7:10 seven days; and 7:11 seventeenth day; and the word night is not found. The Biblical day was counted from sunset to sunset, as it is today in the same region, and was already approximately twelve hours old when Ziusudra's day began. Both of these accounts, Gilgamesh and Biblical, are in complete agreement, for the day the flood began, at the end of Ziusudra's sixth night, and at the beginning of Ziusudra's seventh day, at sunrise, it was still the same Biblical day; for it was the seventeenth Biblical day before sunrise, as it also was the seventeenth Biblical day after sunrise.

Is there a solution to this misnomer? There are several exceptions through which it might be possible to accept the Atra-ḫasīs term "seventh night". The most obvious choices are; if the flood announcement was given to Ziusudra at night, on the night of the 9th day of the 2nd month, which would have to be between sunset and sunrise, and then we counted the following night as the first night; or if the flood announcement was given to Ziusudra at night, on the night of the 10th day of the 2nd month, which would also have to be between sunset and sunrise, and then we counted that same night as the first night, but this would be rather awkward, for the first night would normally be the following night.

In the preceding two Atra-ḫasīs contexts, the term "seventh night" would be correct, for in both, the seven units would be measured in terms of nights, or from night to night. But at the end of the seventh night, at sunrise, it would be the beginning of the eighth day, and this would not agree with either the Gilgamesh or Biblical accounts. Also, the number eight was not considered a mystical number by the ancients, as was the number seven. Therefore, both of the Atra-ḫasīs exceptions which I have given would be highly unlikely choices, for the most common thread and emphasis of all the accounts, is on the word or derivatives of seven, and day, or first shimmer of morning, sunrise, which also implies day.

However, there is another Atra-ḫasīs exception, one of a different nature, another candidate for the solution to this misnomer, and it is my personal favorite choice. At sunrise on the seventh day, as the Sun rose and the sky became lighter, it suddenly and prematurely became so dark, very dense darkness, as I describe in my rendition, that it appeared as night, and was termed the "seventh night", but actually, it was sunrise and the beginning of the seventh day. The clue to solving the misnomer is in my footnote 663, in reference to The Gilgamesh Epic And Old Testament Parallels by Heidel, which reads on page 85, line 106 (And) turned into darkness all that was light. The enigma is solved by simply reading the vivid line 106 as follows; (And) turned into darkness (night) all that was light (day). So it was the seventh day, which suddenly and prematurely became dark, and therefore prematurely became the "seventh night". For what else could it be called? It certainly could not be called the sixth night, which had just passed, or the eighth night, for the eighth day had not yet appeared. It was therefore termed the "seventh night", for the seventh day had already briefly made its appearance, and night follows day. Therefore, the so-called misnomer is not a misnomer after all, it is an accurate description of the unfolding of events at sunrise, the day the flood began, and I forgive the scribe who inscribed the term "seventh night" on the soft clay tablet.

Again, I have prepared the following chronological schedules, Illustrations 19 and 20, for a better understanding of the event.

Illustration 19

HOUR FLOOD BEGAN

↓

	Six Hour Intervals	Sun Rise	12 Mid Day	Sun Set	12 Mid Night	Sun Rise	12 Mid Day	Sun Set	12 Mid Night	Sun Rise	12 Mid Day	Sun Set	12 Mid Night	Sun Rise
*	Clay Tablet Day	16	16	16	16	17	17	17	17	18	18	18	18	19
**	Biblical Day	16	16	17	17	17	17	18	18	18	18	19	19	19
	Todays Day	May 6	May 6	May 6	May 7	May 7	May 7	May 7	May 8	May 8	May 8	May 8	May 9	May 9

*Source: The Gilgamesh Epic And Old Testament Parallels by Heidel, page 84, line 96 As soon as the first shimmer of morning beamed forth (sunrise).

**Source: The Holy Bible, Genesis 7:11 the seventeenth day of the month.

Illustration 20

DAY FLOOD BEGAN

↓

	One Day Intervals	Day 1	Day 2	Day 3	Day 4	Day 5	Day 6	Day 7
*	Day of Announcement 0 Day	1st	2nd	3rd	4th	5th	6th	7th
**	Old Calendar 2nd Month—10th Day	11th	12th	13th	14th	15th	16th	17th
	Todays Calendar April 30	May 1	May 2	May 3	May 4	May 5	May 6	May 7

*Source: Atra-ḫasīs, The Babylonian Story Of The Flood by Lambert and Millard, page 91, line 37 He (God Enki) announced to him (Atra-ḫasīs/Ziusudra) the coming of the flood for the seventh night (premature darkness, actually still seventh day); and The Holy Bible, Genesis 7:4 For yet seven days (in seven days), and 7:10 And it came to pass after seven days, that the waters of the flood were upon the earth.

**Source: The Gilgamesh Epic And Old Testament Parallels by Heidel, page 84, line 96 As soon as the first shimmer of morning beamed forth (sunrise = a new day); and The Holy Bible, Genesis 7:11 in the second month, the seventeenth day of the month.

THE CHRONOLOGY OF THE FLOOD

(According to the old calendar; 30 days per month, 360 days per year)

2nd day of the 2nd month: (Nibiru begins secondary seizure of the Earth. Approximately 14 days before flood begins). Very few light tremors begin.

10th day of the 2nd month: (7 days before flood begins. Today's calendar: April 30). God Enki gives Ziusudra 7 day warning.

17th day of the 2nd month: (Nibiru begins primary seizure of the Earth. Today's calendar: night of May 6th, at sunrise on May 7th, flood begins). Constant severe tremors and volcanic eruptions with lava and thick clouds of smoke, ash, and steam begin, water displacement begins and water is piled into rolling flood waves, violent winds begin, heavy rain begins, very dark clouds and dense darkness, no sunlight, Ziusudra in hexagon shaped reed boat.

24th day of the 2nd month: (Nibiru concludes primary seizure of the Earth. 7 days after flood began). Constant severe tremors and volcanic eruptions cease, very few light tremors continue, flood waves cease, flood waters stop rising, violent winds diminish and become gentle winds, gentle waves, heavy rain diminishes and becomes moderate rain, dark gray clouds and lighter sky, no Sun rays.

9th day of the 3rd month: (Nibiru concludes secondary seizure of the Earth. Approximately 14 days after flood began). Very few light tremors cease, gentle winds and waves and moderate rain continue.

27th day of the 3rd month: (40 days after flood began). Flood waters maintain maximum height, gentle winds and gentle waves continue, moderate rain ceases, normal clouds and sky, uninterrupted Sun rays upon boat, cut window opening out of boat.

17th day of the 7th month: (Nibiru passes perigee of its orbit, its closest distance to the Earth. Approximately 5 months after flood began). Calm winds, calm water, cloudless sky, bright sunshine, three additional window openings cut out of boat, flood waters begin to slowly recede, boat caught and grounded on Mount Nisir.

1st day of the 10th month: (Approximately 7-1/2 months after flood began). Flood waters continue to slowly recede, other mountain tops visible.

12th day of the 11th month: (40 days after other mountain tops are visible). Flood waters continue to slowly recede, Ziusudra sends out a dove, the dove returns to the grounded boat on the same day.

19th day of the 11th month: (7 days later). Flood waters continue to slowly recede, Ziusudra sends out a swallow, the swallow returns to the grounded boat on the same day.

26th day of the 11th month: (7 days later). Flood waters continue to slowly recede, Ziusudra sends out a raven, the raven does not return to the grounded boat.

27th day of the 11th month: (Nibiru begins secondary release of the Earth. This lasts approximately 14 days). Flood waters continue to slowly recede, very few light tremors begin, calm winds increase and become very light winds.

9th day of the 12th month: (Nibiru begins primary release of the Earth. This lasts approximately 7 days). Flood waters continue to slowly recede, constant moderate tremors and distant volcanic activity with thin clouds of smoke, ash, and steam begin, very light winds increase and become mild winds.

16th day of the 12th month: (Nibiru concludes primary release of the Earth. Approximately 10 months after flood began). Flood waters continue to slowly recede, constant moderate tremors and volcanic activity cease, very few light tremors continue, mild winds diminish and become very light winds.

1st day of the 1st month: (Nibiru concludes secondary release of the Earth). Flood waters continue to slowly recede, very few light tremors cease, very light winds diminish and become calm, Earth quiet and stable, Ziusudra opens sealed door, mountain plateau not dry enough to exit.

26th day of the 2nd month: (Nearly 2 months later). Flood waters continue to slowly recede, Ziusudra announces and prepares to depart from boat tomorrow, and to cut new door opening, and to make offerings, and to feast, mountain plateau now dry enough to exit.

27th day of the 2nd month: (12 months and 11 days after flood began, 7 months and 11 days after grounding of boat. Today's calendar: May 17th). Ziusudra and all his people, animals, and birds exit from the hexagon shaped reed boat, make offerings and feast on mountain plateau, God Enki and Goddess Ninhursag and other Gods arrive and partake in feast, God Anu and God Enlil arrive, God Enlil at first angry, but then accepts mankind, God Enlil bestows immortality upon Ziusudra and his wife, and takes them to the abode of the Gods to dwell.

LOCATION OF THE BOAT

Through the years, there have been numerous reports of sightings of pitch coated objects and ice covered sections of rectangular shaped boats upon or near the mountains called Ararat, and each time "Noah's" ark is claimed to have been discovered, as Seagraves describes throughout Search For Noah's Ark, and as the Reader's Digest editors report on page 234 of Secrets Of The Past. What are they, if they are not the hexagon shaped boat of Ziusudra/Noah?

I believe, that in addition to the hexagon shaped boat of Ziusudra, many other objects were also swept away by the huge flood waves, including perhaps nearly entire common rectangular shaped boats, and certainly sections of those boats, and large amounts of miscellaneous debris. Why should the boat of Ziusudra be the only object or boat swept to the north by the huge flood waves, and is every object upon those mountains to be associated with the boat of Ziusudra/Noah? I think not, for the clay tablets state that Mount Nisir, not Mount Ararat, is the landing place of the boat of

Ziusudra. According to the ancient annals of King Ashurnâṣirapal III of Assyria (883-859 B.C.), "Mount Niṣir was a steep mountain, difficult of access, and that it was named Kinipa or Kiniba in the language of the natives," as on lower page 29 through page 32 of The Babylonian Expedition Of The University Of Pennsylvania by Hilprecht, which to my knowledge has not been found and searched, and per Hilprecht "originally was a mountain in the district of the upper courses of the 'Adhaim and Diyâlâ rivers, somewhere between the 35th and 36th degrees latitude;" and is to be associated at or near the approximately 9,000 feet high mountain called Pir 'Omar Gudrûn (or Pir-i-Mukurun), which is located in an area where in the past, the boundaries of Kurdistan and Armenia have overlapped.

In circa 290 B.C., as on page 136 of Atra-ḫasīs, The Babylonian Story Of The Flood by Lambert and Millard, Berossus identified the landing place of the boat of Ziusudra, whom he called Xisuthros in Hellenic, as "the Gordyaean mountains of Armenia;" as Heidel points out on page 250 of The Gilgamesh Epic And Old Testament Parallels "as the mountains of the Gordyaeans, or the Kurds," which are located in central Kurdistan; and as Sollberger denotes on page 27 of The Babylonian Legend Of The Flood, which concurs with the aforementioned. More recently, Syriac and Arabic accounts identify the mountain as Jebel Jûdî, or El-Judi on page 244, Volume I of The Koran Interpreted by Arberry, which is located in southwestern Armenia, which corresponds to central Kurdistan, for again, they once overlapped in this area. The newer translation of the Bible states the boat of Noah grounded "upon the mountains of Ăr´-ă-răt," more correctly identified as Armenia (Urartu) by the older translation, which corresponds to the clay tablets and above mentioned reports, but does not specifically name the mountain within Armenia, and through misinterpretation, having been primarily translated from the original Hebrew, into Hellenic/Greek, and finally into English, with many additional translations occurring more recently, this has been interpreted to mean the highest peak in the area, Mount Massis, or Mount Agridagh, which we now call Mount Ararat (see my footnote 761 and 762).

The clue to the identity of the mountain appears in my footnote 759, which describes the landing place of the boat as a twin-peaked mountain. Mount Ararat does not have twin peaks, but the 16, 945 feet high mountain is locally called Great or Greater Ararat, and has a companion mountain called Little or Lesser Ararat, which reaches 12,877 feet in height. The two Ararat mountains are known as Mother and Child, and extend over 25 linear miles, but are not referred to as twins, or as one mountain with twin peaks, for they are two distinct mountains. Hopefully, when we locate Mount Nisir, it will have twin peaks, and we will have found the resting place of the boat.

Perhaps someday, if we look carefully in the correct place, and for the correct shape of boat, we may also find the answer to this enigma.

LONGEVITY

One of the most challenging and difficult Sumerian enigmas to grasp is the longevity of the ancients: that of thousands of years. Even according to Biblical accounts, beginning in Chapter 5 of Genesis, the ancients lived up to 969 years, as is described, along with interesting views, throughout First Man, Then Adam by Ginsburgh. The Sumerian Priest-King Ziusudra (life of long days) ruled 36,000 years, and was then made immortal, or given an extended life, while the Biblical Noah (respite, delay, prolongation of time/life) lived a total of 950 years, and then he died. These terms are truly fantastic, for the difference between their life spans and ours today is almost incomprehensible.

If only a different means of measuring time is involved, using a space traveler's clock that is independent of any celestial body, like our atomic clock, a "fast clock" or time of the Gods; or perhaps it could be related to, or simply be time dilation, the result of space travel over long distances at great speeds, as is described and suggested on page 56 of Sumerian Mythology by Kramer, and not actual longevity for mankind, it would be more understandable: although both are certainly not exclusively normal Earth time. However, the succession of only one generation, as between Ubar-Tutu and his son Ziusudra, and the simultaneous passing of 28,800 years, seems to refute the theory of time dilation, for certainly many generations should have passed on Earth for mankind during this period of time, while only a few years would have passed for the same space traveler. But if it is a normal Earth time for the Gods, using our present measurement of time, then mankind would actually have these long life spans. After all, if we were genetically creating beings today on some distant planet, would not we want them to be of long term use to us? For before the flood, during the antediluvial period, the Sumerian rulers, nearly all of which, except Ziusudra and his father Ubar-Tutu, were Gods/astronauts/supreme beings, enjoyed a range of reign that was from 43,200 years to 18,600 years; but after the flood, during the postdiluvial period, when the rulers were earthlings/mankind/human beings, the term was 1,200 years, and generally steadily decreased. Was this the result of a rapid advance in their aging process, as I render in the text of this book; or could it be that after the flood, the references to the creation of new categories of people, as described on page 103 of Atra-ḫasīs, The Babylonian Story Of The Flood by Lambert and Millard, possibly implied a reprogramming or altering of the genes of mankind, affecting longevity, as well as the functions of mankind; or did the flood mark the beginning of computing the rulers in mankind's terms, from our perspective, without the use of an external multiplier, or in terrestrial/Earth related years, as opposed to the former astronauts' terms, from their perspective, with the use of an external multiplier, or in cosmic/Shar related years?

Is it possible that it may not be a "fast clock" of the Gods, but may have been a faster rotating Earth? If the ratio between the years of Ziusudra and "Noah" was approximately 36 to 1, could the Earth have been rotating approximately 36 times faster before the flood orbit of Nibiru, and then during the flood orbit, the Earth slowed to its present rotational speed? I doubt it, for as I have previously mentioned, Ziusudra maintained a water clock within the boat, and there is no clay tablet mention of any discrepancy in days after he saw the Sun.

Also, the fact that part of the anger of God Enlil at mankind was due to his lack or deprivation of sleep, ascribed to mankind during 1,200 year intervals, as in my footnote 319 and 328, or 362 and 373, raises a most interesting question. How long does God Enlil sleep? Is there a correlation between his sleep period and that of mankind? Being an extraterrestrial astronaut, it could be possible that his total sleep period lasts proportionately to the cycle of 1 Shar year, 3,600 Earth years, as ours lasts proportionately to the cycle of 1 Earth year, 365 Earth days, which in both terms, the total sleep period to total period ratio would be one to three, or one-third. For mankind, this would be one-third of 365 Earth days, or a total sleep period of 122 Earth days, and for God Enlil, this would be one-third of 3,600 Earth years, or a total sleep period of 1,200 Earth years. Granted we do not enjoy our total sleep period at one time, probably neither does God Enlil, but the clay tablets, by giving 1,200 years as the period, and not identifying by whose terms the years are measured, may be indicating that over the period of 1 Shar year, 3,600 Earth years, or a total cumulative sleep period of one-third of 3,600 Earth years, which is 1,200 Earth years, the sleep of God Enlil was interfered with. As is put forth on the clay tablets, this may only be another way of stating that throughout the course of the 1,200 Earth year sleep period, God Enlil was deprived of sleep.

The other obvious conclusion is that he slept in units of 1,200 Earth years at one time, like a hibernating bear, but this is even more difficult to comprehend and relate to, or compare to the sleep period of mankind, as the total sleep period of the astronauts would be greatly multiplied. Or perhaps he slept in units, either cumulatively or at one time, of 400 Earth years, which would also follow the one-third ratio.

And there is another distinct possibility, that the term "deprived of sleep", simply means that God Enlil had no peace of mind when he was trying to sleep, he was suffering from anxiety at night, worrying about the eroding of his absolute authority and control over mankind, because of the increasing independence demonstrated by mankind.

Although other possibilities exist, in adherence to the clay tablet account, I have rendered 1,200 years, and assume it is in terms of Earth years, as the total time elapsed between his deprivation of sleep, and concurrently the disciplinary measures God Enlil imposed upon mankind. Perhaps we will just have to wait until God Enlil returns, so he can directly inform us exactly how long he sleeps, and solve our enigma.

Time and longevity are slippery concepts to grasp, and until a more satisfactory solution to this question of time and longevity is found, I will have to accept as being correct, although reluctantly and with great reservation and doubt, a 36,000 year rule for Ziusudra, as on page 76 of The Sumerian King List by Jacobsen, and a 950 year life for Noah, as in Genesis 9:29. Furthermore, I admit the two divergent spans of years may be accurate reports of the same span of time, expressed in different terms; perhaps neither being comparable or related to our present Earth years, but representations of time dilation or a similar aberration of time, or another dimension of time unknown to us today.

Additionally, and of considerable interest, is the 36,000 year rule ascribed to Ziusudra, which equals 10 orbits of Nibiru: $36{,}000 \div 3{,}600 = 10$. Thus, confirming that not only was Ziusudra the 10th successive King/Ruler (see my footnote 621), but his rule also extended through a peroid of 10 successive orbits, and his reign expressed in orbit related years would be $3{,}600 \times 10 = 36{,}000$. Therefore, Ziusudra is correctly identified as the 10th King/Ruler who ruled for 10 orbits.

EVOLUTION

Some people believe in the theory of natural selection, which I call arbitrary evolution, mankind very slowly arbitrarily evolving from the apes, either on a direct line, or side by side collaterally. I have a few words regarding both of these theories.

If mankind slowly arbitrarily evolved directly from the apes, orangutan to gorilla to chimpanzee to mankind, there would be no apes present upon the Earth today, for all the apes would have evolved into mankind; and if mankind slowly arbitrarily evolved collaterally with the apes, branching off from the apes during a remote time, mankind would not presently be at the advanced stage we find ourselves in today, for the evidence suggests that a relatively too short period of time has passed for us to slowly evolve within. Why are there still apes, and why did not all apes evolve into mankind, applying the direct line theory; and why has mankind advanced so rapidly, before our time, applying the collateral theory? And if our current intellectual level is due entirely to arbitrary evolution, why, after such a rapid acceleration in brain size, increasing from about 600 cubic centimeters to about 1,300 cubic centimeters in a relatively short period of time, has our brain size not increased over the past few hundred thousand years? And if some apes somehow arbitrarily evolved into human beings, using any theory, why haven't some human beings somehow arbitrarily evolved into "super" human beings? I am looking, but "super" human beings cannot be found anywhere: they simply do not exist. Can we and have we stopped arbitrarily evolving, and why?

I believe the direct line theory is in error, and the collateral theory is only partially correct. Mankind, either through an earlier intervention by the

Gods or other external agents, or independently, may have started slowly arbitrarily evolving from the apes, and then branched off and slowly arbitrarily evolved collaterally to the early Homo erectus stage, as the "creatures" the clay tablets allude to on page 70 of Sumerian Mythology by Kramer, but without the Gods interceding, and through genetic engineering "jumping" mankind forward from the early Homo erectus stage, creating, fashioning, or programming Homo sapiens, as the clay tablets indicate on page 57 through page 63 of Atra-ḫasīs, The Babylonian Story Of The Flood by Lambert and Millard, we would still be near the early Homo erectus stage today, for far not enough time has elapsed for us to slowly arbitrarily evolve within. That is why there are apes today, they are still slowly arbitrarily evolving along a direct line, perhaps only evolving into higher apes, not mankind; but present day mankind did not slowly arbitrarily evolve directly from them, nor did we uninterruptedly slowly arbitrarily evolve collaterally with them. The Gods quickly fashioned mankind, and concurrently through genetic engineering programmed our evolution, and through this programmed genetic development, what I call programmed evolution, as opposed to arbitrary evolution, transformed us into what we are today, as all the ancient clay tablets and most religious sources indicate. In other words, the primary steering factor determining our genetic progress was and is the result of external and deliberate programming, or if I may call it, divine intervention; and the very minor secondary steering factor determining our genetic progress was and is the result of internal and random choice, or the so-called natural selection theory. Thus, normally we have programmed evolution, which allows for very minor arbitrary evolution/alteration, without deviating from the primary steering factor/course/design of the programmed evolution. For I believe the composition and functioning of our genes is far too intricate for them to have exclusively originated and developed out of a primordial mixing of chemicals and gases, or spontaneous creation, as those who advocate the "primordial soup" theory suggest.

Furthermore, it may be possible that we are constantly or at regular intervals being "seeded" with new microscopic life forms floating through the preserving vacuum of space, as in my footnote 40. Therefore, the only logical conclusion I can arrive at is that either directly or indirectly, we/mankind, are the product and result of external intervention and influence. Only thinking along these lines and viewed from this perspective, can we/mankind begin to know and understand who our progenitors were, and are.

Evidence of our programmed evolution, and not arbitrary evolution, can be found within our cranium. Our brain contains three distinct levels of composition and intelligence: the brain of a reptile, the brain of an animal, and the brain of a human being. From one to the next represents a major degree of change. If our brain slowly arbitrarily evolved from the earliest level of composition and intelligence to its present level, then we would not have the three distinct levels, each added and layered onto the former: we would only possess and be aware of one level; one arbitrarily evolved level slowly transformed from the earliest level to the present level. But this is not the case. However, if our brain was "jumped" forward, genetically engineered and improved at specific intervals, adding to and building upon an earlier model, then we would possess and be aware of the different levels, of which we certainly are today. Therefore, I believe that possessing three different levels of intelligence, each within the other, is evidence that the brain of human beings was changed to a major degree, genetically improved, "jumped" forward at three different times by the supreme beings/Gods. Furthermore, "recently", before arriving at our present degree of intelligence within the last major degree change level of intelligence, we were "jumped" forward, genetically advanced to a minor degree, probably not only once, but several times, the most recent example being genetically advancing/altering "Neanderthal" man into "Cro-Magnon" man, mankind as we genetically know ourselves today.

Incidentally, being that the black race still possess physical characteristics/features, especially skin pigmentation, that are very close/similar to the ape/chimpanzee, it may be that the black race was an early creation or prototype for mankind. After genetically engineering/transforming the ape-like creature into an early model of mankind, perhaps the other races represented more refined models, and therefore, later in time and/or procedure, because of the diminished and/or elimination of some of the ape-like features, especially its skin pigmentation. And if this is not the case, then it is possible that the races were genetically engineered/created, or altered/modified/manipulated, at different locations around the Earth, perhaps even at different times, or perhaps because of rivalry between the supreme beings for their own special type/version of mankind, or perhaps by other supreme beings from other extraterrestrial sources, as I further explain in my footnote 194, 745, and 830.

Additionally, if all mankind evolved from a brown (black) skin pigmentation/color ape, why are there four main skin colors among mankind (black, red, yellow, white)? Where did these extremely diverse colors originate from, aside from other obvious different physical characteristics previously alluded to, and to what are they attributable? Why are there only brown (black) skin colored apes: where are the red, yellow, and white skin colored apes?

Also, it is possible that altering the electromagnetic field surrounding the Earth, which could be compared and linked to the electrical aura surrounding each living thing/being, could change/alter mankind, either directly or indirectly. For example, indirectly, a change in the electromagnetic field surrounding the Earth could genetically affect or trigger the weak, by comparison, electrical field/aura surrounding each living being, thus genetically altering that being. Therefore, the past reversals of the polarity of the Earth, may have also triggered changes in living beings, including human beings. There is a recent resurgence of a relatively "new" theory, which attributes sudden changes in the genetic patterns of the arbitrarily evolving creatures of the Earth to repeated catastrophic events. Could the planet Nibiru be the catastrophic agent?

STERILITY

Why are male and female classifications or gender inferences given to the Igigi Gods, if they are sexless surrogate beings, and cannot reproduce?

In Atra-ḫasīs, The Babylonian Story Of The Flood by Lambert and Millard, on page 45, line 47, and page 53, line 173, and page 59, line 210, 225, and line 239, God Wêila, the Igigi God from whom mankind was partially fashioned, is referred to as "his" and "him", which would suggest his and her, or opposite genders. Also, in the same book, on page 45, line 48, and line 4, God Wêila comments, "My brothers," which would suggest a relationship of brothers and sisters, or opposite genders; and on page 49, line 93, 94, 95, and line 96, God Nusku comments, "My lord (God Enlil), sons (the Igigi Gods) are your sons, Why do you fear your own sons," which would suggest a relationship of sons and daughters, or opposite genders. These examples all point toward the same direction: opposite genders.

I believe it can be possible to have a gender, and be unable to reproduce. The gender is called neuter. The Igigi Gods can have a gender, which would be called neuter, or a sexless classification, and still be referred to as male and female, if their reproductive organs are generatively underdeveloped/not sexually functioning when they grow to be adults. This could be achieved by the highest Gods, if the pre-puberty absence of reproductive development in the Igigi Gods is genetically extended through their entire existence. This would create a being that is unfertile, sterile, much the same way pre-puberty children of mankind are today, whose reproductive capabilities are not yet genetically activated, "turned on", which we identify by the gender of male and female, but whose gender is actually neuter. Through this method, it could be possible the highest Gods genetically suspended the reproductive development of the Igigi Gods in their pre-puberty sterile state, before they reached their fertile state, and their reproductive genes remained dormant, and were never "turned on".

It has been known that the reproductive organs of the fetus of mankind are not formed/non-existent during the entire first 6 weeks of gestation, thus the fetus has no sexual designation that is detectable, has a sexless classification; and then after this 6 week stage, a genetic "switch" is triggered, and male or female selection is made. It is very important to note here that the gene/genetic "switch" which selects/determines the male or female reproductive organ development has been recently discovered, and therefore our ability to select the sex of our own children is close in terms of time. This is one "switch", a male or female "selection switch". How close are we to discovering another "switch", the genetic "switch" which is triggered at puberty, to activate/"turn on" the male and female generatively, a male and female "fertility switch". Will we then discover how not to turn this "switch" on? Slowly, we human beings are discovering the "handiwork" and "secrets" of the supreme beings. Incidentally, I believe we humans have a gene which contains a coded record of each person's lineage, a kind of lineage memory bank, and someday, when we become sufficiently technologically advanced, upon its discovery, we will be able to trace back our ancestral lineage and identify whose genes we carry and are composed of: what an interesting learning experience that will be.

Also, the Igigi Gods may be sterile because they are possibly genetically altered higher Gods, or possibly a hybrid, or possibly a hybrid of the highest Gods themselves, as in my footnote 49b, which in any event, was overcome or changed in the sample of genes from God Wêila, by God Enki when he fashioned mankind. But how would it be possible for God Enki to extract generative fluid and then perfect it, from God Wêila, an Igigi God, the lowest rank, a neuter being, who technically has no fertile or active reproductive genes. This may be why a purifying bath was necessary, as in my footnote 186. Perhaps God Enki merely advanced the dormant sample of generative fluid from God Wêila to the active stage, beyond the dormant suspended stage, or genetically "turned the sample on", thus perfecting it.

The genes of Adapa, mankind, may also have been genetically "turned on", in much the same way the generative fluid of God Wêila was, for newly created mankind and the Igigi Gods once shared the same trait: sterility. How could mankind advance from a sterile hybrid state, to a state of fertility? It is possible that a similar procedure to genetically activate the dormant reproductive genes of mankind was employed by God Enki, if not the identical procedure. The reproductive development of mankind, normally activated at puberty, which had been programmed out, or genetically "turned off" by God Enki when he created mankind, was simply programmed in, or genetically "turned on". Or, if newly created mankind was sterile simply because we were hybrid beings, which also could be the state of the Igigi Gods, then this too was overcome or changed by God Enki, and furthermore, the two may be closely linked.

120 YEARS

I agree with the portrayal of most of the events as set forth by Sitchin in The 12th Planet, and in particular, on page 224 through page 229, the 120 years referred to in the Bible under Genesis 6:3 And the LORD said, My spirit shall not always strive with man (shield mankind forever), (having erred) for that he also is flesh (in addition to possessing our Godly spirit/essence, he also possesses human flesh/essence): (And) yet his days shall be (were) an hundred and twenty years. I believe this reference is to the number of Shar years preceding the great flood, for if 1 Shar year equals 3,600 Earth years, then 120 Shar years equals 432,000 Earth years. These

432,000 Earth years measure a period of time and stewardship for the Gods, not for mankind, and spans the period from their landing upon the Earth, to their forced temporary departure with the occurrence of the great flood.

The Biblical account reflects 8,226 years from Adam to Noah and the great flood, whereas Abydenus, who lived between 100 B.C. and 400 A.D., quoting Berossus of circa 290 B.C., states there were ten kings who ruled 120 Shar years, or 432,000 Earth years before the great flood.

If the great flood occurred in 8,800 B.C., we must add 432,000 Earth years to this figure, and we arrive at 440,800 B.C. as the year the Gods decided to come to the Earth, and established their first base, Eridu.

However, there are other interesting interpretations and possibilities. Could 120 years be a reference to galactic years, the age of existence of our Sun, Earth, and of course our solar system, as it orbits around the galactic center every 200 million Earth years, considering one orbit as a galactic year? If the Sun is approximately 5 billion Earth years old, that would equal 25 completed galactic orbits, and our Sun/Earth/solar system has 95 galactic orbits remaining until it expires and ceases to exist as such, or approximately 19 billion Earth years remaining. Or, could 120 years (in terms of years as units), be a reference to the maximum potential cell life of mankind, which now stands at 50 divisions/replications/generations, to be increased to 120; or in other words, a more than doubling of the generations of cells to permit the life expectancy of mankind to be extended to the 180 year range. Or, could the 120 years (in terms of years as years), be a reference to the maximum potential life expectancy of mankind, extended to 120 years, with the existing 50 cell divisions/replications/generations remaining as it is, and not increasing. Or, could 120 years be linked to and be an abbreviation, a variant of the 1,200 year intervals, as I have put forth within 'Longevity'.

ARCHITECTURE

In this book, I have primarily given the Babylonian clay tablet rendition of flavoring details, especially of architecture, which could be a comparatively later version. From the descriptions given, we are led and predisposed to believe in the existence, at the time of the flood, of rectangular houses and rectangular courtyard walls, both constructed of clay bricks, and of rectangular huts, constructed of reed bundles, all of which may be accurate, although changes may have steadily occurred throughout the period rendered. This could lead to some confusion.

The references to a house, door, gate, reed wall, and reed hut serve to indicate and infer the general shape of the structure, as given on page 47, line 69, and page 69, line 375, 380, and line 386, and page 89, line 20, and line 21, and page 123, line 14, and line 15 of Atra-ḫasīs, The Babylonian Story Of The Flood by Lambert and Millard, and as illustrated on page 105, and page 113 of Earliest Civilizations Of The Near East by Mellaart. Even the earlier Sumerian clay tablets were describing bricks, cities, bricklayer, foundation, capitals, and wall, as given on page 141, line 42, 87, and line 92, and page 143, line 151, 152, 153, and line 154 of Atra-ḫasīs, The Babylonian Story Of The Flood by Lambert and Millard. These types of structures were undoubtedly correct for the Sumerian and Babylonian periods, both long after the flood, for they both merely transposed the earlier flood story into their later periods, retaining the main story line, but favoring themselves by injecting their own familiar architecture, which may be accurate, to make the story more understandable and identifiable to their contemporaries.

Therefore, in correctly interpreting the later clay tablets, if their architectural descriptions were not totally accurate, we may be misinterpreting the very earliest clay tablets, and not grasping the true construction and shape of these structures.

But what was the shape of these structures preceding and at the time of the flood? The earliest sources, whether oral or written, which were rendered immediately after the flood, but are now lost to us, except for several Biblical passages, which I give later, undoubtedly inspired the Sumerians and Babylonians, and were probably referring to circular shaped reed huts as houses, like a byre, a type of barn, as illustrated on page 38, and page 39 of The Babylonian Legend Of The Flood by Sollberger, and as illustrated on page 275, Fig. 133a of The 12th Planet by Sitchin, and as illustrated on page 200 of They Wrote On Clay by Chiera, and as described on page 19, 20, 21, 22, 23, Ill. 5, and page 37, 38, Ill. 18, and page 53, 54, and page 55, Ill. 29 of The Earliest Civilizations Of The Near East by Mellaart, and as described on lower page 13 and upper page 14 of The Sumerians by Woolley. These circular huts were probably similar to the primitive huts still in use in remote areas around the world today. However, this does not preclude the possibility that the circular shaped reed huts were actually only byres, shelter/houses for animals, and not the houses of mankind.

It is possible this very early circular design as a house for mankind, if it in fact was a house for mankind, was an attempt to imitate the circular design of the craft of the Gods, their home away from home. For they may have brought large self supporting controlled environment craft with them, which would certainly be practical and functional to serve the astronauts as a base, what mankind eventually called a city.

Why should the Gods build clay brick primary structures, using primitive materials, and methods that were far removed from their advanced technology? They probably did not for themselves, although after they created mankind, when there was a need, they assuredly instructed and probably had mankind build houses for themselves, and build primary clay brick structures for the Gods, probably functional staged communication towers to repress and control early mankind, utilizing one of the most abundant materials in that area, perhaps with their craft as the central feature. Also, the Gods probably dwelt in their large craft-cities, possibly similar to the Duku satellite, as I refer to in my footnote 50, or even in much smaller shuttle craft, all which they took with them when they departed, for we have found no physical traces to this day of their craft, or of the Gods, except ourselves; we only have lingering unusual shaped designs of structures to question as evidence, among which are the circle and the ziggurat, especially the circle, as explained on page 295 of When Time Began by Sitchin, including the circular shaped gufa boats, which could be the last vestiges of the similar legendary hexagon shaped boat of the great flood, as I refer to in my footnote 5 and 531.

Certainly the architecture of the Sumerians and Babylonians was very interesting, for their unique designs in building may be a vestige, an art form, employed by mankind through which to emulate and remember the Gods. For after the flood, probably more than 350 years later, for Noah lived 350 years after the flood, but he is not mentioned further in the present tense, nor in this passage, which presumably takes place after his death, for if he were alive, I am certain he would have been specifically mentioned, mankind returned to Sumer, their home, as is stated in the Bible in Genesis 11:2 And it came to pass (probably long after the waters receded), as they (the descendants of

Noah/Ziusudra) journeyed from the east (from the highlands of the east, part of the so-called fertile crescent: see my footnote 778), that they found (located) a plain (a more favorable topography) in the land of Shĭ´-när (Sumer); and they dwelt there. Furthermore, it can be assumed from the context the art of making bricks was known from before the flood, and could have been practiced 88 somewhat exclusively for the Gods, and possibly only after the flood, did mankind begin to imitate for themselves, with bricks, the primary structures, or towers, of the Gods, which again brought their wrath; but while on the mountains, to the east, immediately after the flood while the waters were receding, they built with stone, as is stated in Genesis 11:3 And they said one to another, Go to (we too), let us (mankind) make brick, and burn them thoroughly. And they had brick for (instead of) stone, and slime had they for morter (mortar).

In any event, it is possible the flavoring details, including architectural designs, were accurately described by the clay tablets, and the confusion may be caused only by our own placement of the renderings within a particular time frame, or not being able to accept the flavoring details in their proper time frame.

Therefore, for continuity of the clay tablet narrative, I adhere to and render the earlier writers descriptions of flavoring details wherever possible, who related to and wrote from their own common experience, which moreover may be totally accurate, but somewhat out of focus to many of us today, although most of the flavoring details are also primarily still within our own common experiences.

Incidentally, after the most recent passing of Nibiru, between 1,600 B.C. and 1,500 B.C., could it have been possible that the style of architecture itself, and explicitly the use of embellished columns, was an attempt by mankind to depict and/or emulate the ascending or descending craft of the Gods/astronauts, with the capital, or uppermost detail of the column being the craft, or the triangular or circular shape of the roof of the temple being the craft, and the columns being the vapor trails of the craft? And at the same time, or taken from the preceding passing of Nibiru, did the Egyptians use their unique obelisk design for the same symbolic meaning, with the pyramidal top representing the craft, and the towering base exemplifying the vapor trail; and could the pyramids themselves be symbolic of the craft of the Gods/astronauts resting upon the Earth after landing?

HARSH CONDITIONS

During the nearly 25,200 Earth years, or nearly 7 Shar years, or nearly 7 orbits immediately preceding the flood orbit, there is a strong possibility the orbital path of Nibiru steadily became closer to the Earth, and its force fields upset the normal rainfall upon the Earth. The effect of this being the diverting of rain toward both polar regions, where it accumulated as snow and ice. During each orbit of this period, this phenomenon grew more intense, causing the central regions of the Earth, including the lands of Sumer, to have sparse rainfall, or none at all, and harsh conditions with drought and famine to exist, as the clay tablets report, and I describe within the text of this book. However, we do not have enough evidence at this time to verify the occurrence of these events.

ICE

Are there any facts that substantiate the abrupt end of the most recent major ice age, which corresponds to the great flood of 8,800 B.C., and occurred as a result of it? On page 103 of Secrets Of The Past, which describes the effects of the flood, and not the cause, scientific evidence is given to support the clay tablet account that the rain was diverted away from mankind, and toward both polar regions. The rain was deposited as snow, which formed ice, and accumulated at the north pole, increasing in height upon the flat ice sheets over the water, for the north pole does not have a solid land foundation. At the south pole, the ice also accumulated to great heights, but upon its solid land foundation.

When the great flood occurred, both polar ice caps, as well as the entire Earth was shaken, and swept over by the receding waters from the equatorial bulge. The south ice cap probably was lifted up from the solid land foundation by the bulging water, after it was shaken loose by the Earth tremors and other disturbances, and the north ice cap was simply lifted up by the bulging water, and then both ice caps melted at the same rate. But only after the displaced flood waters returned to the equator, during the following period of time, after 8,800 B.C., did the fresh water from the melting ice caps begin to leave salinity differences, which we now can measure in core samples of the sediments.

Also, as I have previously mentioned, it could be possible that Nibiru periodically tilts the axis of the Earth, and triggers the ice sheet production process; and similarly, it could be possible that Nibiru is the catalyst for melting the ice caps on Mars, causing the waters to interimly run. During any close or perturbing orbit of Nibiru, and especially the flood orbit between 8,800 B.C. and 8,700 B.C., or approximately 11,000 years ago, the axis of Mars could have been shifted and the waters of Mars could have been pulled to the caps, thus destroying any evidence and/or ending any life/civilization, as we know it, if there was any, that may have existed there. Therefore, the flood orbit of Earth, may also have been a flood orbit for Mars. Perturbations caused by the close passing of Nibiru could have either directly or indirectly caused the waters to flow/run on the surface of Mars for periods of 3,600 years, or multiples thereof; and then trigger/cause a stoppage/locking-up of the waters, for like or long multiple periods, resulting in the unusual topographical features we observe/find on Mars today. A good example of the results a shifted axis could bring, would be if we take notice of the nearly 23-1/2° constant inclination of the axis of the Earth, as it orbits around the Sun. Aside from the obvious question of why is the axis of the Earth tilted, the constant tilt of the Earth during its orbit creates a summer and winter season, accompanied by wide temperature/climate change variations, which causes the same water to thaw/run and then freeze/stop. A result such as this could also be achieved by the "instant" tilting of the axis of a planet, such as Mars, either more or less than the nearly 23-1/2° constant inclination of the axis of the Earth.

Furthermore, in circa 600 B.C., one of the Egyptian priests, as in my footnote 719, a wise man of Sais (Port Said?), informed Solon, a Hellene, who later informed Dropidas, who in circa 430 B.C. informed Critias, who in circa 425 B.C. informed Socrates, who in circa 400 B.C. informed Plato, who subsequently gave his account of Atlantis to us. If the account of Plato is correct, and Atlantis sank "after" 9,600 B.C., and if the last ice age ended during this same period, and if the flood also occurred during this same period, as I believe it did, then all of the events occurred simultaneously, and all were related and triggered by the passing of Nibiru. Therefore, the same agent/force that precipitated the great flood, also sank Atlantis.

CONTINENTAL DRIFT

As we know, the crust of the Earth is divided into large plates, called tectonic plates, with the continents, which are the uppermost part of these plates, being visible above the water level of the oceans. Together, they drift, or are being pushed to and fro around the Earth, dividing and then reuniting again as a single super continent, or so we are told. Why are the continents drifting?

Could the triggering mechanism and continuing driving force of the drifting continents be ascribed to the cleaving/splitting in half of the original water planet Tiamat; the totally cracked upper half of which afterward became the water planet Earth? I believe it could be, if the original super continent, sometimes referred to as Pangaea, which means "all Earth", was located on the upper half of Tiamat, and then after the cleaving of Tiamat, the totally cracked super continent continued to break up and drift toward the lopped off/flattened side of the newly designated Earth, which could originally have been where the Pacific Ocean is today. This phenomenon, the now drifting continents, could have been described as acting like counter weights, or a kind of equalizing factor, to suddenly balance the then lopsided newly created Earth, allowing the Earth time to form a new and properly uniform thickness of crust, and round itself out on the lopped off/flattened side.

Either, after the initial rounding out and balancing of the Earth, the continents continued to drift back and forth to their original location, all on the same side of the originally cleaved planet Tiamat, now called Earth, for they were no longer needed as an equalizing factor on the previously flattened side, and that is what we could be witnessing; or more likely, we could be witnessing the effects of the initial rounding out of the Earth, continuing to this day in the form of the drifting continents.

In other words, if the Earth is the upper half of the split planet Tiamat, then it is possible the continents, as we know them today, could have all fit together, as they appear to, and were originally all a part of one large continent on Tiamat, before the split. After the split, the lopped off side could have become our Pacific Ocean of today, and the single totally cracked large continent, now a part of the newly labeled Earth, continued to break up and drift apart, in order to temporarily balance the Earth, until a new crust could be formed to permanently round out and balance the Earth.

Additionally, depending on the proximity of the Earth in its orbit, a close passing/orbit of Nibiru could jolt the plates of the Earth, causing the plates to very suddenly slide/move to great distances, resulting in both the quick increasing/rising of certain mountains/land masses, such as Machu Picchu, and quick decreasing/submergence of other mountains/land masses, such as Atlantis: "instant" appearance/rising of mountains/land masses, and "instant" disappearance/submergence of other mountains/land masses. Also, some existing land masses that had been raised further during the onset of the passing/orbit, could have returned to their former levels during the conclusion of the event, thus these land masses would not have been flooded, and the indigenous life forms would have survived.

In other words, the major changes in the topography of the Earth, namely, the formation of the mountains, I believe are due to sudden periodic catastrophic events triggered by the forces of an external agent: Nibiru and its host of satellites. Concurrently, the minor corrections or adjustments to the topography of the Earth, occur during the relative "peaceful" interim periods between the major changes. Therefore, I agree with Velikovsky in his Worlds In Collision and Earth In Upheaval, that the primary driving force behind the creation of the mountains occurs when the sections/plates of the Earth are jolted and very suddenly slide/move to great distances, and new mountains are quickly formed, while other lands beneath them are forced downward. Secondary "slow" movement in the plates would then occur when the external force was not present. Also, if the Earth could be slowed, its rotation retarded, while the plates continued to move, as in my footnote 744, the result described could easily be achieved.

AGE OF THE EARTH

From rock samples, the Earth has been estimated to be approximately 4-1/2 billion years old, but that could only indicate the age of Tiamat, and not the age of the newly designated Earth. The age of the oldest rocks do not indicate when the Earth was formed, they only indicate when Tiamat was formed, for they do not tell us when Tiamat was cleaved by the forces of Nibiru, and therefore, when the Earth was actually formed. The question is: how many years ago was the original water planet Tiamat cleaved/split;

and diverse orbit asteroids, comets, and meteoroids formed from the shattered and scattered ten smallest sattelites; and the new water planet Earth, as we know it today, concurrently formed from the totally cracked upper half; and the Asteroid Belt formed from the crushed lower half?

For the answer to this question, we must search for direct preserved evidence of the split: the youngest age of the now solidified molten rocks in the Asteroid Belt; or the youngest age of the now frozen water, the ice within the Asteroid Belt; or the youngest age of the organisms frozen and trapped within the ice in the Asteroid Belt; or with more difficulty and perhaps not as "pure" a sample, due to their increased exposure during their deep space orbits, applying all of the aforementioned to the diverse orbit asteroids, comets, and meteoroids; or the age of the rocks that were cracked or formed, or both, upon the initial break up 200 million years ago of the large super continent Pangaea on Tiamat, or continued break up on newly formed Earth, if this evidence still exists; or geologists must determine the age of the newly formed crust of the Earth, possibly beneath the Pacific Ocean, where the lower half of Tiamat was originally located, which then became the lopped off/flattened side of the newly designated Earth, if this evidence has not already been erased by the eroding actions of the ocean, or covered by subsequent volcanic activity associated with the drifting continents.

Additionally, there is another very obvious possibility of where direct preserved evidence of the split may be found, for if we look up to the sky at night, we view the Moon, which we can clearly see above our head, and mankind/human beings have even walked upon its surface and touched it, and brought samples of its surface material, including rocks, back "home". The evidence could be the oldest age of the still visible craters on the Moon, which could be remnants of "battle" scars from the early period of the Tiamat and Nibiru encounter, and could have been easily caused; either by the smaller debris orbiting among the eleven satellites of Tiamat, shortly before the encounter, as they became erratic in their orbits, as described in Enuma Elish, The Seven Tablets Of Creation by King, on page 67, line 67 As (Marduk/Nibiru) gazed, (Kingu/the Moon) was troubled (became erratic) in his gait (orbit), and line 70 (the ten other satellites of Tiamat) Beheld their leader's (the Moon's) [. . .] (plight), and their sight (orbital path) was troubled (also became erratic); or by smaller debris orbiting among the seven satellites of Nibiru, shortly before the ten satellites of Tiamat were shattered and scattered; or by meteoroids, the smaller debris from the shattering and scattering of the ten satellites of Tiamat into diverse orbits as asteroids and comets, the larger debris: see my interpretations of the initial pass, Illustration 11, line 106 through line 118 on page 332; or by the smaller debris as Tiamat was split in half: see my interpretations of the first orbit, Illustration 12, line 129 through line 132 on page 334.

Furthermore, an interesting thought occurred regarding why the same side of the Moon is always facing toward the Earth, or stating it another way, why does the Moon not rotate on an axis, yet why does there seem to be somewhat even distribution of its craters? It seems obvious to me the Moon once rotated, for it would be far easier to account for the somewhat even distribution of its craters, all over the surface, if the Moon was rotating while being impacted or "bombarded" by the smaller debris coming from one direction during the Tiamat and Nibiru encounter, regardless if the debris was propelled toward the Moon, or the Moon orbited through the debris; than if the Moon was not rotating, and the smaller debris "bombarded" the Moon from all possible directions. For if the Moon had never rotated, all of the craters from the Tiamat and Nibiru encounter would be on the same side, and the opposite side of the Moon would be nearly without craters. It also seems obvious to me the Moon was rotating, probably very rapidly, before the Tiamat and Nibiru encounter, when it was fully charged, being the "exalted" (largest) of the "fuming and raging" host of eleven satellites of Tiamat, for I cannot imagine a "fuming and raging" satellite not rotating, as described in Enuma Elish, The Seven Tablets Of Creation by King, on page 17, line 111 [They prepared for battle] (the encounter), fuming and raging (rotating, as well as orbiting, probably very rapidly; and possibly spouting lava, which could be the cause of some of the very smallest craters on the Moon); and page 19, line 126 After this fashion, huge of stature, [she (Tiamat) made] eleven (monsters/satellites); and line 128 She (Tiamat) exalted (gave dominance over, or gave the power to control or influence the other ten satellites to) Kingu (the Moon, probably because the Moon was the largest of the eleven satellites); in their midst (among the eleven satellites) [she raised] him [to power].

I conclude that it was during the "heated battle" period of the Tiamat and Nibiru encounter, while the Moon was still rotating, that the smaller debris "bombarded" the Moon and caused the craters over its entire surface; and this is confirmation of a once rotating Moon, which adds credence to the scenario given by the clay tablets. I also grant to those of you who doubt my belief, that there are exceptions, and some of the craters on the Moon, as well as on the Earth, were made at a more recent time, after the Tiamat and Nibiru encounter, by meteoroids, the smaller remnants of encounter debris created during the shattering and scattering of the ten satellites of Tiamat into diverse orbits as asteroids and comets, the larger remnants of encounter debris, as the scattered smaller and larger remnant debris orbits around the Sun and back to its place of origination, back "home", which impacted over a period of millions of years; but I am speaking of the vast majority of craters on the Moon, which I believe were made during the same "short" period of time. However, it is also possible that during subsequent and more recent orbits of Nibiru, while orbiting through our solar system near the Asteroid Belt, Nibiru could distort/pull and ricochet some of the various size confined orbit asteroids out of the Asteroid Belt, and these asteroids could "bombard" not only the Moon, but also the Earth and other planets.

But why and when did the Moon stop rotating? Again, I believe it was the result of the Tiamat and Nibiru encounter, but during a later period of the encounter, later than the smaller debris "bombardment" period, and among the possibilities that exist in this encounter period are, with the strongest probability being offered first: during the initial pass of Nibiru, after the smaller debris "bombardment", Nibiru fully neutralized the forces of the Moon, causing its rotation to cease; see my interpretations of the initial pass, Illustration 11, line 120 on page 333: during the first orbit, immediately after Tiamat was cleaved, the newly designated Earth, along with the Moon, together were driven to their present orbit, with the Moon actually being pulled by the forces or larger mass of the Earth, causing the center of mass of the Moon, comprised of its molten core, to somewhat solidify, with the heaviest elements or greater part of the core furthest away from the Earth, or off center, thus establishing what could be referred to as a bary (heavy) center of mass within the Moon, and remaining so to this day; see my interpretations of the first orbit, Illustration 12, line 132 on page 334. Also, there is a weaker candidate earlier in the encounter, as I mention within line 93 on page 331, line 67 and line 68, and I believe the words "His will was destroyed" is a reference to the elimination of the particular force causing the erratic orbit, and although other menacing aspects or forces still existed, the forces of the Moon were partially neutralized, and the words "motions ceased" is a reference to the arresting of the "troubled" (erratic) "gait" (orbit), and I interpret this to mean the erratic orbit of the Moon ceased, and either the Moon stopped orbiting around Tiamat, or its orbit was somewhat corrected; but its rotation did not cease. Furthermore, there are far too many evenly distributed craters over the entire surface of the Moon, which were probably made during the "heated battle" period of the encounter, than I can account for being made prior to that time, during the approach of the two "combatants", Tiamat and Nibiru.

Therefore, taking into consideration all the correlative evidence given above, presently we do not know the age of the Earth, we only know the age of Tiamat, which was created approximately 4-1/2 billion years ago. However, recent discoveries tracing the diverse orbits of comets, and probably asteroids, if we could trace their diverse orbits, the shattered and scattered larger remnants of the ten satellites of Tiamat, back in time, for they also orbit around the Sun, indicate they seem to have originated in the area of the Asteroid Belt, just 4 million years ago. If this information is correct, our question has been answered.

APHELION

What is the furthest distance from the Sun that Nibiru travels in its orbit, or the aphelion of Nibiru? To find a somewhat reasonable answer to this question, we must first find a known comparable, or means by which to measure distance. In this instance, we shall use the aphelion of Pluto, the planet in our solar system with the known widest or most elliptical orbit, and we must assume the velocities of both Nibiru and Pluto are similar, and they may be, for the last act of Nibiru, during the third orbit of the initial encounters with Tiamat/Earth, was to sweep/remove Pluto from its loosened orbit around Saturn, and cast/place it in its present orbit, as I indicate under 'Creation Of Planets, Satellites, And Rings', Illustration 14. Therefore, the velocity of Pluto, and perhaps even its 17 degree different plane of orbit, may be very close to that of Nibiru.

If 1 year on Nibiru equals 3,600 Earth years, and if 1 year on Pluto equals 248.49 Earth years, we divide 3,600 by 248.49 to find the ratio, which is 14.49; or 1 year on Nibiru is 14.49 times as long as 1 year on Pluto. If the aphelion of Pluto is 4,580,000,000 miles, then we multiply this figure by 14.49 to find the aphelion of Nibiru, which could be 6,636,420,000,000 miles. Keep in mind that we are assuming similar velocities, for if they are not similar, the final figure could vary greatly.

MAY NOT BE A PLANET

The planet Nibiru is like a comet in the period of its orbit, which is 3,600 Earth years. However, there is a possibility that the comet-like planet Nibiru may not be a planet at all.

With a moderately more advanced technology than ours, Nibiru may be an artificial satellite, and not a natural celestial body. With an extremely more advanced technology than ours, Nibiru may be only an energy force, like the quantum mechanics photon, but with the ability to change itself into any form the observer could recognize, or from one form to another; or to destroy our visual perception of it and then create a new or the same visual perception of it, or vanish and then reappear, as is allegorically described on page 61, line 19 through line 28 of Enuma Elish, The Seven Tablets Of Creation by King, or as described by the Egyptian accounts of Ra (Nibiru) as Amon-Ra (hidden celestial body) or Aton-Ra (visible celestial body), as I previously mentioned within 'Most Recent Orbit'. If we consider Ra as the name of the celestial body, it becomes apparent the term "hidden" Ra could indicate that near the perihelion of its orbit, when it is near or behind the Sun, the celestial body is temporarily hidden from view as observed from the Earth, thus necessitating a "hidden" and "visible" designation. And if the Sun was being referred to as "bloody star" Ra, as some suggest, I believe it is only because the Sun was being used as a visible symbol for Nibiru, the true Ra/An, perhaps because of its apparent similar large size and/or color to the observer upon Earth, especially during sunrise and sunset: thus, sunset red color. In other words, the Sun could have been worshipped as a daily substitute for Nibiru, which only approaches the Earth and actually appears to earthlings/human beings every 3,600 years.

Additionally, Ra may have been phonetically pronounced differently than how it appears, especially if reversed, or if transposed through metathesis, as commonly was done in transliterating, and could be An (Anu), the Sumerian name for the highest divine being, and also the name for heaven (space), his abode. Furthermore, I believe Ra/An/Anu are found within the Hellenic Ouranos (sky/heaven/celestial): OU-RA-NOS and OUR-AN-OS and OUR-ANOS (see my footnote 48). Therefore, I believe the four names, An, Anu, Ra, and Ouranos, identify the same supreme being, as well as his celestial abode, what we are now referring to by the epithet Nibiru; and did our ancient ancestors, as on page 5 of Enuma Elish,

The Seven Tablets Of Creation by King, initially name our known planet Uranus, Anu (An/Anu/Ra/Ouranos = Uranus), in honor and in remembrance of the supreme being and his home planet An/Anu/Ra/Ouranos = Nibiru; and initially name our known planet Neptune, Nudimmud (an epithet for Ea/Enki), in honor and in remembrance of the supreme being Ea/Enki/Poseidon = Neptune (see my footnote 26). And reflecting this thought, I believe Neptune, whose role it was as the first outer planet to attract Nibiru into our solar system and perfect its orbit into its present orbit, initially was named Nudimmud (Ea/Enki), which means 'He Who Fashions Things', which most certainly God Enki is skilled and noted for, primarily because Neptune "fashioned" a new orbit for Nibiru (see my footnote 64). Also, does Nibiru vibrate on different wave lengths (outside of visible light), or change to different wave lengths, such as infra red (shorter rays) or ultra violet (longer rays) to make it invisible to the human eye, for it has been theorized that only 10% of the matter in the universe is within the visible range of humans, and if this is correct, then 90% would be invisible to us.

The possibility also exists that we are part of a binary/double star system, and the "bloody star"/"brilliant star" (see my footnote 16, 719, and 742), or "dark red"/"afire"/dark companion to our Sun may be Nibiru, the other sun, a dark/sunset red sun/star, which would be correctly identified as a "bloody star"/"brilliant star", the so-called "Nemesis" or "Death Star" (see my footnote 14); and if this is true, the satellites/moons/Winds, as I indicate within 'Winds', may be planets, and not moons.

It is also possible that Nibiru may be the manifestation of a concept far beyond our present comprehension; or is Nibiru simply a red planet, rich in iron oxide, like Mars, and thus its identification with the color red.

However, if Nibiru is not a natural celestial body, and in fact is an artificial satellite or more, we must assume it has guidance/steering capabilities. Therefore, the possibilities are far greater that it is a natural celestial body, for if it were not, the jeopardizing reoccurring encounters with our solar system certainly would have been totally avoided.

NIBIRU

How do I arrive at the name Nibiru? I have applied the name from perhaps the oldest nearly complete clay tablets that have been found. The epithet, which is one of the fifty names given to the celestial body, probably characterizations of manifestations or effects upon the Earth during various stages/periods of its orbit, appears in Enuma Elish, The Seven Tablets Of Creation by King, page 109, line 109 "Let his name be Nibiru, 'the Seizer of the Midst'!"; and in The 12th Planet by Sitchin, page 215 Planet NIBIRU:, and page 218 The Heaven (our solar system) he divides in half (orbits through the center, near the Asteroid Belt) and stands (is visible in its orbit) as Nibiru, Akkad (the land) will overflow with plenty (especially water: see my footnote 14), When Nibiru culminates . . . . (reaches its perigee), and page 222 When standing in (visibly orbiting through) the place of the celestial battle (near the Asteroid Belt): Nibiru (shall be its name)., and upper page 361 It is the god (planet) Nibiru ["Planet of Crossing"].

The celestial body that disturbs the Earth and nearby planets has been called various names throughout the ancient history of our planet, among which are, in order of the commonly accepted oldest name first: Sumerian — Nibiru; Babylonian and Assyrian — Marduk; Egyptian — Ra; and Hellenic — Ouranos. I believe the ancient Sumerians, Babylonians, Hebrews, and others, also had another name for the celestial body, a name that carried a religious connotation, one which we still reverence and exalt today in the same sense in many religions. The name is Lord; as identified by the Biblical Amos in 5:8, one of many descriptions the Bible is rich with, which reads as follows: Seek him that maketh the seven stars (Pleiades) and Ō-rī´-o̲n, and turneth the shadow of death into the morning, and maketh the day dark with (as) night: that calleth for (forth) the waters of the sea, and poureth them out upon the face (surface) of the earth: The LORD is his name.

PROPHECIES

Many religious leaders of today are interpreting Biblical prophecies, and are alerting us to the approaching tribulation period of the so-called 70th week of Daniel, during which time great misery, distress, and instability shall overwhelm, afflict, and burden all mankind. Could there be a correlation between the Biblical tribulation period, and the total period of passing of Nibiru? The length of time, 70 weeks, about 16-1/2 months, or 1-1/3 years, is nearly the same.

I believe there is, for the effects of the force fields of Nibiru, especially the gravitational and electromagnetic, not only pull and cause the fluid parts of the Earth, the molten core, the air, and especially the water, to be shifted or displaced by the forces exerted upon them, but they affect all water, including the water/fluid within our bodies, as described on page 205 of Tides And Pull Of The Moon by Wylie, and especially the water/fluid within our cranium: our brain. This could surely cause all mankind to become somewhat unstable, as described in Volume II of The Koran Interpreted by Arberry, upper page 27 O men, fear your Lord! (etc.), and thou shalt see mankind drunk (unstable, as is the Earth during this period), yet they are not drunk, but God's chastisement is terrible; and perhaps the celestial Lord, Nibiru, shall cause mankind to, as in the Bible in Isaiah 2:4 and Micah 4:3, beat their swords into plowshares, and their spears into pruninghooks.

Ziusudra and his wife were not the only earthlings to have been transported to the abode of the Gods to dwell: taken to Nibiru. Adapa, as described in The Adapa Legend on page 147 of The Babylonian Genesis by Heidel; the Biblical Enoch in Genesis 5:24; and Elijah in II Kings 2:11, also enjoy this privilege. A certain king of Uruk in circa 2,888 B.C., Gilgamesh, whose story is told in The Gilgamesh Epic And Old Testament Parallels by Heidel, embarks on a long journey to the spaceport/launch center, reported to be Baalbek in Lebanon, as I previously mentioned on page 298 in my comments, and possibly ultimately to the abode of the Gods, in search of the secret of immortality, and either communicates with Ziusudra through a device or actually meets Ziusudra, here called Utnapishtim, "the Distant," who dwells in the distance, "at the mouth of the rivers," and his wife. Gilgamesh acquires a secret plant of immortality, "a plant like a thorn, like a rose its thorns will prick thy hands, a wondrous plant, a fragrant plant," to be eaten at a later time during his life, which will return him to his youth, only to lose it as he returns home. But we know that at least five individuals remain at the abode of the Gods; Adapa, Ziusudra and his wife, Enoch, and Elijah.

Additionally, could it be possible that several relatively recent individuals are also at this abode, waiting for the next passing and window period of Nibiru, and their time to return; for there are many human beings either waiting for the appearance of a supreme being, or the return of a supreme being or supreme beings.

Incidentally, in Hellenic "mythology", the bull represented Zeus, and it is very possible the bull also represented and was a symbol for Nibiru, as I believe it was with their contemporaries. And in more recent times, Alexander the Great (356-323 B.C.), sought to emulate the supreme beings/Gods, especially Zeus (Ammon/Enlil), head of the Hellenic Pantheon of Gods. The Gods were nearly always depicted wearing their unique "high technological" headdress, which appeared to human beings as the incorporation and use of horns, one on each side of the headdress. After his Egyptian welcome, and also seeking immortality like Gilgamesh, his visit to the famous temple and oracle of Zeus/Ammon at the Great Oasis (SIWA) in the desert of Lybia and affirmation of his true divine parentage by the God himself, upon which Alexander was deified by the Egyptian priests as a Pharaoh. This confirmed his mother's claim of his lineage to the Gods due to his birth to his human being mother, Olympias, and a supreme being/God father, Zeus/Ammon; and thereafter, Alexander the Great was depicted, especially on his coins, as having horns, one on each side of his head, symbolic of the "horned" headdress of the Gods.

Furthermore, it could be possible that worshipping the bull may be closely linked, or identical to worshipping God Enki, for the horns of the bull, which the bull worshippers used as a symbol of the bull, could represent the crescent shape of the Moon, which is a symbol of God Enki. Therefore, as I describe in my footnote 199, the practice of bull (horns) and crescent Moon (appears as having horns) worship, could have been the result of the symbolic association of horns with supreme beings/Gods; a manifestation of the reverence/esteem of the Gods: even our ancient ancestors had prophetic beliefs, each in their own way.

GODS

Some critics will say the material covered in this book is irreligious, blasphemous, sacrilegious, impious, heretical, or even secular; and they will certainly loudly condemn it. But I believe it serves to confirm our religious beliefs, and we have only been somewhat misinterpreting our religious writings, viewing them through opaque perspectives, during the past 2,000 years or more, for there is an old Hellenic/Greek proverb which states: with whichever teacher you sit, such will be the extent of your learning.

We do not have to lead an ascetic or monastic lifestyle, or renounce, relinquish, or abandon our present widely accepted fundamental religious beliefs, including the belief in the existence of a supreme being, to believe in GODS OF THE UNIVERSE,™ which I envision as a pure supreme religion, a world wide religion: all Earth, all solar system, all galaxy, all universe; GODS OF THE UNIVERSE.™ We must only broaden and supplement our present belief and embrace the belief of supreme beings: a concurrent or simultaneous belief. A belief in one God is simply a belief in the highest God, not the sole God, but the chief God, the head of the pantheon of Gods, the president of the assembly of Gods; or perhaps the highest ranking God upon the Earth, the administrator/commander of the Earth. It is analogous to believing in the singular office of president of a particular country, and also believing in the plurality of offices of the governors of the various states within that country. The word religion itself facilitates the belief in many Gods, for it is defined in most dictionaries as meaning "the belief in God or Gods," which I of course passionately believe is synonymous with aliens/astronauts/Divine Beings/Lofty Ones/supreme beings, for I not only believe in a supreme being, I believe in many of them. There are no supreme beings that are indigenous to Earth, therefore, the supreme beings are extraterrestrials, not from or of the Earth; and I interpret aliens, extraterrestrials, astronauts, supreme beings, deities, or as the ancients called them, dingir/DIN.GIR/Divine Beings, i-lu or ilu/I.LU/Lofty Ones (shepherders), and GODS OF THE UNIVERSE,™ or if you will, Gods, all equivalent in essence and connotation.

Furthermore, no particular religious group has a special privilege, an exclusive relationship, or possesses the franchise on any particular God, or on the Gods, regardless of their claims. The Gods belong to all mankind, or more appropriately stated, all mankind belongs to the Gods.

There are no miracles, just as there is no magic. Miracles are phenomenon beyond the comprehension of the observer. In the past, there were no miracles, for the so-called "miracles" that primitive human beings witnessed at various times upon the Earth were merely manifestations of the high degree of technological advancement of the Gods. In the future, as our own technology advances, and we begin to recognize, identify, and correlate the so-called "miracles" that took place in the past as technological achievements, are we going to continue to call the technological achievements of the Gods "miracles"? I do not believe we will, for we shall have advanced far enough to recognize many of their "miracles" for what they actually are: the achievements of a higher technology.

Moreover, why did not the omnipotent and omnipresent traditional God, which many people envision in their minds eye as floating in the heavens and looking down upon mankind, have simply prevented the flood from occurring through some miracle, or performed a miracle and rescued Ziusudra/Noah from the great flood, without the use of a boat? It was not possible, and that is not what transpired, for the remarkable effects upon the Earth of the passing of their celestial abode are beyond the control and influence of the advanced technology of the Gods. Therefore, Ziusudra/Noah was instructed to build a "practical" high technology hexagon shaped boat for his escape from the huge flood waves and waters. There was no miracle, for it was demonstrated that even the Gods do not possess such omnipotence, there was only high technology.

And an interesting personal observation: if there were only one omnipotent and omniscient God, it would not be necessary for that supreme being to have a specific name, such as the specifically different names which appear in many religious and historical texts throughout the world, other than the general identification of God. For we only apply specific names, for example, to species, in order to differentiate and distinguish and identify between them, which would thereby imply a plurality: when there are more than one. A good example would be the term applied to the single moon of the Earth, called by the general term of moon; and the many moons of Jupiter, which have specific names because of their plurality.

Also, I must point out, the Biblical rendition of the creation and early history of mankind, in the early chapters of Genesis, was probably drawn from the same, older, original clay tablet material, from which I researched this book. For the original script and source of this story, could come from none other the GODS OF THE UNIVERSE,™ our creators, for how else would mankind know of our own creation and early history in such detail.

Additionally, the plural form of the word is used in the following verses in the Bible, for which I provide the underlines, and they agree with the clay tablet account of the creation and early history of mankind as I have rendered, and I believe they are both describing the precise same events: Genesis 1:26 And God said, Let <u>us</u> make man in <u>our</u> image, after <u>our</u> likeness; Genesis 3:5 For God doth know that in the day ye eat thereof, then your eyes shall be opened, and ye shall be as gods, knowing good and evil; Genesis 3:22 And the LORD God said, Behold, the man is become as one of us, to know good and evil: and now, lest he put forth his hand, and take also of the tree of life, and eat, and live for ever; Genesis 6:2 That the sons of God saw the daughters of men that they were fair; and they took them wives of all which they chose; Genesis 6:4 There were giants in the earth in those days; and also after that, when the sons of God came in unto the daughters of men, and they bare children to them, the same became mighty men which were of old, men of renown; Genesis 6:17 And, behold, I, even I do bring a flood of waters upon the earth; and Genesis 11:7 Go to, let us go down, and there confound their language, that they may not understand one another's speech.

POLYTHEISM

In conclusion, generally all of our distant ancestors believed in many Gods, polytheism, and only during the past approximately 2,000 years, a relatively very short time in our history, a time during which there was no window period of the planet Nibiru, has mankind generally believed in one God, monotheism. Have we been deliberately or casually misled to believe in monotheism for the past approximately 2,000 years? Is monotheism a belief in one God, or a belief in one of the many Gods? We accept a plural form of Angel, a plurality of Angels, many Angels, why not accept a plural form of God, a plurality of Gods, many Gods. And as I previously stated within the Introduction, if Angels are also considered Gods/aliens, from the same source, which they certainly are, but a specific category of God/alien, the astronaut category, Anunnaki Gods, as they are called within the clay tablets, for the Anunnaki are also designated as Gods,which is the correct way to view them, then everyone who believes in Angels, whether they realize it or not, already believes in many Gods/aliens, polytheism; for our ancient ancestors had first-hand knowledge, a real life experience by witnessing them and the events surrounding them, and thus a better understanding of who they were, and thereby accurately identifying them as Anunnaki Gods/astronaut aliens, as opposed to Gods/aliens, for all Anunnaki/astronauts are also Gods/aliens, but not all Gods/aliens are Anunnaki/astronauts (see my footnote 44, 45, and 48). A belief in polytheism is analogous to a belief in monotheism, except in polytheism the belief is broadened to include a plurality of supreme beings; a belief in the same genus of supreme being, except a belief in more of them.

It seems strange that in the society of ancient Hellas, where polytheism was acknowledged, that Socrates, who lived approximately 1,100 years after the most recent passing and window period of Nibiru, was convicted of trying to corrupt the youth by introducing new Gods to the already large number of Gods worshipped. Subsequently, in 399 B.C., he was condemned to death by drinking hemlock poison. Why did Socrates postulate that even more Gods, or as I call them, astronauts, could exist? Was he thinking ahead of his time, or did he gain and develop information and then understanding from the window period that was most recent to him?

Is it not time for each of us to open our eyes and mind, and recognize ancient events in the light of our own recent space accomplishments, achieved by our very own astronauts, and therefore reacknowledge polytheism? Our planet is full of people, who like small birds waiting for food with their mouths open, are waiting for religious proof, not prophecies, with which to be satiated. I ask you, "Is not the discovery of the true shape of the boat of the great flood verification of the existence of Gods/astronauts!" And were all of our distant ancestors wrong in believing in polytheism, or did some have exposure to the Gods/astronauts, during a window period, and witness events that shaped their lives and initiated their belief, inspiring our ancestors to venerate them? I believe they did. Again, only time can prove my belief.

Also, within our own experience, we have a somewhat correlative circumstance existing today. For hypothetically, if there were primitive ape-like creatures on our Moon in 1969 A.D., when the United States astronauts landed, because of our advanced technology, our astronauts would have been and would still be worshipped as their Gods. Therefore, since July 20, 1969 A.D., mankind, we the earthlings, to some degree have become Gods of the Moon. Why then, could there not be GODS OF THE UNIVERSE™?

BIBLIOGRAPHY

(Hardback—HB Paperback—PB)

A — Blumrich, Josef F. The Spaceships Of Ezekial. 1974 (PB)

B — Chiera, Edward They Wrote On Clay. 1938 (HB)

C — Gaster, Theodor H. The Dead Sea Scriptures. 1956 (PB)

D — Ginsburgh, Irwin First Man. Then, Adam! 1975 (PB)

E — Heidel, Alexander The Babylonian Genesis. 1942 (PB)

F — Heidel, Alexander The Gilgamesh Epic And Old Testament Parallels. 1946 (PB)

G — Hilprecht, H.V. The Babylonian Expedition Of The University Of Pennsylvania. Series D: Researches And Treatises. 1910 (PB)

H — Jacobsen, Thorkild The Sumerian King List. 1939 (PB)

I — Jacobsen, Thorkild The Treasures Of Darkness. 1976 (PB)

J — King, L.W. Enuma Elish, The Seven Tablets Of Creation. 1902 (HB)

K — Kramer, Samuel Noah Sumerian Mythology. 1944 (PB)

L — Kramer, Samuel Noah The Sumerians. 1963 (PB)

M — Lambert, W. G. and Millard, A. R. Atra-ḫasīs, The Babylonian Story Of The Flood. 1969 (HB)

N — Lloyd, Seton The Art Of The Ancient Near East. 1961 (PB)

O — Mellaart, James Earliest Civilizations Of The Near East. 1965 (PB)

P — Neugebauer, O. The Exact Sciences In Antiquity. 1957 (HB)

Q — Oppenheim, A. Leo Ancient Mesopotamia. 1964 (PB)

R — Reader's Digest Editors	Secrets Of The Past. 1980 (PB)
S — Segraves, Kelly L.	Search For Noah's Ark. 1975 (PB)
T — Sitchin, Zecharia	The 12th Planet. 1976 (HB)
U — Sollberger, Edmond	The Babylonian Legend Of The Flood. 1962 (PB)
V — The Holy Bible	King James Version. (HB)
W—The Koran, Interpreted	Translation by Arthur J. Arberry. 1955 (PB)
X — Velikovsky, Immanuel	Worlds In Collision. 1950 (PB)
Y — Woolley, C. Leonard	The Sumerians. 1965 (PB)
Z — Wylie, Francis E.	Tides And The Pull Of The Moon. 1979 (PB)

SUGGESTED READING ORDER

1 . Sitchin, Zecharia	The 12th Planet. 1976 (HB), and all of his other more recent books; and for his abstract interpretations, which are most interesting, enlightening, and inspiring.
2. Lambert, W.G. and Millard, A.R.	Atra-ḫasīs, The Babylonian Story Of The Flood. 1969 (HB)
3. Heidel, Alexander	The Gilgamesh Epic And Old Testament Parallels. 1946 (PB), especially page 80 through page 88, and page 224 through page 269.
4. Sollberger, Edmond	The Babylonian Legend Of The Flood. 1962 (PB)
5. King, L.W.	Enuma Elish, The Seven Tablets Of Creation. 1902 (HB)
6. Velikovsky, Immanuel	Worlds In Collision. 1950 (PB)

7. And last but certainly not least, for a wide perspective of the general subject matter, I recommend many of the books by Erich von Däniken; whose recognition of evidence, innovative views and theories, inquisitiveness, and intrepid literary efforts have helped to raise our consciousness and stimulate world-wide interest in extraterrestrial visitations to the Earth; and for his resolute acknowledgment of ancient Gods as astronauts.